AF697649

GIS APPLICATIONS IN AGRICULTURE

Volume Four • Conservation Planning

GIS Applications in Agriculture

Series Editor

Francis J. Pierce

Washington State University, Prosser

GIS Applications in Agriculture, *Francis J. Pierce and David Clay*

GIS Applications in Agriculture, Volume Two: Nutrient Management for Energy Efficiency, *David Clay and John F. Shanahan*

GIS Applications in Agriculture, Volume Three: Invasive Species, *Sharon A. Clay*

GIS Applications in Agriculture, Volume Four: Conservation Planning, *Tom Mueller and Gretchen F. Sassenrath*

GIS APPLICATIONS IN AGRICULTURE

Volume Four • Conservation Planning

Edited by

Tom Mueller
John Deere & Company
Urbandale, Iowa, USA

Gretchen F. Sassenrath
Southeast Agricultural Research Center
Kansas State University
Parsons, Kansas, USA

LIBRARY OF CONGRESS
SURPLUS
DUPLICATE

CRC Press
Taylor & Francis Group
Boca Raton London New York

CRC Press is an imprint of the
Taylor & Francis Group, an **informa** business

MATLAB® is a trademark of The MathWorks, Inc. and is used with permission. The MathWorks does not warrant the accuracy of the text or exercises in this book. This book's use or discussion of MATLAB® software or related products does not constitute endorsement or sponsorship by The MathWorks of a particular pedagogical approach or particular use of the MATLAB® software.

Cover Photo Credits: The middle left image on the cover was created by Demetrio Zourarakis for Chapter 2 using LiDAR elevation and terrain analysis techniques. Procedures for creating this figure are described in the on-line appendices for Chapter 2. The middle right hand image of estimates of annual sediment loss were created by Xiuying Wang for Chapter 13 and the procedures for calculating sediment loss with Apex are given in the on-line appendix for this chapter. The lower aerial landscape image was created by Tom Mueller using public imagery and terrain data using methods and procedures described in the on-line appendices for Chapter 2.

CRC Press
Taylor & Francis Group
6000 Broken Sound Parkway NW, Suite 300
Boca Raton, FL 33487-2742

© 2015 by Taylor & Francis Group, LLC
CRC Press is an imprint of Taylor & Francis Group, an Informa business

No claim to original U.S. Government works

Printed on acid-free paper
Version Date: 20150107

International Standard Book Number-13: 978-1-4398-6722-8 (Hardback)

This book contains information obtained from authentic and highly regarded sources. Reasonable efforts have been made to publish reliable data and information, but the author and publisher cannot assume responsibility for the validity of all materials or the consequences of their use. The authors and publishers have attempted to trace the copyright holders of all material reproduced in this publication and apologize to copyright holders if permission to publish in this form has not been obtained. If any copyright material has not been acknowledged please write and let us know so we may rectify in any future reprint.

Except as permitted under U.S. Copyright Law, no part of this book may be reprinted, reproduced, transmitted, or utilized in any form by any electronic, mechanical, or other means, now known or hereafter invented, including photocopying, microfilming, and recording, or in any information storage or retrieval system, without written permission from the publishers.

For permission to photocopy or use material electronically from this work, please access www.copyright.com (http://www.copyright.com/) or contact the Copyright Clearance Center, Inc. (CCC), 222 Rosewood Drive, Danvers, MA 01923, 978-750-8400. CCC is a not-for-profit organization that provides licenses and registration for a variety of users. For organizations that have been granted a photocopy license by the CCC, a separate system of payment has been arranged.

Trademark Notice: Product or corporate names may be trademarks or registered trademarks, and are used only for identification and explanation without intent to infringe.

Library of Congress Cataloging-in-Publication Data

GIS applications in agriculture. Volume 4, Conservation planning / editors, Tom Mueller and Gretchen F. Sassenrath.
pages cm. -- (GIS applications in agriculture)
Includes bibliographical references and index.
ISBN 978-1-4398-6722-8 (alk. paper)
1. Agricultural conservation--Planning. 2. Conservation of natural resources--Planning. 3. Geographic information systems. I. Mueller, Tom (Thomas G.), 1967- editor. II. Sassenrath, Gretchen F., editor. III. Title: Conservation planning. IV. Series: GIS applications in agriculture ; v. 4.

S604.5.G57 2015
631.4'5--dc23 2014035374

Visit the Taylor & Francis Web site at
http://www.taylorandfrancis.com

and the CRC Press Web site at
http://www.crcpress.com

I dedicate this book to my wonderful wife and incredible parents for their love and for sticking with me through thick and thin.

Tom Mueller

This is dedicated to my parents, who, even though they have no idea what this is about, will buy a copy, place it in the middle of the coffee table, and make everyone who visits the house read it. This is also dedicated to my two wonderful children. They won't understand it either, but at least they won't keep a copy on the coffee table and assault visitors with it.

Gretchen F. Sassenrath

Contents

Series Preface

GIS Applications in Agriculture, Volume Four: Conservation Planning, edited by Tom Mueller and Gretchen F. Sassenrath, is the fourth volume in the book series GIS Applications in Agriculture, which is designed to enhance the application and use of geographic information systems (GISs) in agriculture by providing detailed GIS applications that are useful to scientists, educators, students, consultants, and farmers. The first volume, *GIS Applications in Agriculture*, edited by Francis J. Pierce and David Clay, was published by CRC Press in 2007. The second volume, *GIS Applications in Agriculture: Nutrient Management for Improved Energy Efficiency*, edited by David Clay and John Shanahan, and the third volume, *GIS Applications in Agriculture: Invasive Species*, edited by Sharon Clay, were published by CRC Press in 2011. While the newest book in this series, the idea of a book on conservation planning using GIS was identified in 2007 when the book series began. Intuitively, conservation planning through GIS applications should appeal to all conservationists who clearly understand that a key to achieving soil and water conservation is rooted in an understanding of the spatial and temporal variation in both soil and water resources and natural and human-induced forces that affect the quality and quantity of those resources.

GIS Applications in Agriculture, Volume Four: Conservation Planning presents 14 chapters that address various aspects of soil and water conservation planning, including the availability and use of technical data sources, application of models and indices, information systems, targeting conservation programs, example applications, and decision making. Most of the chapters include exercises that allow the reader to recreate the GIS application for personal understanding, for use in either formal or informal education or training, or for direct application to situations that he or she encounters. While writing a meaningful chapter or editing a book on conservation planning alone is an achievement, including detailed applications that are tested and ready for the reader to use is quite another accomplishment. The book took a great deal of effort and time for all involved. I thank the authors who contributed excellent individual chapters with detailed examples when possible and the book editors, Dr. Tom Mueller and Dr. Gretchen F. Sassenrath, for organizing, collating, and editing the book and for their due diligence in testing the exercises to ensure that they provide useful applications of GIS in agriculture. I also thank Randy Brehm, CRC Press editor, for this volume and the staff at CRC Press for their hard work in making this volume possible.

It is my pleasure to serve as a series editor for this book series GIS Applications in Agriculture. I thank those who long ago recognized the need for publishing applications of GIS in agriculture, including Max Crandall, Pierre C. Robert, Harold Reitz, Jr., and Matthew Yen, who joined me in numerous conversations more than a decade ago about this important need. It has been more than 10 years since the untimely death of Pierre C. Robert in December 2003. I know Pierre would have been proud of this latest volume with a focus on what he loved—conservation planning and the use of soil information systems.

This book series can and will continue but only with the dedication and willingness of people to serve as an editor and who can bring together a team of authors to create a new, topical volume. If anyone wishes to prepare a volume in this series, please contact CRC Press.

Francis J. Pierce
Series Editor

Preface

Conservation planning is a process that involves making management and land use decisions based on careful analysis of the natural limitations of land to protect soil and water resources. Sound and effective conservation planning is becoming more important than ever with anticipated population growth, increasing demand for feedstocks for biofuels, dwindling freshwater resources, increasing land degradation, and increasing demand for meat products in the developing world, with additional complications associated with potential changing climate.

This book focuses on the use of spatial technologies for agricultural and natural resource conservation. The text provides a compilation of approaches from a broad range of topics developed by leading researchers working at the intersection of conservation and geospatial sciences and technologies. Nonpoint source pollution from agricultural practices is a serious problem across much of the United States. Targeting conservation practices to the most critical landscapes and soils will provide the greatest benefit to preserving natural resources while reducing economic expenses.

Rapid and substantial progress has been made in site-specific management of agriculture and natural resources. Much of the information has been presented in scientific publications, documenting the problems arising from current production methodologies. This book is unique in that it combines detailed information and background on conservation problems with the nuts and bolts of conservation planning, including step-by-step guidelines on developing solutions through conservation planning and implementation. Each chapter includes a detailed background on the specific topic, with case studies describing the design and implementation of the solution. The reader is then guided through online step-by-step exercises to gain experience in executing the conservation practice. Substantial online data and modeling are available that can be immediately implemented or modified to suit the users' needs. The exercises are written in such a way as to be amenable for classroom use as well as detailed enough for self-instruction by a highly motivated conservation professional.

This book will fill a need at both the university level for teaching graduate and advanced undergraduates and practicing professionals in agriculture and natural resource conservation. The book is an excellent resource for background information, along with detailed instructions for designing unique solutions to conservation problems. Through the use of case studies, the authors give direct examples of the application of the technology to solving conservation problems in the field. The chapters with the associated exercises are appropriate for direct incorporation into student instruction and can be expanded for student research projects. The information in this book will provide tools for conservation agencies to help them target critical areas for the implementation of conservation practices.

Tutorials and datasets are available through the CRC Press website at http://www.crcpress.com/prod/isbn/9781439867228. These tutorials expand the information presented in the chapters, allowing the student to work through data acquisition and analyses. The exercises have been tested with different versions of ArcGIS, generally version 10.0 or higher.

This book will help fill the dual requirement of addressing conservation of natural resources while maintaining agricultural productivity. Conservation professionals are being continually challenged to address problems in natural resources without negatively impacting agricultural production. The tools and resources provided in this book will assist in that effort.

Acknowledgments

The editors express their appreciation to Jason Corbitt for his assistance in reviewing the exercises in the book and to Marla Sexton for her assistance with formatting references. The editors thank the staff at Taylor & Francis for their exceptional support and service, and unimaginable patience.

We wish to express our appreciation to Scott Shearer for his early contributions to the development of this book.

Editors

Tom Mueller, PhD, is an agronomic data researcher with the Decision Science and Modeling team at the Intelligent Solutions Group, John Deere and Company in Urbandale, Iowa. His work there focuses on data analytics and introducing new technologies into the MyJohnDeere platform. Prior to joining Deere, Dr. Mueller was a faculty member at the University of Kentucky for 14 years in the Department of Plant and Soil Sciences in Lexington, Kentucky, where he taught and conducted research on site-specific crop management and precision conservation. He earned his BS in 1990 and MS in 1994 from Purdue University. He completed his PhD work at Michigan State University in 1998. Dr. Mueller is currently the leader of the Precision Agriculture community of interest for the American Society of Agronomy. He has served on the editorial boards for the *Agronomy Journal*, *Crop Science*, and the *Journal of Soil and Water Conservation*. He was the Geospatial Sciences and Technology Working Group Committee chair for the University of Kentucky. Dr. Mueller has a passion for using geospatial tools and technologies for conservation planning and natural resource management.

Gretchen F. Sassenrath, PhD, is an associate professor of agronomy at the Kansas State University Southeast Agricultural Research Center in Parsons, Kansas. She earned her BA degree from Oberlin College, MS degree in biophysics from the University of Illinois, and PhD degree in plant physiology from the University of Illinois. After completing postdoctoral work at Michigan State University and the University of California–Davis, Dr. Sassenrath joined the research team at the USDA Agricultural Research Service Crop Simulation Research Unit in Starkville, Mississippi. She transferred to the Application and Production Technology Research Unit in Stoneville, Mississippi, in 1997, where she continued her research on cotton production and cropping systems. She led research examining the status of farming systems in the United States and the future of agriculture. Dr. Sassenrath accepted the position at Kansas State University in 2013 and has established a research program on integrated crop and animal production systems that improves agronomic productivity, enhance soil and water resources, and increases economic return for producers. Dr. Sassenrath is dedicated to improving the long-term sustainability of farming for both farmers and consumers. She has published on a range of scientific topics and served as a technical editor for the *Journal of Cotton Science* in its early years. She served as the chair of the regional project Water Management in Humid Areas, and the Beltwide Cotton Production Physiology Conference.

Contributors

James Adams
NAI Isaac Commercial Real Estate Services
Lexington, Kentucky

James C. Ascough II
Agricultural Systems Research Unit
Agricultural Research Service
United States Department of Agriculture
Fort Collins, Colorado

Stephen D. Austin
Landscape Architecture Program
School of Design and Construction
Washington State University
Pullman, Washington

Rafael Mazza Barbieri
Athenas Agricultural Consulting and Laboratory
Cidade Jardim
São Paulo, Brazil

J. Bean
National Laboratory for Agriculture and the Environment
Agricultural Research Service
United States Department of Agriculture
Ames, Iowa

Shannon Belmont
Department of Environment and Society, Watershed Sciences
Utah State University
Logan, Utah

D. Terrance Booth
High Plains Grassland Research Station
Agricultural Research Service
United States Department of Agriculture
Cheyenne, Wyoming

Derya Özgöç Çağlar
Investment Support Office
Ankara Regional Development Agency
Ankara, Turkey

Dennis L. Corwin
Water Reuse and Remediation Research Unit
U.S. Salinity Laboratory
Agricultural Research Service
United States Department of Agriculture
Riverside, California

Samuel E. Cox
Bureau of Land Management
United States Department of the Interior
Cheyenne, Wyoming

Seth M. Dabney
Watershed Physical Processes Research Unit
National Sedimentation Laboratory
Agricultural Research Service
United States Department of Agriculture
Oxford, Mississippi

Jorge A. Delgado
Soil Plant Nutrient Research Unit
Agricultural Research Service
United States Department of Agriculture
Fort Collins, Colorado

Carl R. Dillon
Department of Agricultural Economics
University of Kentucky
Lexington, Kentucky

Michael G. Dosskey
National Agroforestry Center
United States Forest Service
United States Department of Agriculture
Lincoln, Nebraska

Gabriel D. Faleiros
São Paulo State University
São Paulo, Brazil

Richard L. Farnsworth
Natural Resources Conservation Service
United States Department of Agriculture
Beltsville, Maryland

Eduardo M. Gianello
Field Production
Monsanto Company
São Paulo, Brazil

David E. James
National Laboratory for Agriculture and the Environment
Agricultural Research Service
United States Department of Agriculture
Ames, Iowa

Todd Kellerman
National Agroforestry Center
United States Forest Service
United States Department of Agriculture
Lincoln, Nebraska

Jill A. Kostel
The Wetlands Initiative
Chicago, Illinois

Brian D. Lee
Department of Landscape Architecture
College of Agriculture, Food and Environment
University of Kentucky
Lexington, Kentucky

Joe D. Luck
Biological Systems Engineering Department
University of Nebraska
Lincoln, Nebraska

Blazan Mijatovic
Department of Plant and Soil Sciences
University of Kentucky
Lexington, Kentucky

David J. Mulla
Department of Soil, Water and Climate
University of Minnesota
St. Paul, Minnesota

Surendran Neelakantan
Department of Computer Science
University of Kentucky
Lexington, Kentucky

Joel Nelson
Department of Soil, Water and Climate
University of Minnesota
St. Paul, Minnesota

Naresh Pai
Stone Environmental, Inc.
Montpelier, Vermont

Eduardo A. Rienzi
Department of Plant and Soil Sciences
University of Kentucky
Lexington, Kentucky

Marcos Rodrigues
Campus Ciências Agrárias
Universidade Federal do Vale do São Francisco
Petrolina, Brazil

Dharmendra Saraswat
Department of Biological and Agricultural Engineering
University of Arkansas
Fayetteville, Arkansas

Jeanne M. Schneider
Great Plains Agroclimate and Natural Resources Research Unit
Agricultural Research Service
United States Department of Agriculture
El Reno, Oklahoma

Jordan M. Shockley
Agricultural Economist
Independent Consultant
Houston, Texas

Mark D. Tomer
National Laboratory for Agriculture and the Environment
Agricultural Research Service
United States Department of Agriculture
Ames, Iowa

Pushpa Tuppad
Department of Environmental Engineering
Sri Jayachamarajendra College of Engineering
Mysore, Karnataka, India

Dalmo A.N. Vieira
Watershed Physical Processes Research Unit
National Sedimentation Laboratory
Agricultural Research Service
United States Department of Agriculture
Jonesboro, Arkansas

Xiuying Wang
Blackland Research and Extension Center
Texas AgriLife Research
Texas A&M University System
Temple, Texas

Jimmy R. Williams
Blackland Research and Extension Center
Texas AgriLife Research
Texas A&M University System
Temple, Texas

Daniel C. Yoder
Biosystems Engineering and Soil Science
University of Tennessee
Knoxville, Tennessee

Demetrio Zourarakis
Division of Geographic Information
Frankfort, Kentucky

1 Geospatial Technologies for Conservation Planning

An Approach to Build More Sustainable Cropping Systems

Gretchen F. Sassenrath, Tom Mueller, and Jeanne M. Schneider

CONTENTS

EXECUTIVE SUMMARY

Current agricultural production systems must adapt to meet the increasing demands for more economically and environmentally sustainable cropping systems. The application of precision agricultural technologies and geospatial and environmental modeling for conservation planning can aid in this transition. To achieve this, conservation systems must address complicated, nonlinear, multidimensional, and cross-disciplinary problems. Such application of the physical and biological science of agroecosystems is further complicated by ever-evolving societal and fiscal realities. New conservation planning technologies and tools can be used to design unique, dynamic solutions that address various social, political, economic, environmental, and agronomic production goals. An overview of the techniques, tools, and analytical methods presented in this book is described here.

KEYWORDS

Conservation planning, conservation systems, sustainable agronomic systems, targeted conservation

1.1 INTRODUCTION

Over the last 100 years, agriculture has made many strides in the development of more sustainable production practices both in the United States and worldwide, but there are still great needs and opportunities for further progress. Current production methods externalize environmental costs to the detriment of the natural resource base that sustains agricultural systems. These externalities also increase the costs of production. Agriculture must increase productive capacity to provide rapidly escalating societal needs for food, feed, fuel, and fiber while simultaneously adapting to the impacts of climate change, population growth, dwindling freshwater resources, impaired soil health, greater demand for meat products in the developing world, and increasing land degradation. To achieve these goals, we need a dynamic analysis approach to conservation planning that allows us to more realistically evaluate management practices, understanding that adaptation will necessarily be a continuous process. Here, the word "dynamic" is used in the sense of "constant change," with an additional nod to the complexity and nonlinearity of many of the physical and agroecological processes involved. This book offers tools that will facilitate such a dynamic approach to conservation planning, detailing a number of useful observational, computational, modeling, and analysis approaches on the "physical" side of the problem. These can then be linked with comparable tools from the "living" side of the problem, incorporating considerations such as the importance of microbial contributions to soil function, competitive plant pressures, and pest adaptations under a changing climate to develop more robust management tools. Each use of such a dynamic system would employ the relevant components for a given location and conservation planning application. But first, the physical aspects, the "bones" of the system under analysis, need to be well defined and represented, covering hydrology, soils, elevation, extant erosion, land use, and other fundamental aspects of the agroecosystem. The addition of wildlife surveys, land use, and agricultural production histories, along with social and economic analysis tools, literally "lays the groundwork."

1.1.1 Thoughts on Tools and Approaches for Conservation Planning

Conservation planning is a process that involves making management and land use decisions based on careful analysis of the natural limitations of both landscape and resident agroecological systems in order to protect soil and water resources. Increasingly, the diverse goals of the agronomic community for maintaining and improving productive capacity must be coordinated with nonfarm valuation of landscapes for aesthetics or recreation (Sassenrath et al. 2010). Effective conservation planning should be made within a place-specific socioecological system that is adaptable, locally focused, and controlled; integrates multiple priorities across temporal and spatial scales; and is continuous (Ostrom et al. 2007). Precision (targeted) conservation has the potential to greatly improve productive capacity while addressing concerns for natural resource degradation (Delgado et al. 2011) if sufficiently and accurately framed with detailed and accurate data and objectives (Stokes and Morrison 2008).

Advances in the physical study of the agroecosystem have enhanced the knowledge of topography and its relationship to water and soil erosion. The knowledge of physical structure and erosive processes has furthered our understanding of the magnitude and extent of soil and nutrient loss, as well as factors contributing to those losses. Remote sensing and aerial imagery provide large-scale monitoring of the physical world. Elevation maps and light detection and ranging (LiDAR) provide detail on topographic changes across landscapes. Updated soil databases (Feng et al. 2009), now available through the Internet, improve the knowledge base for decision making, as do continuous sensors for measuring soil properties in situ, such as electrical conductivity.

1.1.2 Tools for Conservation Planning

Modeling tools greatly expand the potential to analyze and explore complex, interconnected, and highly dynamic environmental systems. Analytical tools and models have greatly improved and include those that examine soil erosion (Revised Universal Soil Loss Equation [RUSLE]) and soil and water resources (Soil and Water Assessment Tool [SWAT]). Other models can be used to explore nitrogen (Nitrogen Losses and Environmental Assessment Package [NLEAP]) or to examine the mapping of electrical conductivity using EC_e Sampling, Assessment, and Prediction (ESAP) software. Still, other models allow an examination of the entire agronomic system (Agricultural Policy/Environmental Extender [APEX]) for watershed-level planning.

Incorporating geospatial analysis with these new technologies is possible with geographic information systems (GIS) such as ArcGIS. Recent advances in geospatial techniques now provide land managers with tools and resources to conserve soil and water resources more efficiently than has ever been possible. These technologies include differential and survey-grade Global Positioning Systems (GPSs), GPS guidance technologies, map-based application control, GIS, Internet mapping technologies, remote sensing, erosion modeling software, terrain analysis software, and 2-D erosion and environmental modeling software. Further automation of geospatial analysis is possible through ModelBuilder. Sharing powerful networks through cloud computing further expands the speed of analysis and accessibility, facilitating an integration of knowledge and goals to develop answers that can direct production and inform public policy makers. Most significantly, incorporating expenses and potential income allows the determination of the economic impacts of implementing conservation practices, enabling the landowner to determine the feasibility of implementation and potential payoff.

1.1.3 Observational and Analytical Approaches

In Chapter 2, Mueller et al. (2015) use US Department of Agriculture National Agricultural Imagery Program (USDA NAIP) imagery and US Geological Service digital elevation models (DEMs) to identify eroded concentrated water flow pathways in agricultural fields and poorly vegetated grassed waterways from publically available data. Erosive processes in these areas could be improved by targeted conservation practices including landscape modifications such as grassed waterways, wooded and herbaceous riparian buffers, and filter strips, as shown in Figure 1.1. They also use terrain and economic modeling techniques to determine the profitability of cropping the steeper slopes in agricultural fields through a slope-and-yield map analysis. Four online tutorials teach students how to download available data from a range of sources and analyze the vegetation and the topography of land for conservation planning.

In Chapter 3, ravine morphometry is employed in the Minnesota River Basin to delineate the impact of the upland contributing area and ravine size on water discharge and sediment loading (Belmont et al. 2015). Using GIS tools to analyze the area, slope, relief, and stream power index values, they conclude that increased volumes of water discharged from larger upland contributing areas into larger ravines could potentially deliver more sediment to the Minnesota River Basin, making them ideal candidates for targeted conservation practices such as cover crop or perennial grass plantings to reduce water discharge and sediment loading to the ravine. In the tutorial, students analyze LiDAR DEM data and aerial photography to identify ravines using Spatial Analyst in ArcGIS.

Chapter 4 describes a fine-detail topographic analysis of a 6200-ha watershed analysis performed by Tomer et al. (2015) using LiDAR data. The enhanced elevation data available using LiDAR provide the necessary details required to identify sites that are appropriate for implementation of conservation practices such as wetlands,

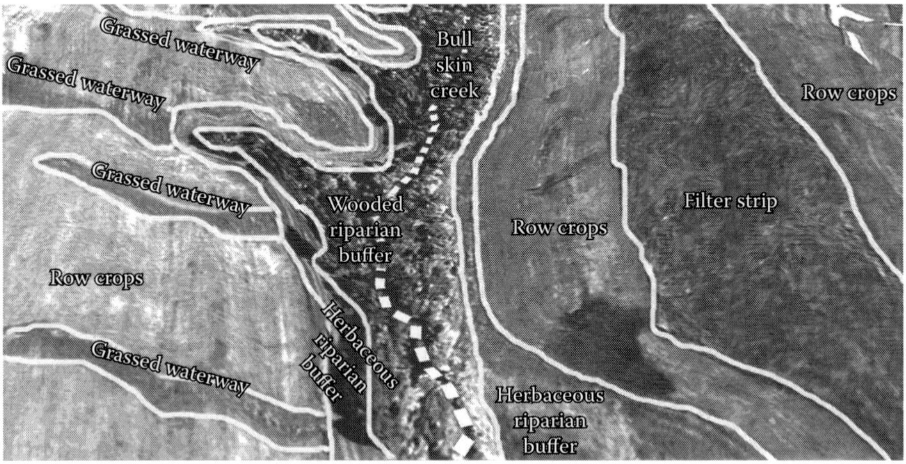

FIGURE 1.1 Conservation practices in an agricultural landscape in Kentucky.

two-stage ditches, and grassed waterways, with the intent to improve downstream water quality. They emphasize the critical role that landowners play in establishing conservation practices, recognizing that effective conservation must address both technical and social issues. The exercise teaches students how to process DEM data to evaluate individual fields for hydrologic routing.

In Chapter 5, Dabney et al. (2015) use GIS tools and high-resolution topography elevation data to expand the current model of soil erosion, the Revised Universal Soil Loss Equation—Version 2 (RUSLE2), to account for concentrated flow erosion. By linking the current version of RUSLE2 with a process-based channel erosion model, conservation planning for agricultural fields will be improved by taking into account the often substantial concentrated flow erosion. The exercise explores GIS-based terrain analysis to define hillslope profiles to estimate erosion using RUSLE2. The students will determine flow directions and define channel networks to visualize surface drainage within the watershed. An erosion simulation is defined from these geometric data using soils and agronomic management data.

The focus of Chapter 6 is the design and assessment of filter strips, which are useful conservation practices that reduce sediment and other pollutants carried off of agricultural fields in an overland runoff. In this chapter, Dosskey et al. (2015) present a GIS procedure for designing filter strips for implementation at agricultural field margins. The model uses terrain analysis to design a constant, user-selected level of trapping efficiency, reducing the loss of sediment and sediment-bound pollutants from nonuniform agricultural fields. The design speed is enhanced by incorporating the design model into GIS. In the online tutorial, students will use AgBufferBuilder in ArcGIS to design and assess vegetative filter strips around agricultural fields. The core AgBufferBuilder model uses terrain analysis in the GIS to produce designs and estimate performance.

In Chapter 7, Ascough and Delgado (2015) present an improved NLEAP that allows producers to improve nitrogen use efficiency and reduce nitrogen losses from agricultural systems using a GIS approach. NLEAP was originally released about two decades ago and has recently been revised in 2010 (Shaffer et al. 1991, 2010; Delgado et al. 1998a,b, 2010). The more user-friendly GIS version of the nitrogen management tool interactively displays 3-D geographic information (Ascough and Delgado 2015). Using embedded GIS-integrated technologies and statistical analysis, the user can also visually display nitrogen losses within the agroecosystem. Alternative management practices and nitrogen savings can also be analyzed using the Nitrogen Trading Tool, allowing determination of the economic impact of management or potential trading credits for future air or water quality markets. The students will use NLEAP GIS 5.0 in the exercises to explore nitrogen flows in agricultural fields. The Nitrogen Trading Tool is also demonstrated.

Soil characteristics within agricultural fields are usually highly variable and, when coupled with increasing salinity due to irrigation, pose a challenging problem for crop production. This variability has important implications for the development of accurate, site-specific management protocols for nutrients and irrigation. In many areas, salinity and spatial variability of salinity severely impair crop production. In Chapter 8, Corwin (2015) describes the spatial tools and analytical techniques to map the spatiotemporal distribution of salinity within agricultural fields. Geospatial measurement and analytical protocols provide field-scale information on soil salinity, water content, and soil texture and can be used to delineate site-specific management units for irrigation. These tools are important for optimal use of limited water resources for crop production and limit soil degradation due to excessive salinization. In the exercise, Corwin presents a thorough discussion of electrical conductivity measurements in soils. The EC_e Sampling, Assessment, and Prediction (ESAP) software is introduced, and the student is trained on its use for analyzing data and performing soil surveys. Relationships between soil characteristics and crop yield can be analyzed and mapped in either ESAP or ArcGIS.

Chapter 9 confronts the challenges of developing realistic management scenarios for rangelands. Booth and Cox (2015) present a new approach to adequate assessment of vast expanses of rangelands that combines ground surveys, remote imagery from high-altitude or satellite systems, and GIS to create sufficient scale and detail to monitor plant and animal distributions over large landscapes. The authors also present hands-on examples of applications of rangeland surveys and the associated development of landscape-scale management protocols. In the exercise, Booth and Cox provide instructions on using aerial surveys to develop conservation management plans for rangelands. The student is guided through the development of an overall conservation program and uses ArcGIS to create a sampling grid. The student is then instructed on the use of remote imagery to answer questions of habitat and distribution.

1.1.4 Modeling Systems for Landscape Conservation

Chapter 10 addresses the tension created by competition for limited land resources among agronomic productive value, provision of ecosystem services, and urban development. Lee et al. (2015) describe the use of a modified Land Evaluation and Site Assessment expert system approach to delineate land value based on different possible uses. Spatial data layers and weighting factors are then used to develop valuation algorithms. The strength of this method is its ability to incorporate input from multiple, and often conflicting, stakeholders. The resultant Landscape Tension Index can be further valued and modified for stakeholder input through an iterative Delphi Process. In the exercise, students use Spatial Analyst and ModelBuilder in ArcGIS to develop a suitability model. The student can change parameters to develop a new weighted overlay output, and use the approach described in the text to create the Landscape Tension Index.

In Chapter 11, another approach that successfully incorporates stakeholder values is detailed by Özgöç Çağlar and Farnsworth (2015). The Spatial Multiple-Criteria Decision Analysis (SMCDA) model, developed in ModelBuilder in ArcGIS, uses decision rules based on the Environmental Quality Incentives Program to identify and rank areas for conservation program implementation. The SMCDA further selects optimal regions for funding based on environmental impact scores. The exercise details the development of the model using ModelBuilder and demonstrates how decision rules can be modified and adapted to other regions and priorities.

The SWAT software is useful in exploring water conservation issues within watersheds. In Chapter 12, Pai and Saraswat (2015) demonstrate a Land Use Update module that expands SWAT2009, allowing dynamic updating of land use during simulation runs. This allows more accurate assessment of watersheds, especially for those in rapidly changing environments such as urban and periurban. The tutorial guides the user through the use of SWAT and the Land Use Change module with various land-use/land cover options.

The APEX model, described in Chapter 13, integrates multiple environmental factors into a systems model, allowing conservation planning targeted to critical areas for agricultural water resources and other environmental problems (Wang et al. 2015). APEX simulates the long-term sustainability of land management by modeling the flow of water, nutrients, sediments, and pesticides across landscapes and can account for changes in soil (erosion, sediment loss, nutrient flows), water (amount and quality), as well as plants, weather, and pests, greatly improving conservation efficacy. The online tutorial introduces students to the APEX system and performs a simulation of a watershed.

A significant factor that influences the implementation and long-term commitment to conservation programs is the economic cost or return on investment. In Chapter 14, Dillon et al. (2015) present a spatial economics decision-making guide for selecting land for Conservation Reserve Program (CRP) enrollment. Agronomic production data (yield maps) and economic analysis tools are employed within a GIS context to maximize profitability. The case study demonstrates the advantages of using precision agricultural tools and the importance of a detailed, flexible decision-making tool that encourages adoption of conservation systems. Their precision strategy of selecting land for enrollment in CRPs led to environmental improvements over traditional strategies. The tutorial allows students to determine potential land for CRP enrollment and net return.

1.2 CONCLUSIONS

The advances in observational, analytical, and modeling tools described in this volume give us an opportunity for transitioning current agricultural production systems to sustainable systems through greater conservation effectiveness. Using the conservation tools presented in this book, the agriculturalist can develop alternative practices based on an improved knowledge base of environmental functionality, such as erosive processes or sediment and nutrient runoff. Through integrative research at multiple levels of abstraction, GIS tools and models allow agroecosystems to be redesigned based on a more complete understanding of ecological processes and relationships. And most importantly, the simulation tools allow greater connectivity across disciplines to integrate multiple goals within agroecosystems, encouraging input from nonagronomic entities and moving toward a culture of sustainability (Gliessman 2010). Collaborative planning allows an integration of the entire system: production and consumption, urban and rural (Sassenrath et al. 2010). The strength of an integrated modeling approach is its ability to incorporate social aspects into the conversation. This is critical to developing realistic conservation plans that garner committed, long-term support (Ikerd 1997).

Improved knowledge-based and analytical tools for developing conservation methods are critical to address complex social, political, and environmental goals while enhancing the sustainability of the agricultural production system (Sassenrath et al. 2008). Greater implementation of conservation tools and practices has the potential to reduce reliance on nonrenewable resources and chemical controls, reduce waste by improved tracking of resources through the system to identify areas of better utilization or reuse, and provide realistic solutions to "wicked" problems. Critical to the success of these conservation technologies is their implementation in a format that is easily and readily accessible to the end user, such as web-based applications.

ACKNOWLEDGMENT

This chapter is contribution no. 15-006-B from the Kansas Agricultural Experiment Station.

REFERENCES

Ascough, II, J.C., and J.A. Delgado. 2015. Modeling landscape-scale nitrogen management for conservation. In *GIS Applications in Agriculture, Volume Four: Conservation Planning*, eds. T.G. Mueller, and G.F. Sassenrath, pp. 99–115. Taylor and Francis Group, Boca Raton, FL.

Belmont, S., D.J. Mulla, and J. Nelson. 2015. Identification and analysis of ravines in the Minnesota River Basin with geographic information system. In *GIS Applications in Agriculture, Volume Four: Conservation Planning*, eds. T.G. Mueller, and G.F. Sassenrath, pp. 37–51. Taylor and Francis Group, Boca Raton, FL.

Booth, D.T., and S.E. Cox. 2015. Resource management in rangeland. In *GIS Applications in Agriculture, Volume Four: Conservation Planning*, eds. T.G. Mueller, and G.F. Sassenrath, pp. 147–165. Taylor and Francis Group, Boca Raton, FL.

Corwin, D.L. 2015. Use of advanced information technologies for water conservation on salt-affected soils. In *GIS Applications in Agriculture, Volume Four: Conservation Planning*, eds. T.G. Mueller, and G.F. Sassenrath, pp. 117–145. Taylor and Francis Group, Boca Raton, FL.

Dabney, S.M., D.C. Yoder, and D.A.N. Vieira. 2015. Erosion modeling in 2-D with Revised Universal Loss Equation-Version 2: A tool for conservation planning. In *GIS Applications in Agriculture, Volume Four: Conservation Planning*, eds. T.G. Mueller, and G.F. Sassenrath, pp. 69–82. Taylor and Francis Group, Boca Raton, FL.

Delgado, J.A., R.F. Follett, J.L. Sharkoff, M.K. Brodahl, and M.J. Shaffer. 1998a. NLEAP facts about nitrogen management. *Journal of Soil and Water Conservation* 53:332–337.

Delgado, J.A., P.M. Gagliardi, D. Neer, and M.J. Shaffer. 2010. *Nitrogen Loss and Environmental Assessment Package with GIS Capabilities (NLEAP GIS 4.2) User Guide*. USDA Agricultural Research Service, Soil Plant Nutrient Research Unit, Fort Collins, CO. Available at http://www.ars.usda.gov/SP2UserFiles/ad_hoc/54020700NitrogenTools/NLEAP_GIS_4_2_Manual_Nov_29_2010.pdf (accessed July 14, 2014).

Delgado, J.A., R. Khosla, and T. Mueller. 2011. Recent advances in precision (target) conservation. *Journal of Soil and Water Conservation* 66:167A–170A.

Delgado, J.A., M.J. Shaffer, and M.K. Brodahl. 1998b. New NLEAP for shallow- and deep-rooted crop rotations. *Journal of Soil and Water Conservation* 53:338–340.

Dillon, C.R., J.M. Shockley, and J.D. Luck. 2015. Spatial economics decision-making guide for conservation reserve program enrollment. In *GIS Applications in Agriculture, Volume Four: Conservation Planning*, eds. T.G. Mueller, and G.F. Sassenrath, pp. 233–244. Taylor and Francis Group, Boca Raton, FL.

Dosskey, M.G., S. Neelakantan, T. Mueller, T. Kellerman, and E.A. Rienzi. 2015. Application of geographic information system and terrain analysis for designing filter strips. In *GIS Applications in Agriculture, Volume Four: Conservation Planning*, eds. T.G. Mueller, and G.F. Sassenrath, pp. 83–97. Taylor and Francis Group, Boca Raton, FL.

Feng, G., B. Sharratt, J. Vaughan, and B. Lamb. 2009. A multiscale database of soil properties for regional environmental quality modeling in the western United States. *Journal of Soil and Water Conservation* 64:363–373.

Gliessman, S.R. 2010. The framework for conversion. In *The Conversion to Sustainable Agriculture*, eds. S.R. Gliessman, and M. Rosemeyer, pp. 3–14. CRC Press, Boca Raton, FL.

Ikerd, J.E. 1997. Assessing the health of agroecosystems: A socioeconomic perspective. Presented at the First International Ecosystem Health and Medicine Symposium, Ottawa. Available at http://web.missouri.edu/ikerdj/papers/Otta-ssp.htm (accessed July 11, 2014).

Lee, B.D., J. Adams, and S.D. Austin. 2015. Landscape tension index. In *GIS Applications in Agriculture, Volume Four: Conservation Planning*, eds. T.G. Mueller, and G.F. Sassenrath, pp. 167–182. Taylor and Francis Group, Boca Raton, FL.

Mueller, T.G., D. Zourarakis, G.F. Sassenrath, B. Mijatovic, C. Dillon, E. Gianello, R. Barbieri, M. Rodrigues, E.A. Rienzi, and G.D. Faleiros. 2015. Soil surveys, vegetation indices, and topographic analysis for conservation planning. In *GIS Applications in Agriculture, Volume Four: Conservation Planning*, eds. T.G. Mueller, and G.F. Sassenrath, pp. 11–35. Taylor and Francis Group, Boca Raton, FL.

Ostrom, E., M.A. Janssen, and J.M. Anderies. 2007. Going beyond panaceas. *Proceedings of the National Academy of Sciences of the United States of America* 104:15176–15178.

Özgöç Çağlar, D., and R.L. Farnsworth. 2015. Prioritizing land with geographic information system for Environmental Quality Incentives Program funding. In *GIS Applications in Agriculture, Volume Four: Conservation Planning*, eds. T.G. Mueller, and G.F. Sassenrath, pp. 183–196. Taylor and Francis Group, Boca Raton, FL.

Pai, N., and D. Saraswat. 2015. Integrating land use change influences in watershed model simulations. In *GIS Applications in Agriculture, Volume Four: Conservation Planning*, eds. T.G. Mueller, and G.F. Sassenrath, pp. 197–210. Taylor and Francis Group, Boca Raton, FL.

Sassenrath, G.F., P. Heilman, E. Luschei, G. Bennett, G. Fitzgerald, P. Klesius, W. Tracy, and J.R. Williford. 2008. Technology, complexity and change in agricultural production systems. *Renewable Agriculture and Food Systems* 23:285–295.

Sassenrath, G.F., J.D. Wiener, J. Hendrickson, J. Schneider, and D. Archer. 2010. Achieving effective landscape conservation: Evolving demands, adaptive metrics. In *Managing Agricultural Landscapes for Environmental Quality: Achieving More Effective Conservation*, eds. P. Nowak, and M. Schnepf, pp. 107–120. Soil and Water Conservation Society, Ankeny, IA.

Shaffer, M.J., J.A. Delgado, C. Gross, R.F. Follett, and P. Gagliardi. 2010. Simulation processes for the Nitrogen Loss and Environmental Assessment Package (NLEAP). In *Advances in Nitrogen Management for Water Quality*, eds. J.A. Delgado, and R.F. Follett, pp. 361–372. Soil and Water Conservation Society, Ankeny, IA.

Shaffer, M.J., A.D. Halvorson, and F.J. Pierce. 1991. Nitrate Leaching and Economic Analysis Package (NLEAP): Model description and application. In *Managing Nitrogen for Groundwater Quality and Farm Profitability*, eds. R.F. Follett, D.R. Keeney, and R.M. Cruse, pp. 285–322. Soil Science Society of America, Madison, WI.

Stokes, D., and P. Morrison. 2008. GIS-based conservation planning: A powerful tool to be used with caution. *Conservation Magazine*. University of Washington. Available at http://conservationmagazine.org/2008/07/gis-based-conservation-planning-a-powerful-tool-to-be-used-with-caution/ (accessed July 9, 2014).

Tomer, M.D., D.E. James, J.A. Kostel, and J. Bean. 2015. Siting multiple conservation practices in a tile-drained watershed using LiDAR topographic data. In *GIS Applications in Agriculture, Volume Four: Conservation Planning*, eds. T.G. Mueller, and G.F. Sassenrath, pp. 53–67. Taylor and Francis Group, Boca Raton, FL.

Wang, X., P. Tuppad, and J. Williams. 2015. Combining landscape segmentation and agroecosystem simulation model. In *GIS Applications in Agriculture, Volume Four: Conservation Planning*, eds. T.G. Mueller, and G.F. Sassenrath, pp. 211–232. Taylor and Francis Group, Boca Raton, FL.

2 Soil Surveys, Vegetation Indices, and Topographic Analysis for Conservation Planning

Tom Mueller, Demetrio Zourarakis, Gretchen F. Sassenrath, Blazan Mijatovic, Carl R. Dillon, Eduardo M. Gianello, Rafael Mazza Barbieri, Marcos Rodrigues, Eduardo A. Rienzi, and Gabriel D. Faleiros

CONTENTS

EXECUTIVE SUMMARY

Powerful computers and geographic information systems (GISs) allow analysts to more effectively utilize digital data resources such as soil survey maps, remotely sensed imagery, and terrain data for conservation planning. The objectives of this chapter are to (1) demonstrate analyses of digital data for conservation assessment and (2) provide an assignment and detailed instructions for readers to analyze publically available geospatial data using GIS and automated modeling techniques. Research studies demonstrate that (1) eroded concentrated water flow pathways in agricultural fields requiring grassed waterways can be identified with early-spring high-resolution (i.e., 0.3-m) imagery and light detection and ranging (LiDAR)–derived terrain maps (i.e., hillshade and terrain attribute maps); (2) vegetation indices calculated from the National Agricultural Imagery Program (NAIP) imagery are effective tools for detecting poorly vegetated grassed waterways requiring replanting; and (3) terrain and economic modeling techniques can be used for slope and yield map analysis for determining the profitability of cropping steeper slopes. Four appendices are provided with step-by-step instructions for (1) obtaining publically available USDA NAIP imagery and US Geological Service (USGS) digital elevation models (DEMs); (2) calculating vegetation indices with USDA NAIP imagery; (3) creating terrain attribute, contour, and hillshade maps from USGS DEM data; and (4) conducting elevation models from public point cloud LiDAR elevation data.

KEYWORDS

Elevation modeling, remote imagery, soil survey maps, terrain attributes

2.1 INTRODUCTION

Achieving global food security amid increasing challenges of human population growth, diminishing water resources, soil degradation, and climate change will require effective and efficient conservation planning (Delgado et al. 2011). Land resource assessment is fundamental to conservation planning and can be enhanced with geospatial analysis of digital soil, imagery, and terrain data. Specifically, these datasets provide planners a better understanding of soil factors that limit land use and management, the ability to assess the vigor and health of vegetation in grassed fields and grassed waterways, and the capacity to identify areas where soil erosion is likely to or has already occurred.

There are two primary goals for this chapter. The first is to demonstrate how digital data could be analyzed for conservation assessment. The second is to provide readers, students, and instructors an assignment with detailed instructions for analyzing publically available geospatial data that utilize geographic information system (GIS) and automated modeling techniques. The remainder of this introduction will provide background information about digital soil data, remote sensing-derived vegetation indices, and digital terrain analysis.

2.1.1 Digital Soil Maps

Soils throughout the world have been mapped based on detailed soil profile observations using local, regional, or global soil classification systems. Extensive laboratory, field, and statistical research results have been used to develop detailed land use and management interpretations for mapping units. Traditionally, surveys were disseminated through paper maps, but now, they are often distributed electronically. Currently, soil maps in the United States can be visualized through the Web Soil Survey (http://websoilsurvey.nrcs.usda.gov) where soil data can be downloaded in tabular and GIS formats. The Soil Data Viewer (http://soils.usda.gov/sdv/) can be used with Microsoft Access to view the tabular data and with ArcGIS to view the spatial data.

Land capability classifications and subclassifications (Klingebiel and Montgomery 1961) are provided for USDA soil map units. Eight capability classes indicate suitability for specific land uses. As the class number increases, greater conservation efforts or more restrictions on land use are recommended to protect the soil from degradation. Soils belonging to classes I through IV are suitable for cultivated crops, pasture, rangeland, forestry, and wildlife. Very intensive cultivation practices are acceptable for class I soils, but only limited cultivation practices should be used for class IV soils. Classes V through VII are not appropriate for crops but can be used for pasture, rangeland, forestry, and wildlife. Class V allows intensive grazing, whereas only limited grazing is acceptable for class VII soils. Class VIII soils are only suitable for wildlife, recreation, and aesthetic purposes. Land capability subclasses indicate the nature of the primary limitation for land use and management. Planners may consider erosion control practices such as conservation tillage and contour farming for soils that are susceptible to erosion (subclass e). Other measures may be appropriate for soils that are limited by water (subclass w), such as the installation of drainage systems. Soils with root zone limitations (subclass s) such as shallow, water-limited soils may benefit from reduced seeding rates, whereas salt-affected soils may benefit from reclamation practices.

After determining the land capability classes and subclasses, conservation planners should examine the soil survey datasets for additional information that may describe the nature of the limitations. For example, details are provided about soil physical (e.g., available water storage, depth to restrictive layers) and chemical (e.g., cation exchange capacity, pH, salinity) properties. Nonirrigated and in some cases irrigated yield estimates for different crops are useful for assessing the relative productivity of soils on a farm.

Conservation planners must consider that county soil surveys in the United States were intended for coarse-scale planning (US Department of Agriculture–National Resources Conservation Service [USDA-NRCS] 1993). Map unit boundaries are not exact, and soil properties can vary considerably within a taxonomic class. Therefore, the most effective conservation planning decisions also consider additional spatial data (e.g., aerial imagery, terrain information) and utilize field site assessments. Tracts of land have different capabilities that make them more suitable for some uses than others. While some limitations can be overcome with sufficient financial resources, considering the natural limitations of the land can save money.

2.1.2 Remote Sensing and Vegetation Indices

Remotely sensed imagery can help assess problems or potential problems with erosion and investigate plant growth in crop fields and vegetative conservation structures. However, interpreting imagery can be challenging because spectral reflectance is impacted by many factors, including vegetation density, the concentration of pigments in plant material (e.g., chlorophyll, carotenoids), canopy structure, soil properties (e.g., water, oxidized iron, carbon content), solar intensity, angle of the sun, atmospheric factors, and optics of the remote-sensing platform.

Vegetative indices have been developed to quantify plant biomass in remotely sensed imagery in a way that accounts for variation in soil properties. The normalized difference vegetation index (NDVI) (Rouse et al. 1974) is one of the most commonly used vegetation indices and is expressed as follows:

$$\text{NDVI} = \frac{(\text{NIR} - \text{Red})}{(\text{NIR} + \text{Red})} \tag{2.1}$$

where NIR and Red represent the image value of spectral reflectance in the near-infrared (~>725 nm) and red (~600–725 nm) regions of the spectrum, respectively. Red reflectance is used because chlorophyll absorbs strongly in this region of the spectrum. Chlorophyll and carotenoids both absorb blue light making this region of the spectrum less than ideal for estimating vegetation. Red and near-infrared reflectances are both impacted by soil properties, including soil moisture, organic carbon, and so on; consequently, near-infrared reflectance is used to correct for the impact of soil property variability in the calculation of NDVI.

Potentially, NDVI could be useful for examining the condition of vegetation in grassed waterways because this index is correlated ($r^2 = 0.68$) with tall fescue biomass (Flynn et al. 2008). Tall fescue is a recommended species for grassed waterways in some areas of the United States and throughout the world. Maps of NDVI have been significantly related to corn (Huang et al. 2009; Yin and McClure 2013) and soybean (Thenkabail et al. 1994) biomass. NDVI can be a useful tool for assessing past crop management because areas with poor ground cover will be more susceptible to erosion early in the spring. Lower NDVI values may also be indicative of past erosion because degraded soils generally have properties that are less favorable to plant growth (e.g., higher bulk density, lower carbon content, lower water storage capacity, lower water aggregate stability) and lower crop productivity (Mokma and Sietz 2002). Despite its potential, the use of NDVI can be problematic because it is affected by spatial variation of surface cover (Vrieling 2006); specifically, the relationship between Red and NIR absorption shifts as plant densities increase, known as the tasseled cap effect (Jackson and Huete 1991). To adjust for this shift, Huete (1988) developed a modified index, the soil-adjusted vegetation index (SAVI), calculated as

$$\text{SAVI} = (1 + L)\frac{(\text{NIR} - \text{Red})}{(\text{NIR} + \text{Red} + L)}, \tag{2.2}$$

where L is the correction factor for vegetation ranging from 0 (high vegetative cover) to 1 (low vegetative cover). The green normalized difference vegetation index

(GNDVI) developed by Gitelson et al. (1996) is sometimes useful because NDVI is not as sensitive to chlorophyll at higher concentrations. GNDVI is defined as

$$GNDVI = \frac{(NIR - Green)}{(NIR + Green)}, \tag{2.3}$$

where Green represents spectral reflectance image values in the green (~500–600 nm) region of the spectrum.

2.1.3 Terrain Analysis

Soil topographic data are essential for conservation planning because topography influences soil formation (Jenny 1941) and is a major driver of soil processes such as erosion, water dynamics, and nitrogen processes that impact soil conservation. Topographic quadrangle maps have been created with stereo orthophotogrammetry throughout the world and by organizations such as the US Geological Survey (USGS). The USGS has used these topographic maps to create 10- and 30-m digital elevation models (DEMs) in a grid or raster format. Readers can find step-by-step instructions on the National Map website (http://nationalmap.gov/viewer.html) for downloading and using these products.

Elevation data products can also be created using LiDAR. This technology logs the timing and geometry of up to 400,000 light pulses per second emitted toward the ground from a rapidly swinging laser that is mounted on aircraft or satellite; it also records the return times of pulses from the landscape surface using sensitive detectors. These are then used to determine the precise location and orientation of the remote-sensing platform using real-time kinematic (RTK) Global Positioning Systems (GPS) and onboard gyroscopes. The data are processed to create highly accurate, spatially intensive point cloud elevation data. The point cloud data can be converted to high-spatial-resolution raster DEM formats using techniques and various GIS procedures. Step-by-step instructions are provided in the online exercises with this chapter for downloading point cloud LiDAR datasets for some areas in the United States and then creating a raster DEM using a binning technique and inverse distance weighted interpolation.

Digital contour maps are available from the USGS (http://nationalmap.gov/viewer.html) and can also be created from DEMs using GIS software. The orientation of contour lines can be used to visualize landscape geometry. Closely spaced contour lines indicate a steeper slope gradient than contours that are farther apart. The shapes of the contour lines describe landform geometry. Ephemeral gullies, gullies, and ravines form in the areas that are indicated by perpendicular, "V"-shaped contours, with the bottom of the "V" pointing upslope. Ravines are larger than gullies, and gullies are larger than ephemeral gullies. Ephemeral gullies are small enough so they can be repaired with normal tillage operations but will tend to reform in the same location each year. Ridges generally occur when a linear section of land, in the direction that is perpendicular to the slope gradient, is convex in shape. These are indicated by "U"-shaped contours where the bottom of the "U" points downslope.

Small circular contours with no other contour lines within can indicate depressions or hill peaks. The circles on topographic contours indicate depressions with hash marks in these circles pointed inward.

Hillshading digitally mimics natural light from the sun and shadows on gray-scale elevation maps, which very effectively communicate landscape geometry to viewers. When hillshade maps are created with detailed LiDAR elevation models, gullies (Perroy et al. 2010) and detailed topographic differences are clearly visible.

Slope gradient, the first derivative of elevation, is critical because areas with high slope gradients are much more susceptible to soil erosion (Fox and Bryan 2000) and consequently have reduced crop productivity (Kravchenko and Bullock 2000; Kitchen et al. 2003). Areas of higher slope could be targeted for site-specific conservation practices such as planting of winter cover crops (Langdale et al. 1991), making these fields eligible in the United States for payments through the USDA-NRCS Conservation Steward Program (CSP) and Environmental Quality Incentives Program (EQIP). When production costs are high and grain prices are low, removing sloping areas from crop production could be economically advantageous and could also increase the environmental protection of the watershed. Sloping areas with limited economic return from crop production could be determined by combining terrain, yield maps, and partial budgets and performing a simple break-even analysis.

Landscape curvature values represent the second derivative of elevation. Profile curvature is measured along the steepest portion of the slope direction. Water accelerates along slopes with convex profile curvature and decelerates along those that have concave shapes. Plan curvature is measured along a contour line that intersects a slope. Water will concentrate in areas where plan curvature is concave and will disperse where it is convex. General curvature is the total curvature of the landscape, not just in the direction down the slope or along the contour. Different GIS programs use different conventions for calculating profile, plan, and general curvature, so readers should examine software manuals when comparing results from different studies.

The upslope catchment area is also referred to as the specific catchment area. For each point in a landscape surface, this terrain attribute estimates the area of the landscape that drains into it. To calculate the specific catchment using GIS, depressions must generally first be removed from the digital elevation so that when flow direction is calculated, all cells in the DEM will eventually flow into other cells that flow off of the raster surface. Next, flow direction is calculated, which estimates the direction of flow for each point in the DEM. Some software programs use the eight nearest neighbors to calculate flow direction in one of the eight different directions (D8 method), whereas other software programs use sophisticated multidirectional flow models. For example, the D∞ method (Tarboton 1997) estimates flow direction in an almost infinite number of directions. The D∞ method is considered a multidirectional model because flow from a cell is directed into multiple cells. The flow accumulation procedures tally the number of upslope cells or partial cells that flow into each cell according to the flow direction raster. Finally, flow direction is calculated by multiplying the number of upslope cells by the area of each cell. For example, the area of each grid cell for a 9.145-m (30-ft) raster is $\simeq 83.6\ m^2 \simeq 900\ ft^2$.

The topographic wetness index (Beven and Kirkby 1979) indicates the capacity of land to hold water and is defined as

$$\text{Topographic wetness index } = \ln\left(\frac{\text{Specific catchment area}}{\tan\beta}\right), \tag{2.4}$$

where the specific catchment area (SCA) is the upslope contributing area per unit contour length, estimated using a flow accumulation algorithm; β is the slope angle; $\tan \beta$ is the fractional slope; and $\tan \beta * 100$ is the percent slope. The topographic wetness index is greater in areas when flow accumulation is large, such as low flat areas.

2.2 MATERIALS AND METHODS

The case study was conducted in an area in Shelby County, Kentucky (38°20′N, 85°11′W), located in the Outer Bluegrass physiographic region of Kentucky. The focus area was a 50-ha field in a no-till corn (*Zea mays* L.), wheat (*Triticum aestivum* L.), and double-crop soybean (*Glycine max* (L.) Merr.) or corn and full-season soybean rotation for more than 30 years. The extent of the GIS data also includes fields from an adjacent farm that has been intensively tilled with a moldboard plow for more than 20 years and in a corn, soybean, and burley tobacco (*Nicotiana tabacum* L.) rotation. The soils in this region developed primarily from limestone residuum overlain with pedisediment from limestone-weathered materials and loess (Soil Conservation Service 1980).

Soil survey spatial and tabular data were obtained and downloaded through the Web Soil Survey (http://websoilsurvey.nrcs.usda.gov). Maps and tables were created with the Web Soil Survey, Soil Data Viewer (http://soils.usda.gov/sdv/), ArcGIS 10.1 (ESRI, Redlands, CA), and Microsoft Access.

Orthocorrected 1-m-resolution NAIP images were captured on June 20, 2010 and June 18, 2012 and were obtained from the USGS National Map website. Bands 1, 2, 3, and 4 corresponded to red, green, blue, and near-infrared spectral reflectance. Using these bands, NDVI, SAVI, and GNDVI were calculated with Equations 2.1, 2.2, and 2.3.

Orthocorrected 0.3-m-resolution Kentucky Aerial Photogrammetry and Elevation Data Program (KYAPED) imagery from circa March and May 2012 was obtained from the Kentucky Division of Geographic Information at the following website: http://kyraster.ky.gov/arcgis/rest/services/ImageServices/. Although the exact date of the imagery was not recorded, there was relatively little vegetative cover indicating that it was likely captured in March 2012.

A LiDAR-derived 1.52-m (5-ft) DEM was obtained from the KYAPED, and a USGS 10-m DEM was obtained from the National Map. Flow direction and flow accumulation were calculated with TauDEM. The DEM had substantial noise associated with vehicle traffic in the field, which is problematic for flow modeling. To reduce the noise, the data were contoured at a 2-m contour interval, and the "TopoToRaster" command was used to create a smoothed DEM from the contour map (Pike et al. 2012). Slope gradient was calculated, and depressions in the DEM were filled. Spatial Analyst was used to calculate flow direction (D8 method) and flow accumulation in order to estimate the upslope contributing area.

Grassed waterways had been established by the producers for the entire farm. The basis for identifying where these waterways should be placed was visible evidence of concentrated flow erosion, as determined after being trained by the NRCS. The senior author of this paper and the producer visited the field in 2009 with an NRCS conservationist for validation. The NRCS conservationist determined that the producer did use proper techniques to identify the concentrated flow pathways in the field.

Corn (2010) and soybean (2012) yield data were obtained from the producer who used a single calibrated combine for harvest in both years. Point yield data were removed from areas where the header was not engaged and from those that were flagged as "not mappable." Histograms of distance traveled, grain flow, grain moisture, and corn yield were visually assessed and were used to identify outliers for removal. ArcGIS Geospatial Analyst was used to model semivariograms, and the data were interpolated with kriging using the five closest points with a variable search radius with a maximum distance of 7.6 m.

The cleaned corn and soybean point yield data were visually assessed. Areas within the field that were long straight combine passes were included in the analysis. Areas near turn rows and grassed waterways and zones where NDVI values had different patterns than the rest of the field were excluded. The interpolated slope and yield values for these areas were determined using standard GIS techniques.

Linear regression was used to relate slope and yield in the form

$$Y = B_0 + B_1 \times S, \tag{2.5}$$

where Y is the crop yield, B_0 is the intercept, and B_1 is the regression coefficient that indicates the incremental change in yield per unit of slope gradient change. A break-even slope was estimated by assuming that total revenue was equal to total costs or more specifically

$$P \times Y - C = 0, \tag{2.6}$$

where P is the crop price and C is the cost of production. Next, Equations 2.5 and 2.6 were combined to give

$$S_{BE} = \frac{(C - P \times B_0)}{P \times B_1}, \tag{2.7}$$

where S_{BE} represents the break-even slope gradient. To examine this relationship graphically, corn prices between \$0/Mg and \$200/Mg (\$0/Bu to \$12.54/Bu), as well as several different production costs (i.e., \$382/ha, \$1000/ha, \$1618/ha, and \$2336/ha), were considered. The \$1618/ha (\$655/ac) cost was utilized because this was the sum of the fixed plus variable costs provided in the University of Kentucky Corn and Soybean Budget spreadsheet template for central Kentucky for no-till corn in 2013 (Halich 2013).

2.3 CASE STUDY

Most of the soils in this field belonged to capability classes II and III (Figure 2.1, Table 2.1), indicating moderate and severe limitations for cultivated crops, respectively. The most severe limitations are for the two soils (LoC and NhC) with slopes ranging from 6% to 12%. The land capability subclass for most soils is erosion (subclass "e") except for the Nolin soil, which is primarily limited (class IV) by water (subclass "w"). The producers managing this field have been addressing the erosion limitations with conservation tillage, grassed waterways, vegetative filters, and rye grass planted on steep slopes. The excess water limitation for the Nolin soil was addressed by the producer by removing this area from production. With the exception of the Nolin soil, the Shelbyville soil has the greatest available water storage; and as a result, these areas are expected to have greater yields. All soils are well drained but the Nicholson soil, which has a restrictive layer (i.e., fragipan) at 66 cm that can result in a perched water table, particularly in the spring. Consequently, the producer uses delayed nitrogen fertilizer applications to improve crop nitrogen use efficiency and to conserve water resources.

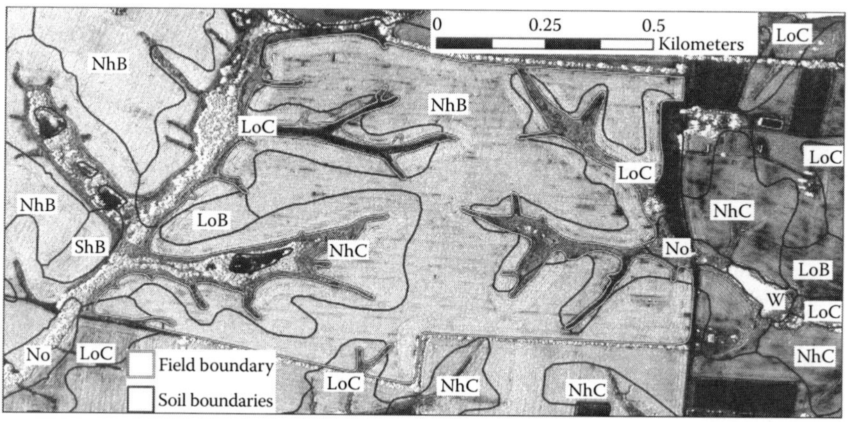

FIGURE 2.1 Soil survey boundaries. The background image is an inverted black-and-white orthoimagery that was collected in 2010 (leaf-on, NAIP imagery). The names of the map unit symbols are as follows: LoB is a Lowell silt loam (2% to 6% slope), LoC is a Lowell silt loam (6% to 12% slope), NhB is a Nicholson silt loam (2% to 6% slope), NhC is a Nicholson silt loam (6% to 12% slope), No is a Nolin silt loam (0 to 2% slope), and ShB is a Shelbyville silt loam (2% to 6% slope). More details about the properties of each soil are shown in Table 2.1. (Data obtained from the Web Soil Survey.)

TABLE 2.1
Map Unit Symbols, Names, and Properties and NRCS Crop Yield Estimates

Map Unit Symbol	Map Unit Name	Slope %	Capability		Available Water Supply in the Top 1-m (cm)	Depth Fragipan (cm)	Corn Yield (Mg/ha)	Soybean Yield (ha)
			Class	Subclass				
DAM	Dam, large earthen							
LoB	Lowell silt loam	2 to 6	2	e	16.7		8.0	2.7
LoC	Lowell silt loam	6 to 12	3	e	16.4		7.5	2.8
NhB	Nicholson silt loam	2 to 6	2	e	16.9	66	8.0	3.4
NhC	Nicholson silt loam	6 to 12	3	e	16.9	66	6.6	3.0
No	Nolin silt loam	0 to 2	2	w	21.0		9.4	3.7
ShB	Shelbyville silt loam	2 to 6	2	e	20.0		8.5	3.4
W	Water							

Note: The spatial distribution of each soil map unit is displayed in Figure 2.1. Soil data were obtained from the Web Soil Survey.

Many features were clearly visible in the black-and-white 1-m NAIP late-June imagery (Figure 2.2a) including trees (Point A), water (Point B), and roads (Point C). For example, grassed waterways with denser vegetation had darker image values (Point D) than those that had been treated with herbicides (Point E). Upon closer examination, it was possible to see a soil that is scoured by erosion within a waterway that may require attention (Point F). With higher-resolution (0.3 m) imagery from earlier in the year (Figure 2.2b), these erosion features could be more easily observed (Point G). However, the condition of the vegetation (Points H and I) was not as clearly detectable this early in the spring. Simple imagery can be helpful for examining vegetative practices for soil conservation and erosion patterns in agricultural fields.

In case studies involving yield map analysis, it is critical to examine remotely sensed imagery for patterns that may compromise analyses. Two areas in the NDVI maps for 2010 and 2012 (Figure 2.3) have low values of NDVI, indicating poorer vegetative growth (the lighter-colored areas surrounding Points A and B). These patterns do not appear to be related to soils or terrain but are likely due to differences

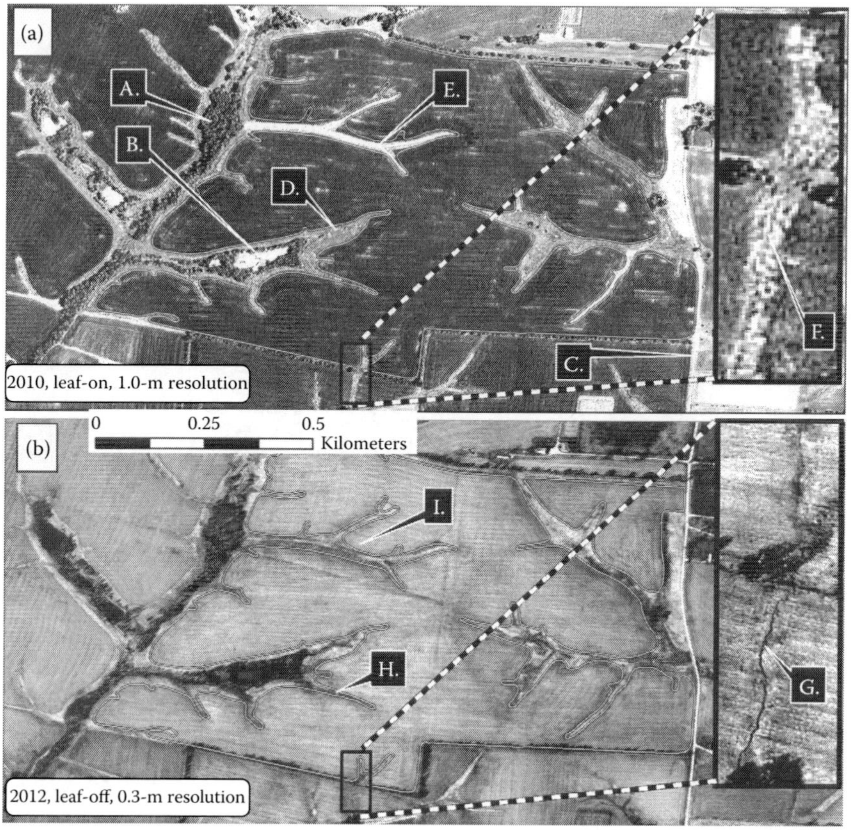

FIGURE 2.2 (a) Orthocorrected leaf-on 1-m resolution NAIP imagery collected on June 20, 2010 showing (A) trees, (B) water, (C) roads, (D) dense vegetation, (E) vegetation treated with herbicides, and (F) soil scouring. (b) Orthocorrected leaf-off 0.3-m resolution KYAPED imagery collected on an unknown date between March and May 2012 showing (G) soil scouring, (H) dense vegetation, and (I) vegetation treated with herbicides. The images were converted to a black-and-white format for display purposes by summing the values of the red and green with one-fourth the value of the blue band.

in management (e.g., different nutrient or herbicide rates, different planting dates or varieties). Planners can use these maps to help producers monitor vegetative cover early in the growing season to assess how well cropping practices are protecting the soil from erosion. In some cases, fields may be poorly vegetated and have low NDVI values because of past erosion that has occurred. For example, a soybean field from a neighboring farm can be seen in the eastern part of the images in Figure 2.3 (labeled "IT soybeans"). This field has sloping ground and has been intensively tilled for many years. The highly variable vegetation patterns have consistently occurred in this field because of shallow top soil associated with years of erosion.

Vegetative index values are substantially lower in sprayed-out waterways (Points C and D) than those with more vegetative cover (Points E and F), as denoted in

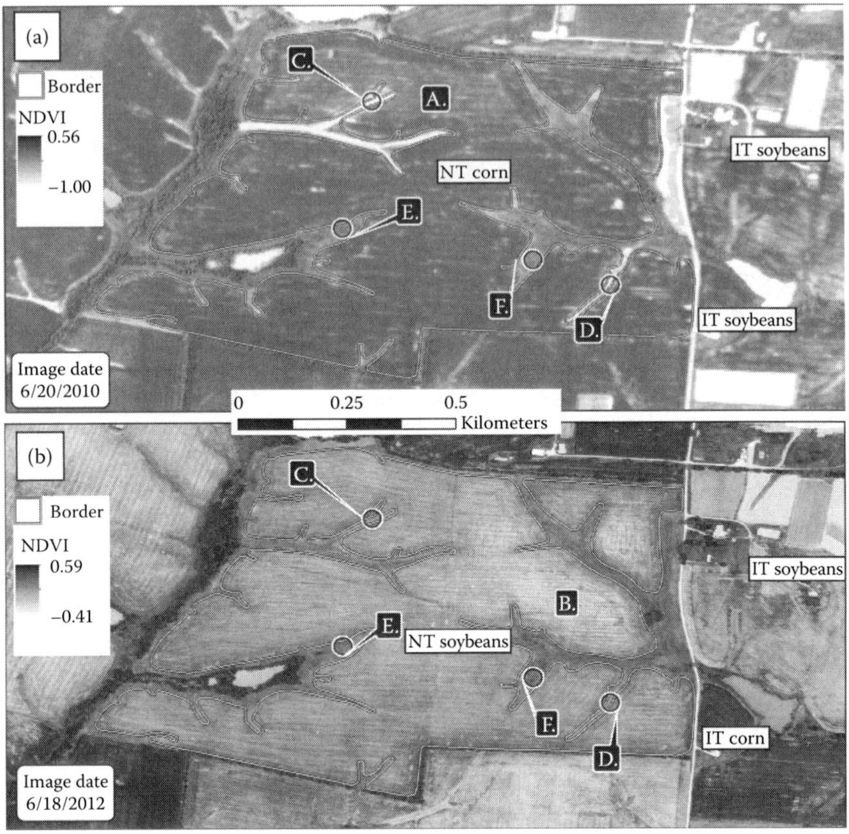

FIGURE 2.3 NDVI calculated from a 1-m resolution NAIP imagery. The top image (a) was collected on June 20, 2010 and the lower image (b) was collected on June 18, 2012. The abbreviations NT and IT denote no-tillage and intensive tillage, respectively. Areas around points A and B indicate poor vegetative growth of the crop; C and D are "sprayed-out" waterways in which herbicide overspray has reduced vegetative coverage; E and F are waterways with greater vegetative coverage.

Figure 2.3 and Table 2.2. Planners could use the freely available four-band USDA NAIP imagery that is published biennially to help producers monitor grassed waterways and potentially other conservation structures (e.g., vegetative buffers).

The contour map (Figure 2.4) shows "V"-shaped contours (Points A and B) in areas where grassed waterways have been installed to control ephemeral gully erosion, whereas "U"-shaped contours occur where convex slopes gradually increase down slope (Points C and D). Wider contour spacing occurs on flatter ridges (Point F) in the center of the field, and smaller spacing indicates greater slopes that are closer to the waterways (Point E) and some field edges (Point G). Figure 2.4 demonstrates the utility of combining contour data with remote-sensing imagery for providing better insights for conservation management.

TABLE. 2.2
Vegetation Index Values for Sprayed-Out and Better-Vegetated Grassed Waterways for Locations C through F in Figure 2.3

Condition of Grassed Waterway	Location in Figure 2.3	2010			2012		
		NDVI	SAVI	GNDVI	NDVI	SAVI	GNDVI
Sprayed-out	Point C	0.005	0.010	0.034	0.116	0.231	0.033
Sprayed-out	Point D	0.090	0.180	0.092	0.093	0.188	0.025
Vegetated	Point E	0.289	0.577	0.173	0.221	0.453	0.073
Vegetated	Point F	0.275	0.548	0.167	0.224	0.455	0.075

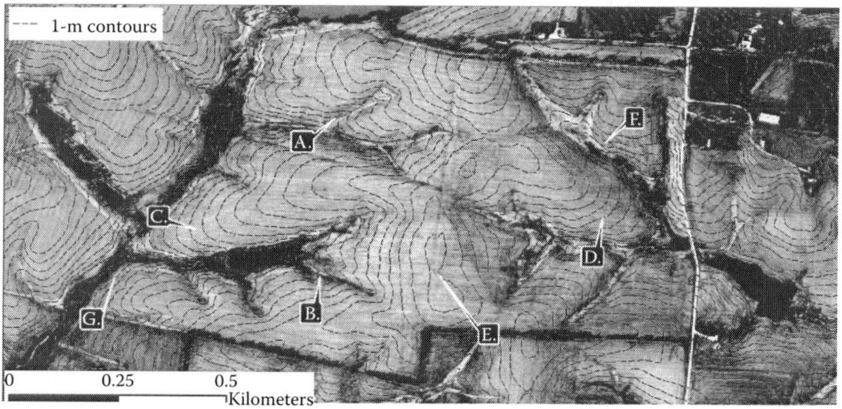

FIGURE 2.4 Elevation contours (1 m) calculated from a 1-m LiDAR DEM. The background black-and-white imagery is a leaf-off 0.3-m resolution image. Points A and B indicate "V"-shaped contours with grassed waterways. Points C and D are "U"-shaped contours with gradually increasing convex down slope. Point E is a flat ridge near the center of the field. The smaller spacings between contours indicate greater slopes such as the areas along some of the waterways (e.g., Point F) and close to field edges (e.g., Point G).

Hillshade maps created from LiDAR data (Figure 2.5) are outstanding tools for soil conservation, enhancing features and patterns to make them more visible. Stream cuts appear deeper and darker (Point A) than those due to ephemeral gully erosion (Points B and C) where erosion control practices are needed. Planners could use these maps as a tool to identify grassed waterways that need repair, as evident in sprayed-out waterways with clear erosion cuts (Point D) that are not apparent in the better-vegetated grassed waterways (Point E). Surface water appears flat, as can be seen at Point F. Historical management regions within fields are useful for zone sampling in agricultural fields and old fence lines, clearly delineated in a hillshade map (Points G and H), which can help define these areas. Conservation structures can be seen in imagery such as the terraces in the field to the east (Point I). Straight lines (Points J) occurring throughout the study area are small ruts resulting from

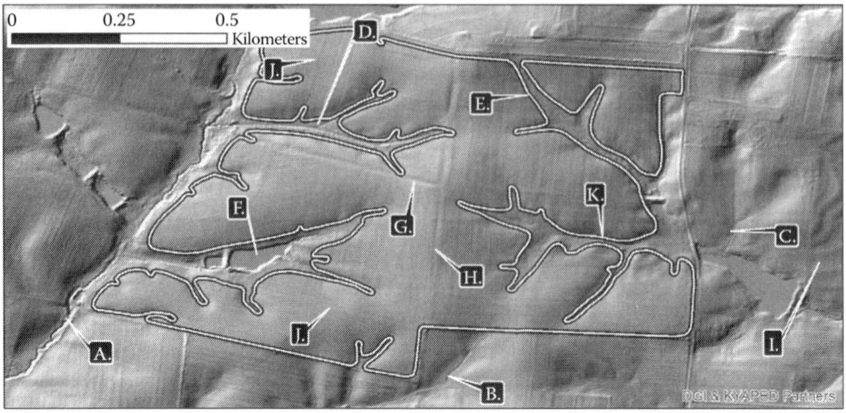

FIGURE 2.5 Hillshade image (5 m) obtained from the KYAPED. The source of the elevation data used to create the hillshade map was LiDAR. Points of interest are as follows: stream cuts (A), ephemeral gully erosion (B and C), erosion cuts on sprayed-out waterways (D), grassed water with adequate vegetation (E), surface water (F), old fence lines (G and H), terraces (I), small ruts from heavy farm machinery movement (J), and encroachment of farming activities on conservation structure (K).

movement of heavy farm equipment that traverses the field. These lines can be problematic for molding overland flow across the landscape, as described in the methods section, because water can be routed preferentially along rather than downhill during a rainfall event. Areas where farming activates potentially encroach very close to conservation structures can be identified and flagged for field scouting (Point K).

The producer targeted cereal rye planting for use as a cover crop on steeper slopes, as is evident from the good correspondence between the LiDAR-derived slope maps (Figure 2.6, Points A and B) and the areas with rye (Points C and D). Farmers and conservation planners could use digital slope maps to help farmers much more efficiently plan targeted planting of cover crops for control erosion.

In a production agriculture environment involving thousands of hectares of farmland, visual assessment of terrain steepness is difficult to achieve perfectly. Consequently, some sloping regions (Point E) were not planted with rye cover crop (Point F). Conservation professionals could provide producers with highly accurate slope maps to help them more accurately and time-efficiently identify vulnerable erosion-prone slopes in fields.

Darker, more negative areas in the plan and profile curvature maps (Figure 2.7) indicate areas with greater potential for erosion. Runoff tends to accelerate in areas with negative profile values and converge in areas with negative plan curvature values. Profile curvature values (Figure 2.7a) are convex and more negative next to ponds (Point A), along the edges of the old field borders (Point B), adjacent to streams (Point C), and the contours of terraces designed to slow runoff (Point E). Conservation efforts should focus in these areas by building structures such as bank protection or terrace maintenance. Plan curvature (Figure 2.7b) can help identify optimal placement of grassed waterways because the values are more negative along erosion channels (Points F and G).

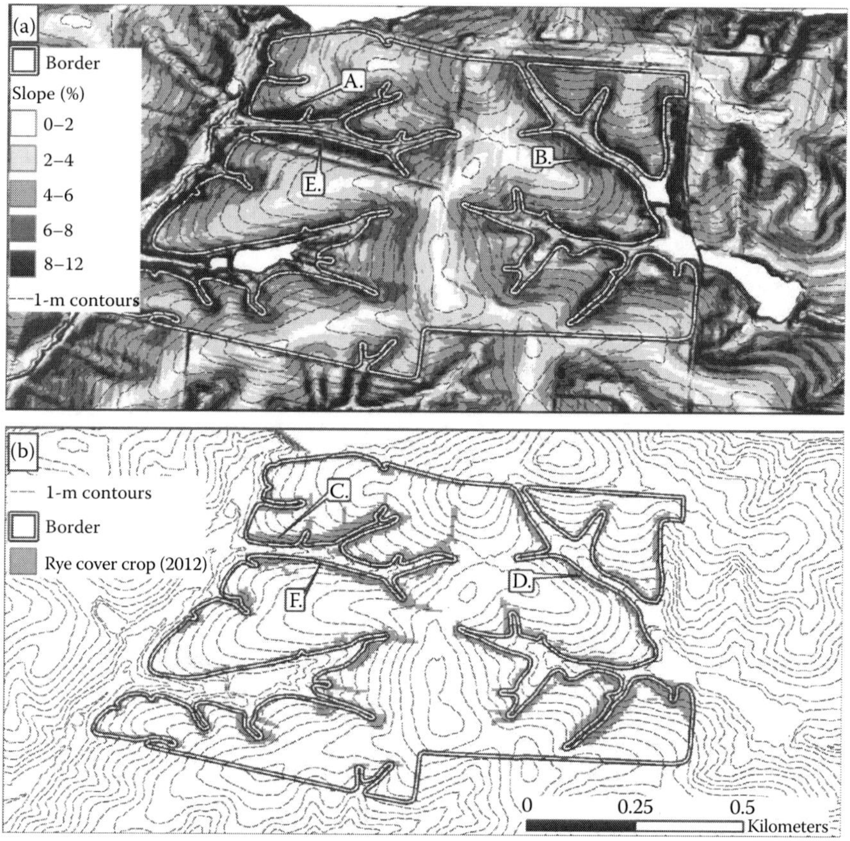

FIGURE 2.6 Maps of (a) slope and (b) 1-m contours overlain and areas that were planted to rye cover crop in 2012. Both maps also include elevation contours and the field boundary. The slope, the upslope area, and the contours were calculated from 1-m LiDAR-derived DEMs. Areas of interest include the following: areas of steep slope (A and B), areas planted to rye (C and D), sloping region (E), and not planted to rye cover crop (F).

Plan curvature is just one factor in determining the potential for ephemeral gully erosion. For example, an area with larger negative profile curvature values (Point H) did not require a grassed waterway because the runoff and sediment load from upslope was not large enough to justify the use of a grassed waterway. Substantial noise in plan and profile curvature maps (Figure 2.7, Points J and K) makes interpretation challenging, particularly when the maps are derived from LiDAR data. Better filtering techniques for DEM creation will help conservationists prepare more easily utilizable maps for planning.

The association between upslope contributing area maps and grassed waterways was remarkable (Figure 2.8, Points A and B) because the dimension and placement of grassed waterways depend on the volume of runoff requiring evacuation from upslope areas. The correspondence was not perfect (Points C and D), but this would not likely be a great problem because the design of a grassed waterway involves a

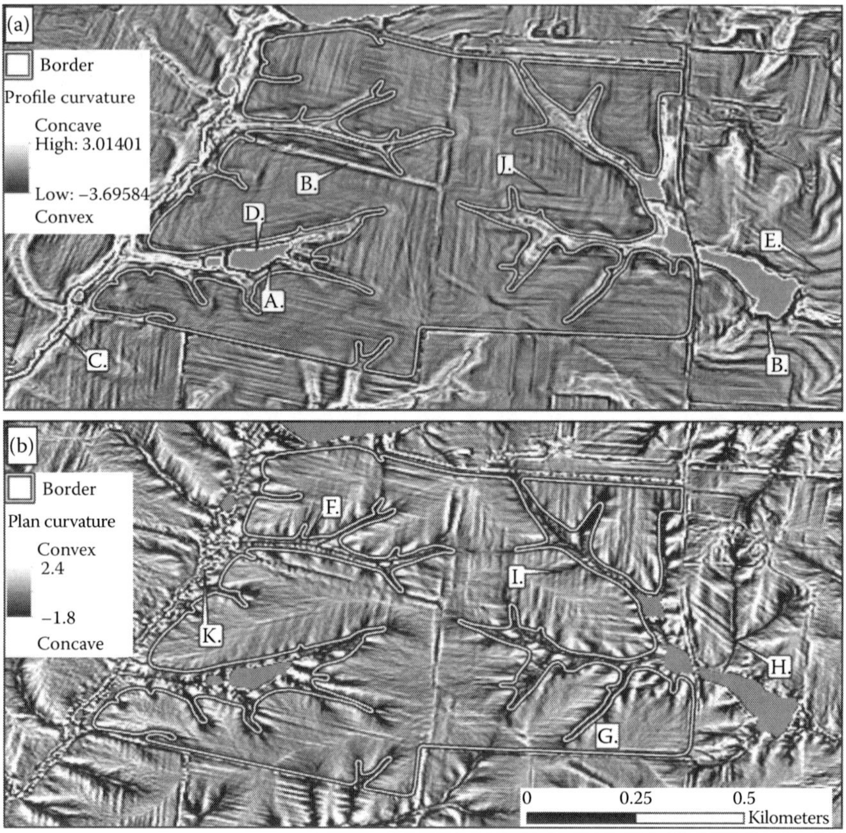

FIGURE 2.7 Maps of (a) profile and (b) plan curvature with field border. Points of interest include the following: ponds (A); old field borders (B); adjacent to streams (C); terrace contours (E); and erosion channels (F and G); area with larger negative profile curvature value (H); noise in plan and profile curvature maps (J and K).

field visit for verification. In some streams, a highly specific catchment area map appeared to be discontinuous (Point E) for two reasons. First, a multidirectional flow model was used, and catchment area values decreased as streams widened downstream and flow became spread out over grid cells. Secondly, those raster values with flow-accumulation values that exceed a threshold value were depicted (rather than actual area values); so the areas where the streams widened and narrowed that also had contributing area values nearly equal to the threshold appeared to be broken.

A proof-of-concept website for the entire state of Kentucky (Neelakantan et al. 2013) used these same techniques (i.e., TauDEM software with D∞ flow direction algorithm) to calculate flow accumulation; however, there were key differences. First, they used a supercomputer rather than a personal computer, establishing that these procedures can be conducted not only at the field scale but also at a statewide scale or finer. Additionally, they calculated terrain attributes using 10-m USGS data rather than LiDAR elevation data because USGS data provide useful information

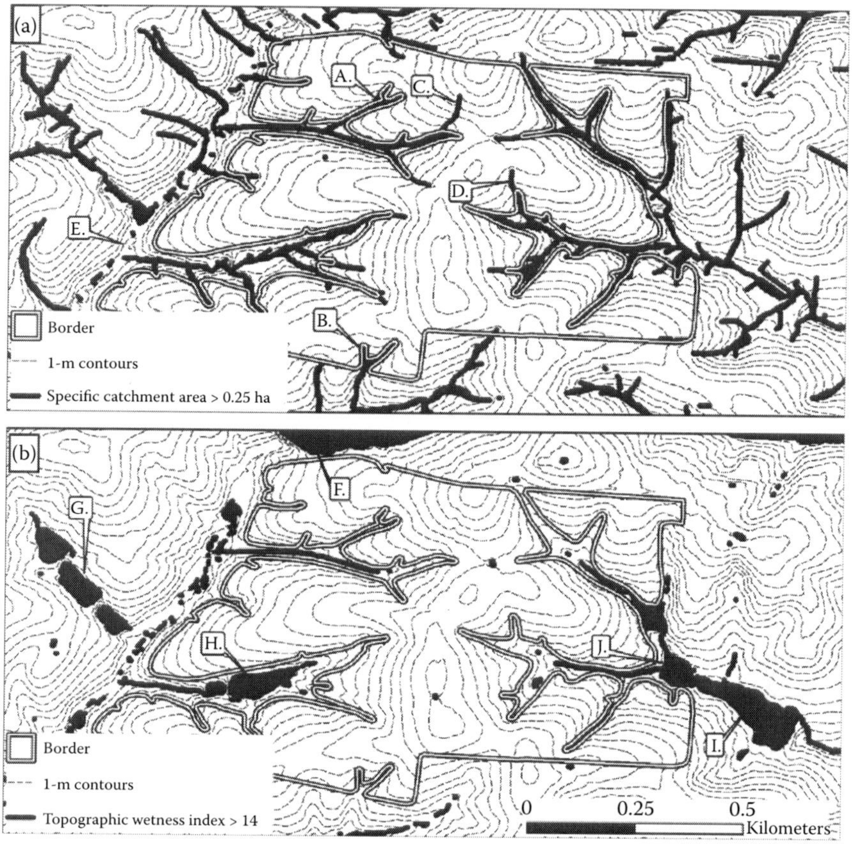

FIGURE 2.8 Maps of (a) specific catchment area > 0.25 ha and (b) topographic wetness index > 14. Both maps have overlays of 1-m contours and field borders. The association between upslope contributing area maps and grassed waterways is good at points A and B but imperfect at C and D; a discontinuous high specific catchment area is seen at point E. Points F, G, H, and I are ponds; high specific catchment area values at point J is a depressional bottom land soil (i.e., Nolin soil series).

about the presence or absence of eroded channels, even though they do not precisely locate them nor as accurately define their boundaries (Pike et al. 2010, 2012). This website also used open-source hybrid-cloud architecture to display these data, demonstrating that computer techniques exist that can be used to disseminate conservation planning information to wide audiences.

The topographic wetness threshold map related well with ponds (Figure 2.8, Points F, G, H, and I) and soils with water limitations (Nolin, Point J). Conservation planners could use upslope contributing area threshold maps to identify areas that are prone to concentrated flow where waterways may be needed. Different thresholds could be used to create topographic wetness maps, identifying soils with low infiltration rates that are prone to ponding. These areas would likely benefit from split nitrogen applications to reduce denitrification losses and could be targeted for tile drainage. It will be challenging for planners and GIS analysts to identify the best site-specific thresholds for these maps because concentrated flow and saturation depend on factors that vary spatially including saturated hydraulic conductivity, texture, internal drainage, the presence of restrictive soil layers, and vegetation. Furthermore, the choice of terrain modeling procedures (e.g., flow direction model chosen, use of filters) can impact the absolute values of terrain attributes. To establish the best site-specific thresholds, planners should view terrain attribute data with leaf-off (bare soil) imagery and examine the terrain data in the field with producers and local soil science experts.

2.3.1 Slope–Crop Yield Relationships

Slope gradient was significantly and negatively associated with corn yield in 2010 but not with soybean yield in 2012. This slope–corn relationship determined for 2010 (Figure 2.9) was

$$\text{Corn yield} = 13.9 - 0.486 \times \text{slope gradient}. \tag{2.8}$$

This equation is expressed in the form of Equation 2.5 where $B_0 = 13.9$ and $B_1 = -0.486$. The relationship between slope and yield was substantial with a 0.486 Mg (7.75 bu) corn yield reduction per unit percent slope gradient. The correlation coefficient ($r = -0.43$) indicated that the slope only explained approximately 18% of the variability in corn grain yield, which was similar to that found by others. For example, an Indiana and Illinois corn-and-soybean study found significant correlations between yield and slope in 8 out of 28 site-years ranging between −0.10 and −0.42 (Kravchenko and Bullock 2000). In a corn, soybeans, wheat, and milo study, there were significant correlations in eight out of nine site-years that ranged between

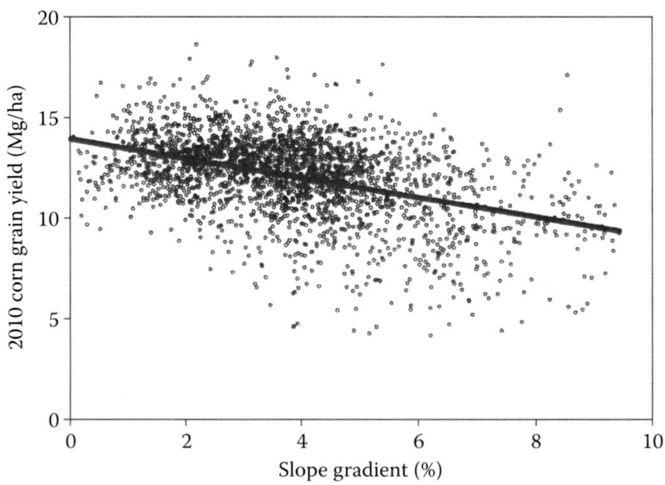

FIGURE 2.9 The relationship between interpolated (7.6-m grids) corn grain yield-and-slope values in 2010. The regression analysis for the line is given by Equation 2.8. The slope values were calculated from LiDAR 1-m DEMs obtained from the KYAPED.

−0.04 and −0.46 (Kitchen et al. 2003). Correlations in yield studies are often low, and biologically significant factors are often found to be statistically nonsignificant because yield is a function of numerous factors that vary spatially and temporally (e.g., nutrients, pH, water, pests, and soil structure). Moreover, yield monitor data can be considerably noisy due to both yield monitor and GPS errors.

Slope gradient often has a negative relationship with yield because more sloping soils can have truncated soil profiles, poorer soil structure, reduced water storage, and less plant-available nutrients, resulting in less vigorous plant growth. Farming eroded soils is a concern for soil conservation because with reduced plant growth, less plant biomass is returned to soils, resulting in less organic matter formation. With less organic matter, plant biomass, and residues for protection, soils become more susceptible to erosion.

While several conservation practices such as conservation tillage, contouring, terracing, or cover crops could be considered for sloping fields, it may be more profitable to remove steeper areas from production depending on the strength and consistency of the yield–slope relationship over time, the future price of corn, and future variable and fixed costs. The break-even slope for the 2010 corn yield in this study was given as

$$S_{\mathrm{BE}} = \frac{(C - P \times 13.9)}{P \times -0.486}, \tag{2.9}$$

This relationship depicted graphically for different crop prices and costs of production (Figure 2.10) demonstrates that the slope gradient above, which is no longer profitable to grow crops, would have increased with greater grain prices and lower

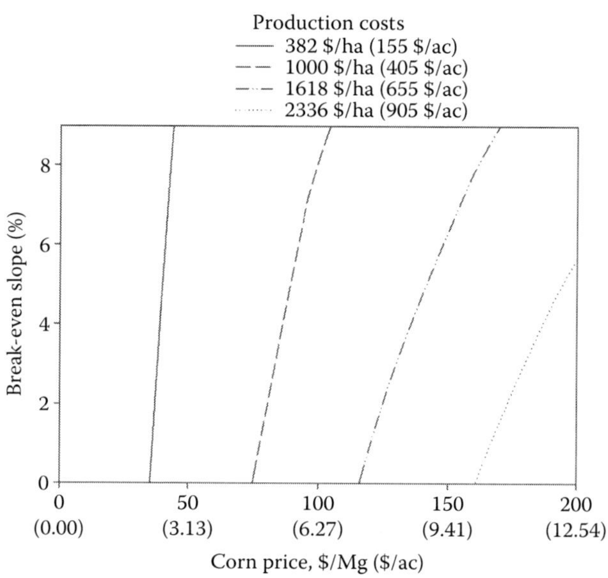

FIGURE 2.10 Break-even slope versus price of corn at four production costs as described in the legend.

production costs. For example, it would not have been economical to farm slopes greater than 6% for this field in 2010 if the cost of production were more than $1618/ha ($655/ac) or the price of corn dropped below $153.72/ Mg ($9.64/bu). If this relationship was understood across multiple fields and sites, farmers could potentially use it for land use decisions. Furthermore, such a model could help producers determine how much to reduce input costs on steeper land to make these areas more profitable. Growers could also use marketing tools such as futures to reduce the uncertainty of S_{BE} estimates.

2.4 SUMMARY AND CONCLUSIONS

This chapter demonstrated how the geospatial analysis of soil, imagery, and terrain data could be used for conservation planning. Specifically, soil surveys should be used to assess land potential and limitations by creating GIS maps that display these land assessments (e.g., land capability class and subclasses, soil properties). Vegetation indices from publically available NAIP imagery can be utilized to evaluate the status of vegetation in grassed waterways and across agricultural fields. High-resolution imagery and LiDAR-derived hillshade, plan curvature, and specific catchment area maps can be used as tools to assist in determining where grassed waterways in agricultural fields could be placed to control ephemeral gully erosion in agricultural fields. Profile curvature maps can help identify sensitive areas near streams, ponds, and along terraces where conservation buffers may be required. Accurate slope gradient maps could be used to determine where to place cover crops within fields. Slope maps can also be used in combination with yield maps, and regression and economic modeling can be used to determine which slopes are too steep to farm economically. The terrain datasets used for these analyses were obtained with LiDAR, which is not yet available in many areas of the United States and the world. Lesser-quality DEMs are available for many areas including USGS DEMs. Readers will be able to compare the LiDAR results from this chapter with the USGS dataset results used to create the step-by-step exercise provided the online exercise Appendix 2.3.

2.5 HANDS-ON STEP-BY-STEP GUIDES PROVIDED IN ONLINE APPENDICES

Several online exercises have been prepared to help students obtain and analyze public terrain data and remote-sensing imagery. Online Appendices 2.1 through 2.4 can be found on the CRC website at http://www.crcpress.com/prod/isbn/9781439867228.

2.5.1 Obtaining Digital Elevation and Imagery from the National Map (Online Appendix 2.1)

This appendix provides step-by-step instructions for readers to obtain 10-m DEMS and NAIP imagery from the USGS National Map web portal. These datasets will be required to complete the exercises in online Appendices 2.2 and 2.3.

2.5.2 Vegetation Indices Calculated with Four-Band NAIP Imagery (Online Appendix 2.2)

This appendix provides step-by-step instructions for automating the calculation of vegetation indices (NDVI, SNDVI, and GNDVI) from NAIP imagery using ArcGIS. The emphases of this and online Appendix 2.3 are on automating GIS tasks with ArcGIS ModelBuilder. Example diagrams from online Appendix 2.2 are provided in Figure 2.11. The top diagram in this figure is a screenshot from ArcMap illustrating some of the detailed steps that required to create maps of vegetation indices using ArcGIS. The bottom diagram is a screenshot from ModelBuilder, which is used to illustrate the complex tasks required to automate the calculation of vegetation indices. A screenshot of example output that readers can create by following the instructions provided in online Appendix 2.2 is given in Figure 2.12. This figure was created by combining multiple maps in the layout view of ArcGIS.

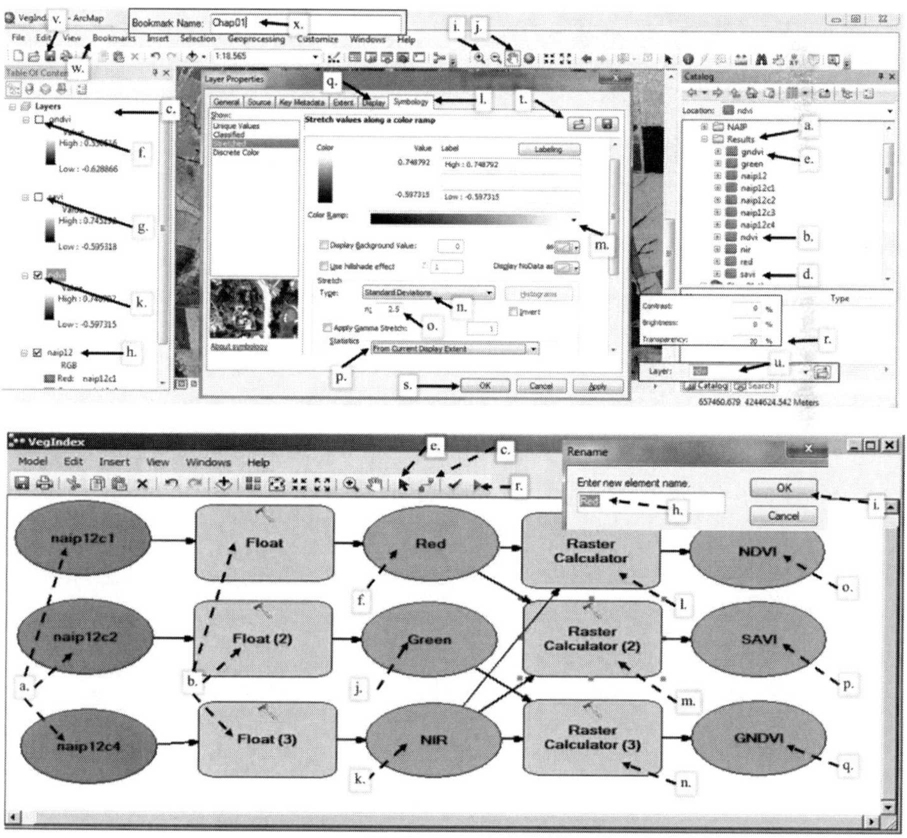

FIGURE 2.11 Screenshots from online Appendix 2.2 used to help describe for readers the steps that can be used to create maps of vegetation indices using ArcGIS. The top image is a screenshot from ArcMap and the bottom from ModelBuilder. The inset annotations in both figures are referenced in the appendix instructions.

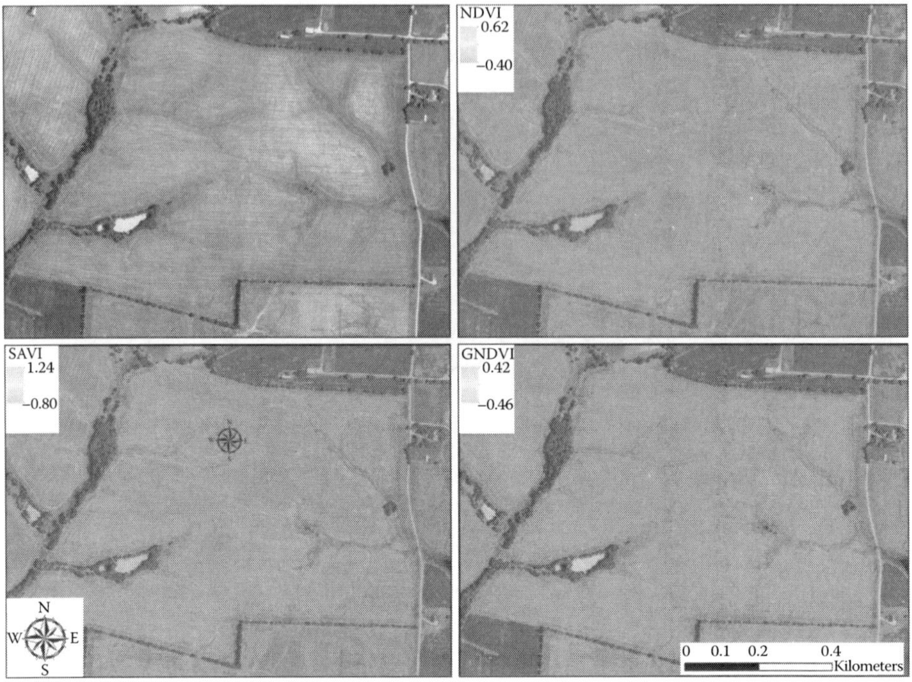

FIGURE 2.12 **(See color insert.)** Example output from three different methods of calculating vegetative index (NDVI, SAVI, and GNDVI) using USGS NAIP imagery in ArcGIS. Detailed instructions are given in the online exercise.

2.5.3 Terrain Analysis with USGS DEMs (Online Appendix 2.3)

This appendix provides step-by-step instructions for automating the creation of hillshade, contour, and terrain attributes (i.e., slope, aspect, curvature, flow-accumulation threshold, and topographic wetness threshold) maps from the publically available 10-m USGS elevation data. A screenshot from this appendix of the automated ModelBuilder analyses used to calculate terrain attributes is displayed in Figure 2.13. This analysis uses tools from ArcGIS and third-party software TauDEM.

2.5.4 Terrain Analysis with LiDAR Data (Online Appendix 2.4)

This appendix provides step-by-step instructions for readers to download LiDAR point clouds from a USGS web portal and to create DEMs, digital surface maps, and intensity maps. A screenshot of a 3-D rendering of a LiDAR-derived hillshade map overlain by elevation contours from online Appendix 2.4 is presented in Figure 2.14.

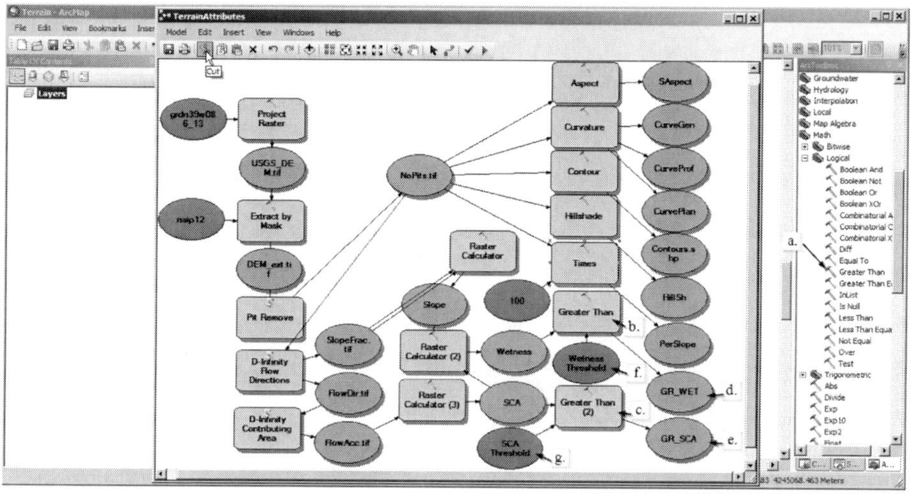

FIGURE 2.13 Screenshots from online Appendix 2.3 used to help describe for readers the steps that can be used to conduct terrain analysis using ArcGIS and TauDEM. The inset annotations are referenced in the appendix instructions.

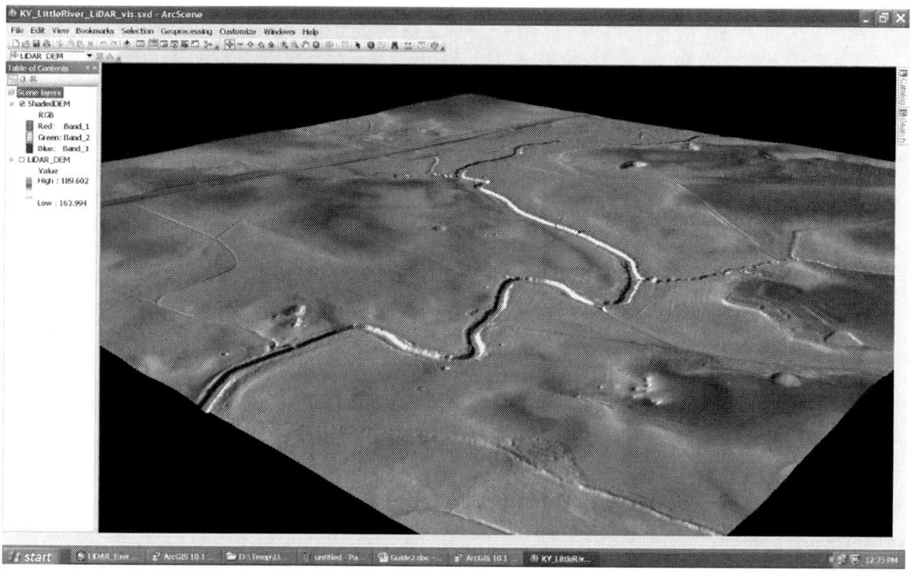

FIGURE 2.14 **(See color insert.)** 3-D rendering of a LiDAR-derived hillshade map overlain by elevation contours calculated from the USGS DEM data using ArcGIS. Detailed instructions are given in the online exercise.

2.6 EXAMPLE: CONSERVATION ASSESSMENT ANALYSIS REPORT

To reinforce the concepts described in this chapter, readers and classroom students should create a conservation assessment report for an area of interest following the hands-on exercises provided in the appendices. Paper copies or digital images should be exported from ArcGIS for the report.

1. First, identify an area of interest for a location in the United States that is not larger than 100 ha. Describe the location and explain the reason why this area was chosen. Download the 10-m USGS DEMs and the 1-m NAIP four-band imagery (see online Appendix 2.1 for detailed instructions). If these datasets are not available for your area of interest, find another area where they are available. Include in your report information about your study location (e.g., why you picked this area, unique attributes, etc.) and a table with the names and file locations of the datasets. Update this table as you learn more about your data (e.g., projections).
2. Visit the Web Soil Survey (http://websoilsurvey.sc.egov.usda.gov), and find the link "How to use Web Soil Survey." Download and study this information. Use these instructions to find and download a shapefile of the soils in your area of interest. Include in your report a soil map created with ArcGIS, a table with land capability class and subclass data and other relevant information (e.g., drainage class, available water storage), and a description of conservation interpretations of the table and maps.
3. Estimate vegetative indices with four-band NAIP imagery according to the instructions provided in online Appendix 2.2. Include in your report copies and descriptions of color imagery and vegetative index maps.
4. Create terrain attributes (i.e., slope, aspect, plan and profile curvature, specific catchment area threshold maps, and topographic wetness threshold), hillshade, and topographic contour maps (see online Appendix 2.3 for instructions). Include in your report copies of each map, and describe conservation interpretations for each.
5. If LiDAR data are available for the area of interest, advanced students should follow the instructions provided in online Appendix 2.4 for downloading point cloud data and creating maps. In your report, include 3-D hillshade and elevation intensity maps and conservation interpretations for each.
6. Finally, readers should summarize the information in the report and draw conclusions about the potential to use geospatial data for conservation planning and management in their area of interest.

ACKNOWLEDGMENTS

The authors especially thank Jim, Mike, and Seth Ellis and Don Griffin from Worth and Dee Ellis Farms for providing the datasets used in the case study. We also express our appreciation to the Brazilian government for funding the students who worked on this project. We express our gratitude to the University of Kentucky, College of

Agriculture, for supporting this work. Finally, we acknowledge the Conservation Innovation Grants program grant funding through the Kentucky NRCS office.

REFERENCES

Beven, K.J., and M.J. Kirkby. 1979. A physically based, variable contributing area model of basin hydrology/Un modèle à base physique de zone d'appel variable de l'hydrologie du bassin versant. *Hydrological Sciences Bulletin* 24:43–69.

Delgado, J.A., R. Khosla, and T.G. Mueller. 2011. Recent advances in precision (target) conservation. *Journal of Soil and Water Conservation* 66:167A–170A.

Flynn, E.S., C.T. Dougherty, and O. Wendroth. 2008. Assessment of pasture biomass with the normalized difference vegetation index from active ground-based sensors. *Agronomy Journal* 100:114–121.

Fox, D.M., and R.B. Bryan. 2000. The relationship of soil loss by interrill erosion to slope gradient. *Catena* 38:211–222.

Gitelson, A.A., Y.J. Kaufman, and M.N. Merzlyak. 1996. Use of a green channel in remote sensing of global vegetation from EOS-MODIS. *Remote Sensing of Environment* 58:289–298.

Halich, G. 2013. *Corn and Soybean Budgets (Central KY 2013)*. University of Kentucky, Lexington, KY. Available at http://www2.ca.uky.edu/agecon/index.php?p=29 (accessed May 8, 2014).

Huang, J., D. Chen, and M.H. Cosh. 2009. Sub-pixel reflectance unmixing in estimating vegetation water content and dry biomass of corn and soybeans cropland using normalized difference water index (NDWI) from satellites. *International Journal of Remote Sensing* 30:2075–2104.

Huete, A.R. 1988. A soil-adjusted vegetation index (SAVI). *Remote Sensing of Environment* 25:295–309.

Jackson, R.D., and A.R. Huete. 1991. Interpreting vegetation indices. *Preventive Veterinary Medicine* 11:185–200.

Jenny, H. 1941. *Factors of Soil Formation*. McGraw-Hill, New York.

Kitchen, N.R., S.T. Drummond, E.D. Lund, K.A. Sudduth, and G.W. Buchleiter. 2003. Soil electrical conductivity and topography related to yield for three contrasting soil–crop systems. *Agronomy Journal* 95:483–495.

Klingebiel, A.A., and P.H. Montgomery. 1961. *Land-Capability Classification. Agriculture Handbook No. 210*. Soil Conservation Service USDA, Washington DC. Available at http://www.nrcs.usda.gov/Internet/FSE_DOCUMENTS/nrcs142p2_052290.pdf (accessed May 8, 2014).

Kravchenko, A.N., and D.G. Bullock. 2000. Correlation of corn and soybean grain yield with topography and soil properties. *Agronomy Journal* 92:75–83.

Langdale, G.W., R.L. Blevins, D.L. Karlen, D.K. McCool, M.A. Nearing, E.L. Skidmore, A.W. Thomas, D.D. Tyler, and J.R. Williams. 1991. Cover crop effects on soil erosion by wind and water. In *Cover Crops for Clean Water*, ed. W.L. Hargrove, 15–22. Soil and Water Conservation Society, Ankeny, IA.

Mokma, D.L., and M.A. Sietz. 2002. Effects of soil erosion on corn yields on Marlette soils in south-central Michigan. *Journal of Soil and Water Conservation* 47:325–327.

Neelakantan, S., C. Bumgardner, T.G. Mueller, B. Mijatovic, G.D. Faleiros, M. Ferguson, and S. Crabtree. 2013. *Kentucky Data Cloud for Soil Management*. University of Kentucky, Lexington, KY. Available at http://128.163.190.221/cig/ (accessed July 8, 2014).

Perroy, R.L., B. Bookhagen, G.P. Asner, and O.A. Chadwick. 2010. Comparison of gully erosion estimates using airborne and ground-based LiDAR on Santa Cruz Island, California. *Geomorphology* 118:288–300.

Pike, A.C., T.G. Mueller, E.A. Rienzi, S. Neelakantan, B. Mijatovic, A.D. Karathanasis, and M. Rodrigues. 2012. Terrain analysis for locating erosion channels: Assessing LiDAR data and flow direction algorithm. In *Research on Soil Erosion*, ed. D. Godone, pp. 45–63. InTech, Rijeka, Croatia. Available at http://www.intechopen.com/books/research-on-soil-erosion/terrain-analysis-for-locating-erosion-channels-assessing-lidar-data-and-flow-direction-algorithm (accessed November 15, 2014).

Pike, A.C., T.G. Mueller, A. Schörgendorfer, J.D. Luck, S.A. Shearer, and A.D. Karathanasis. 2010. Locating eroded waterways with United States Geological Survey elevation data. *Agronomy Journal* 102:1269–1273.

Rouse, J.W., R.H. Haas, J.A. Schell, and D.W. Deering. 1974. Monitoring vegetation systems in the Great Plains with ERTS. In *Proceedings Earth Resources Technology Satellite*, eds. S.C. Freden, E.P. Mercanti, and M.A. Becker, pp. 309–317. Third Earth Resources Technology Satellite-1 Symposium-Volume I: Technical Presentations. NASA SP-351, Washington, DC.

Soil Conservation Service. 1980. *Soil Survey of Shelby County, Kentucky*. USDA, NRCS, Fort Worth, TX.

Tarboton, D.G. 1997. A new method for determination of flow directions and upslope areas in grid digital elevation models. *Water Resources Research* 33(2):309–319.

Thenkabail, P.S., A.D. Ward, and J.G. Lyon. 1994. Landsat-5 Thematic Mapper models of soybean and corn crop characteristics. *International Journal of Remote Sensing* 15:49–61.

USDA-NRCS. 1993. *Soil Survey Manual. Agriculture Handbook No. 18*. U.S. Government Printing Office, Washington, DC.

Vrieling, A. 2006. Satellite remote sensing for water erosion assessment: A review. *Catena* 65:2–18.

Yin, X., and M.A. McClure. 2013. Relationship of corn yield, biomass, and leaf nitrogen with normalized difference vegetation index and plant height. *Agronomy Journal* 105:1005–1016.

3 Identification and Analysis of Ravines in the Minnesota River Basin with Geographic Information System

Shannon Belmont, David J. Mulla, and Joel Nelson*

CONTENTS

* Contact author: mulla003@umn.edu.

EXECUTIVE SUMMARY

Ravines are common features within the Minnesota River Basin (MRB) and are the natural product of a landscape adjusting to disequilibrium caused by a regional baselevel fall 11,500 years before present. Overall sediment contribution by ravines to the Minnesota River (MNR) must be understood to mitigate sediment loading to the MNR and focus remediation strategies. This study defines the morphometry of MRB ravines and their relationship with characteristics of the upland contributing area (UCA) upslope of the ravine. Boundaries for 70 study ravines in the MRB and their associated UCA were digitized. Primary and secondary attributes such as area, slope, relief, and stream power index (SPI) values were estimated for each ravine and their associated UCA. Ravine area, volume, and relief were strongly correlated with ravine slope, length, and SPI. Ravine width-to-depth ratios were negatively correlated with the UCA mean slope. Ravine area was strongly correlated with UCA, suggesting that increasing volumes of water discharged from the UCA result in larger ravines that potentially deliver more sediment to the MRB than smaller ravines. Ravines with larger UCA could be targeted for conservation practices such as cover crops or perennial grass plantings to reduce discharge of water and sediment loading to the ravine.

KEYWORDS

Erosion, ravine, sediment loss, soil conservation

ABBREVIATIONS

CTI: compound topographic index; DEM: digital elevation model; DOQ: digital orthoquad; MRB: Minnesota River Basin; SPI: stream power index; UCA: upland contributing area

3.1 INTRODUCTION

The Minnesota River (MNR) covers 44,000 km^2 in Minnesota, with small portions in North and South Dakota. Less than a million people live in the Minnesota River Basin (MRB), and approximately 75% of its area is in cultivated annual row crops (Mulla and Sekely 2009). The MNR is listed by the American Rivers Council as 1 of the 20 most polluted rivers in the United States (Kober et al. 2008). The Minnesota Pollution Control Agency (MPCA) has listed 18 reaches of the MNR as impaired for turbidity (MPCA 2005) under Section 303d of the Clean Water Act (EPA 2009). In addition, the MNR contributes over 80% of the sediment that enters Lake Pepin (Kelley and Nater 2000; Sekely et al. 2002; Kelley et al. 2006; Engstrom et al. 2009). Lake Pepin is a natural lake in the Mississippi River downstream of Minneapolis and St. Paul, and is impaired for both sediment and phosphorus. These identified impairments have launched a multifaceted effort to understand the hydrologic and geomorphic processes at work in the MRB in order to identify the sediment sources within the watershed and focus on efforts to reduce turbidity. Ravines are common features within the MRB and are the natural product of a landscape that is adjusting to disequilibrium caused by a regional decline of the baselevel 11,500 years bp (Sidorchuk 2006). Overall sediment contribution by ravines to the MNR must be understood to mitigate sediment loading to the MNR and to focus on remediation strategies.

For the purposes of this chapter, we follow the definition by Bates and Jackson (1984) whereby the term *ravine* refers to a small, narrow, and deep depression that is smaller than a valley and larger than a gully. Gully or ravine initiation is the result of nonequilibrium conditions in a landscape (Simon and Darby 1999). The main external drivers of gully or ravine formation include the following: tectonic uplift; baselevel lowering; changes in hydrology; loss of vegetation; and/or anthropogenic disturbances to the watershed such as increased impervious surfaces, infiltration reductions due to soil compaction from machinery, or conversion to turf grass lawns (Schumm 1999; CCMR 2004). Baselevel lowering at the end of the last glaciation is the primary factor that controls ravine initiation in the MRB. More recent anthropogenic alterations to the hydrology, including installation of tile drain outlets in ravines, may be driving a new era of dynamic geomorphic instability.

Ravine size is one of the most important attributes to quantify for the purposes of understanding ravine sediment production. There are many ways to evaluate ravine size, including planform area (2-D), actual surface area (3-D), volume, perimeter, extent of branching, longest channel length, and total channel length. There are also relationships that combine attributes to quantify ravine size, such as drainage density (total channel length divided by area) and the longest channel length divided by the area (which estimates width).

The potential energy gradient, or severity of incision, is a second characteristic of high interest. The incision extent may relate to the age of the ravine, the location within the MRB, the proximity to the baselevel fall, the current erosional forces at work within the ravine, the local gradient surrounding the ravine, the erosional stream power coming from the upland contributing area (UCA), or simply the regolith of the ravine. Incision can be analyzed by several different factors, such as volume, slope, the ratio of 2-D area to volume, manually derived ravine relief or elevation change from the uplands surrounding the ravine to the channel elevation nearest the mouth of the ravine, and automatically derived ravine relief (maximum minus minimum elevations from within the ravine-digitized boundary).

3.2 OBJECTIVES

Ravines are common features within the MRB due to the baselevel fall and the resulting disequilibrium of the landscape. In order to mitigate sediment loading to the MNR and to focus on remediation strategies, the main sediment sources to the river must be understood and quantitatively constrained. There is a large effort underway to identify and apportion the potential contribution of sediment from eroding streambanks, failing bluffs, incising gullies and ravines, topsoil erosion from the upland agricultural fields, and in-channel processes to create a sediment budget for the system as a whole. The objective of this chapter is to document methods for the characterization of ravine morphometry in order to begin to understand their contribution to sediment pollution in the MRB.

3.3 MATERIALS AND METHODS

3.3.1 Ravines Selected for Study

The MRB is an area of approximately 4.4 Mha composed of 12 major subwatersheds that are located in the lower half of the state of Minnesota in the northern United States (Figure 3.1). To understand the dynamics of ravine formation and its impact on sediment loss to the Minnesota River, 70 ravines were selected for study based on the field. This is approximately 1% of the total number of ravines in the MRB by area. The initial selection process consisted of creating an evenly spaced grid over the entire MRB using ArcGIS (ESRI 1994), with each grid cell representing 1/64 of a quadrangle. Where possible, one ravine was selected per grid cell, allowing even distribution of study ravines over the entire watershed. Field validation of geographic information system (GIS) predictions was done in June 2007 by a team that visited each of the 70 selected ravines; measured ravine width and depth and angle of side slope; and took notes on incoming tile lines, culverts, and the overall vegetative state. Field validation confirmed that 90% of the selected ravines existed.

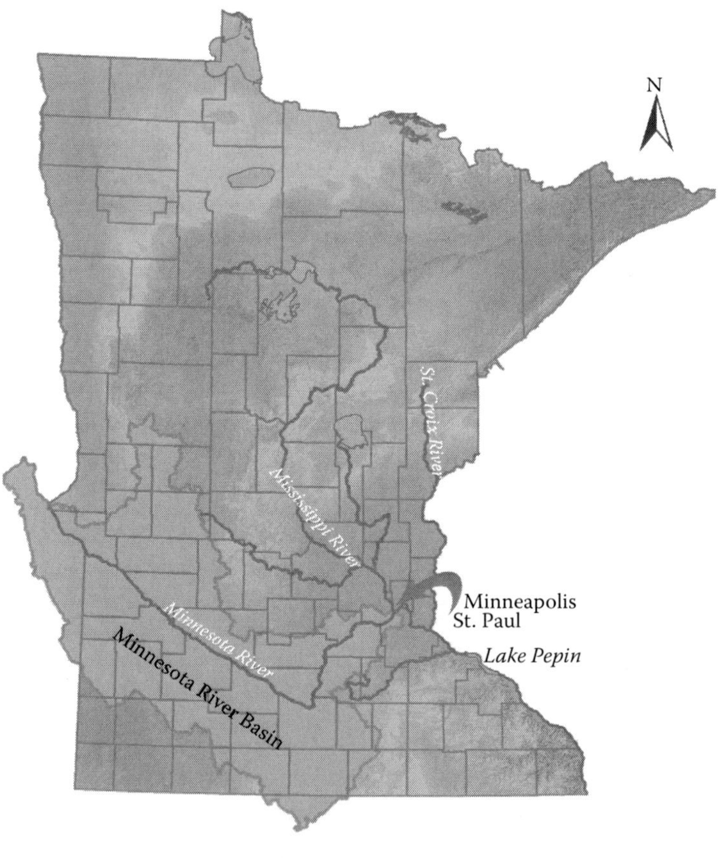

FIGURE 3.1 Study location map showing MRB in Minnesota.

3.3.2 Ravine Morphometry

3.3.2.1 Digitizing Ravines and Delineating Watersheds

Ravine boundaries in the MRB were defined with 3-m DOQs using the USGS 30-m DEM (USGS National Elevation Dataset 2012) as a semitransparent drape over the landscape to help define specific areas where forested bluffs obscured the direction and terminus of ravine features. All 70 study ravines were digitized using this methodology.

Pour point outlets for each ravine were chosen manually using the DOQ overlain with a semitransparent flow accumulation grid calculated from the 30-m DEM. The point file was required as an input for the ArcGIS 9.3 watershed delineation process described by Balco (2001), which was used to delineate the catchment area or the UCA for each ravine. The watershed delineation script also required a DEM of the region and a flow direction grid, which was calculated from the 30-m DEM using the D8 algorithm that is utilized by ArcGIS. The resulting delineated watersheds included the ravine area. The digitized ravine polygon shapefile was used to clip out the ravine portion of DEMs so that attribute extraction and analysis could be done on the ravines and their contributing areas separately.

3.3.2.2 Terrain Attribute Extraction

Approximately 130 characteristics were collected for the ravines and their corresponding catchment areas using ArcGIS and datasets from the Minnesota DNR (2012), the Natural Resources Conservation Service's (NRCS) Soil Survey Geographic Database (USDA-NRCS 2014), and shapefiles for agroecoregions of the MRB (Hatch et al. 2001).

Plan and profile curvature were calculated for the ravines and for each ravine's center channel. Channel slopes were estimated by averaging the cell slope values that are calculated for the grid of the isolated centerline cells of each ravine. Ravine width was estimated by dividing the ravine area by the length of the ravine, estimated by the length of the longest channel. Using flow accumulation to isolate the ravine centerline cells, a series of calculations were made to determine the channel slope, stream power index (SPI), and plan and profile curvature to document the character of the main ravine channel. A manually extracted ravine relief attribute was calculated as the difference in the general elevation taken from the uplands surrounding the ravine to the channel or valley bottom nearest the ravine outlet. This metric was needed because the ravine pour point was not always placed at the valley bottom but was sometimes set back a distance from the receiving drainage channel.

Secondary terrain indices are calculated from linear combinations of two or more primary attributes and often provide a better opportunity to describe landscape processes than the patterns described by primary terrain attributes (Wilson and Gallant 2000). Two such indices, SPI and the compound topographic index (CTI), were calculated. Both combine the slope measured in each cell with the area contributing flow to that cell and are calculated as

$$\text{CTI} = \ln(A_s/\tan \text{slope}) \tag{3.1}$$

$$\text{SPI} = A_s * \tan \text{slope} \tag{3.2}$$

where A_s is the specific catchment area, and SPI and CTI are hydrologically based indices that characterize water movement on a landscape. They are derived from a primary attribute slope and a specific catchment area (Moore et al. 1991, 1993). Stream power increases as the gradient and the amount of water contributed by the upland areas both increase. CTI is also referred to as the steady-state wetness index (Moore et al. 1991) or the topographic wetness index (Beven and Kirkby 1979). It expresses the potential for a landscape to be wet in addition to helping to describe water movement on the landscape (Yang et al. 2007).

The specific catchment area is determined from the upslope area contributing to each cell in the grid. The UCA is the product of the number of cells draining through each cell and the cell area. The specific catchment area is the UCA divided by the width of the cell (Tarboton 1997). This is equivalent to flow accumulation in ArcGIS (ESRI 1994) or simply the number of cells contributing to any other cell. Characteristics of the UCAs upslope of ravines were calculated and/or measured using techniques similar to those that are used for the analysis of ravines.

3.3.3 Statistical Analysis

Ravine and upland attribute data were used to calculate Pearson correlation (*r*) coefficients and to create scatterplots and boxplots in Statistica, version 8.0 (StatSoft, Inc. 2008) and Excel (Microsoft 2003).

Attribute relationships were screened using scatterplots and Pearson correlation coefficient matrices for 91 ravine and 41 UCA characteristics. The results were analyzed to determine which attributes had the most influence on each other and on ravine size in particular. After screening, several attributes were dropped as insignificant, and a smaller, more manageable set of attributes was carried forward for further analysis. Descriptive statistics, scatterplots, and Pearson correlation (*r*) coefficients were calculated for the smaller, focused set of attributes: 19 for the ravines and 9 for the UCA (Table 3.1).

TABLE 3.1
Ravine and UCA Attributes

Ravine Attributes		UCA Attributes
2-D area	% area slope >30%	2-D area
3-D area	Slope mean	Longest drainage path
Volume	SPI mean	Relief
Volume/2-D area ratio	Longest channel length	Slope range
Manual relief	Total channel length	Slope mean
Automatic relief	Ksat	Ksat
Width estimate	% area C_{pro} nonzero	SPI mean
W/D ratio	Channel slope mean	CTI mean
Drainage density	Channel SPI mean	% area CTI 9–13
% area slope <15%		

3.4 RESULTS AND DISCUSSION

3.4.1 Digitized Ravines and Delineated Watersheds

Planform areas of the 70 test ravines were digitized from the 30-m DOQ with a semi-transparent overlay of the DEM to clarify feature boundaries (Figure 3.2). UCAs were also delineated for each of the test ravines (Figure 3.2). In general, UCAs were larger than ravines. The predominant land use in the ravines was forest, whereas the predominant land use in UCAs was typically a corn-and-soybean crop rotation.

3.4.2 Ravine Primary Attributes

Ravine 3-D areas range from 7745 to 1,267,816 m^2, with a mean of 171,453 m^2 and a median of 88,558 m^2 (Table 3.2). Ravine volumes range from 35,240 to 25,629,150 m^3 with a mean of 2,803,300 m^3 and a median of 1,057,230 m^3. Manual ravine relief is the difference in elevation from the uplands surrounding the head of the ravine to the channel elevation that is nearest the ravine mouth. This is an attribute manually obtained from the 30-m DEM. The values range from 12 to 71 m with a mean of

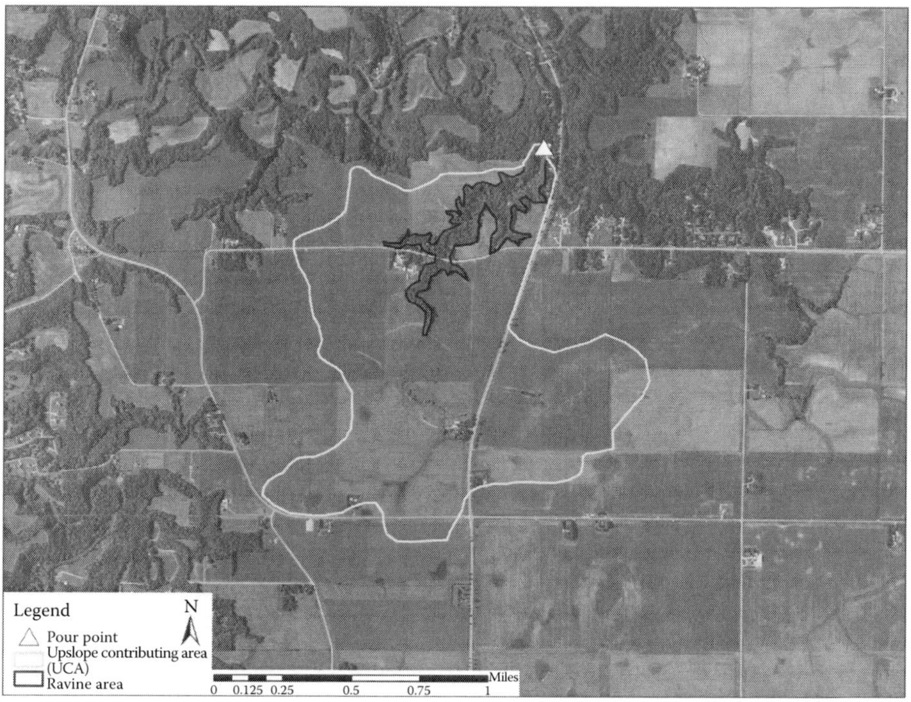

FIGURE 3.2 Pour point, digitized ravine boundary, and UCA for an example ravine draped over a DOQ photo.

TABLE 3.2
Summary of Descriptive Statistics for Primary Ravine Attributes

	3-D Area m^2	Volume m^3	Auto Relief m	Manual Relief m	Longest Channel m	Total Channel m	Mean Slope %
Mean	171,453	2,803,301	34	36	834	1614	16
Median	88,558	1,057,229	35	38	613	910	17
Standard deviation	269,181	5,107,111	16	15	763	2229	5
Minimum	7514	35,237	7	12	160	160	7
Maximum	1,267,816	25,629,151	68	71	3965	11,975	31
Confidence level (95.0%)	64,184	1,217,747	4	4	182	532	1

36 m and a median of 38 m. Auto ravine relief values, calculated automatically in ArcMap by subtracting the minimum elevation from the maximum elevation within the ravine, range from 7 to 68 m, with a mean of 34 m and a median of 35 m. The longest channel length and the total channel length were both measured manually within each ravine using ArcMap measurement tools. The minimum value for the longest channel length was 160 m, and the maximum value was 3965 m with a mean of 834 m and a median of 613 m. The minimum total channel length within the ravines was 160 m, and the maximum total length was 11,975 m, with a mean of 1614 m and a median of 910 m. The slope for each ravine was estimated by averaging the slope values calculated for each cell within the ravine-digitized boundary. Mean ravine slope values range from 7% to 31%, with a mean of 16% and a median of 17%.

The primary attributes of ravines indicate that ravines are, as expected, steep and long and have high relief from top to bottom. Ravines are relatively small (mean area 17 ha), but because of active channel erosion, they are considered as important sources of sediment to the MRB.

3.4.3 Ravine Secondary Attributes

Ravine width estimates range from 35 to 440 m, with a mean of 145 m and a median of 134 m (Table 3.3). Ravine drainage density, which is the total channel length of the ravine divided by the ravine area (multiplied by 100 for reporting purposes), ranges from 0.62 to 2.89 m/m^2, with a mean of 1.21 m/m^2 and a median of 1.12 m/m^2. Ravine mean SPI values range from 0.10 to 240, with a mean of 25.5 and a median of 11.2.

SPI values typically indicate the erosive power of runoff (Galzki et al. 2011). Farmed upland fields in the MRB are classified as being critical source areas of sediment if they have SPI values that range between 0.1 and 10 (Galzki et al. 2011). In comparison, the SPI values for ravines are much larger than those for critical source areas in agricultural fields, indicating a much higher potential for runoff and sediment production per unit area.

TABLE 3.3
Summary of Descriptive Statistics for Secondary Ravine Attributes

	Width m	Drainage Density (*100) m^2/m	Mean SPI
Mean	145.0	1.2	25.5
Median	134.0	1.1	11.2
Standard deviation	74.0	0.5	47.8
Minimum	35.0	0.6	0.1
Maximum	439.0	2.9	239.5
Confidence level (95.0%)	18.0	0.1	11.4

3.4.4 Ravine Descriptive Attributes

The following descriptive attributes assist in understanding the overall character of the ravines. The percent ravine area with a slope less than 15% was calculated for each of the 70 study ravines. Values range from 16% to 100%, with a mean of 55% and a median of 48% (Table 3.4). The percent ravine area with slopes greater than 30% ranged from 0% (there were many ravines with no cells steeper than 30%) to 48%, with a mean of 12% and a median of 9%. The percent area of the ravines with nonzero profile curvature (C_{pro}) ranged from essentially 0% to 47%, with a mean of 20% and a median of 1.4%. The ratio of ravine volume to the 2-D planform area, which describes the vertical downcutting within the ravine, ranged from 3 to 25 m, with a mean of 13 m and a median of 13 m.

TABLE 3.4
Summary of Descriptive Statistics for Ravine Attributes

	Area Slope <15% %	Area Slope >30% %	Area C_{pro} Nonzero %	Volume/2-D Area Ratio m^3/m^2	Channel Slope Mean %	Channel SPI Mean	W/D Ratio
Mean	55	12	20	12.97	10.9	96.0	4.67
Median	48	09	18	12.80	11.0	33.5	4.37
Standard deviation	23	11	11	5.30	3.8	206.2	1.86
Minimum	16	0	0	3.06	0.5	2.0	2.34
Maximum	100	49	47	24.93	20.5	960.5	11.93
Confidence level (95.0%)	5	3	3	1.26	0.9	49.2	0.46

Note: C_{pro} is the profile curvature; SPI is the stream power index; W/D is the width to depth ratio.

Mean channel slopes range from 1% to 21%, with mean and median values of 11%. Ravine channel mean SPI values range from 2 to 961, with a mean of 96 and a median of 34.

3.4.5 Pearson Correlation Coefficients for Ravine Primary and Secondary Attributes

Ravine 3-D area and volume are both correlated to ravine relief (area $r = 0.77$ and volume $r = 0.86$; see Table 3.5) and ravine mean slope (area $r = 0.48$ and volume $r = 0.62$). Ravine area correlates, as it should, to ravine volume ($r = 0.97$). The longest channel length and total channel length, both autocorrelated, have strong r values (area $r = 0.97$ and volume $r = 0.93$ for the longest channel; area $r = 0.97$ and volume $r = 0.92$ for the total channel).

Ravine mean SPI is strongly correlated with both the ravine 3-D area and the ravine volume ($r = 0.66$ and $r = 0.61$, respectively). Ravine drainage density is inversely correlated to the ravine area and volume ($r = -0.65$ and -0.69, respectively). The ravine width, as estimated by the 2-D area divided by the length of the longest stream channel, is strongly correlated to ravine size ($r = 0.92$ area and $r = 0.90$ volume).

3.4.6 Manual Ravine Relief—Primary Attributes

Manually extracted ravine relief measures the change in elevation from the uplands surrounding the ravine head to the channel bottom nearest the ravine mouth. The manual ravine relief measurement correlates strongly with the ravine size attributes ($r = 0.65$ with area, $r = 0.76$ with volume; see Table 3.6). The relationship is stronger

TABLE 3.5
Summary of Pearson Correlation (*r*) Coefficients (with Statistical Significance) for Ravine Area and Volume and Ravine Primary and Secondary Attributes

	Pearson *r* Values	
Ravine Attribute	**3-D Area**	**Volume**
3-D area		*0.97
Volume	*0.97	
Auto relief	*0.77	*0.86
Slope mean	*0.48	*0.62
Longest channel length	*0.97	*0.93
Total channel length	*0.97	*0.92
SPI mean	*0.66	*0.61
Drainage density	*–0.65	*–0.69
Width	*0.92	*0.90

*$p < 0.01$.

TABLE 3.6
Summary of Pearson Correlation (*r*) Coefficients (with Statistical Significance) for Manual Ravine Relief and Ravine Primary Attributes

	r Values
Ravine Attribute	**Manual Relief**
3-D area	**0.65
Volume	**0.76
Volume/2-D area	**0.87
Slope mean	**0.83
Channel slope	0.10
Channel SPI	*0.25

$^*p < 0.05$; $^{**}p < 0.01$.

for the ratio of ravine volume to the 2-D area ($r = 0.87$), which reflects the extent of incision in the ravine. Manual ravine relief is strongly correlated ($r = 0.83$) with the ravine mean slope. In contrast, the ravine relief and the mean SPI are not strongly correlated ($r = 0.25$).

3.4.7 UCA Primary Attributes

UCAs range from 14,386 to 34,309,900 m^2, with a mean of 2,540,416 m^2 and a median of 402,158 m^2 (Table 3.7). Catchment area relief ranges from 3 to 68 m, with a mean of 22 m and a median of 19 m. The longest drainage path length was measured manually for each UCA following the flow lines that are calculated by ArcGIS. The minimum

TABLE 3.7
Summary of Descriptive Statistics for UCA Primary Attributes

	UCA Area m^2	UCA Relief m	UCA Longest Drainage Path m	UCA Slope Mean %
Mean	2,540,416	22	2290	2.76
Median	402,158	19	1332	2.43
Standard deviation	7,323,034	13	3078	1.51
Minimum	14,386	3	305	0.80
Maximum	34,309,900	68	16,132	8.58
Confidence level (95.0%)	1,746,330	3	734	0.36

drainage path length is 305 m and the maximum drainage path length is 16,132 m, with a mean of 2290 and a median of 1332 m. Mean slope values for the contributing areas range from 0.80% to 8.6% with a mean average slope of 2.8% and a median of 2.4%.

In comparison with ravines, the UCAs are much larger (245 ha vs. 17 ha). The UCAs also have much flatter slopes and overall relief than ravines, as expected. Ravines should not be considered as geomorphic features that are unconnected to the surrounding landscape. Indeed, the upslope contributing area has a strong influence on the physical characteristics of their associated ravines, as shown in the following.

3.4.8 UCA Secondary and Descriptive Attributes

The mean SPI values were calculated from the grid and were averaged within each UCA-delineated boundary. The mean stream power values for the catchment areas range from 0 to 2.3, with a mean of 0.27 and a median of 0.11 (Table 3.8). Mean UCA CTI or wetness index values range from 2.7 to 6.7, with a mean of 4.9 and a median of 4.8. The CTI is an attribute that is intrinsically log-transformed, so the values are typically normally distributed. The descriptive attribute "slope range" was calculated for each UCA to try to identify the catchment areas with a relatively higher relief, a detail that might be lost when only the catchment area averages are analyzed. The minimum range in slope was 4.9% and the maximum range was 51.5%, with a mean range of 26.4% and a median of 25.0% (Table 3.8).

3.4.9 Influence of UCA Characteristics on Ravine Morphometry

Two integral questions for this study were as follows: What factors drive ravine morphometry? Is ravine size influenced by the area, the mean slope, or the SPI of their UCA? Figure 3.3 shows that the UCA has a significant effect on ravine area ($r = 0.74$). In contrast, the UCA was less strongly correlated with ravine volume ($r = 0.65$) and has no effect on ravine mean slope or ravine relief ($r = 0.00$ and $r = 0.00$, respectively).

TABLE 3.8
Summary of Descriptive Statistics for UCA Secondary and Descriptive Attributes

	UCA SPI Mean	UCA CTI Mean	UCA Slope Range %
Mean	0.28	4.87	26.4
Median	0.11	4.78	25.0
Standard deviation	0.47	0.84	10.6
Minimum	0.00	2.72	4.9
Maximum	2.26	6.67	51.5
Confidence level (95.0%)	0.11	0.20	2.5

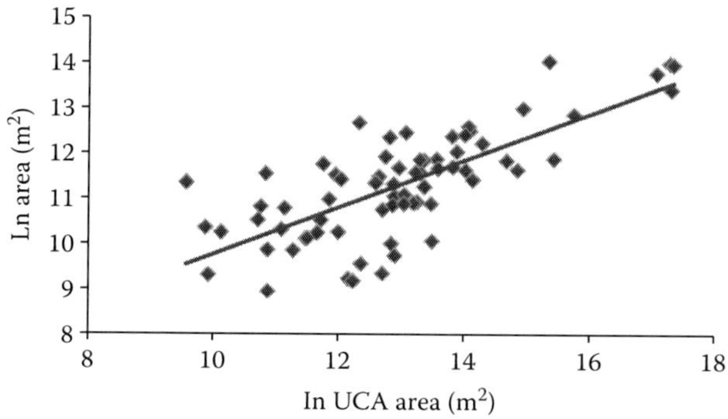

FIGURE 3.3 UCA plotted against ravine 3-D area ($r = 0.74$).

UCA SPI has only a small effect on ravine area and volume ($r = 0.39$ and $r = 0.35$, respectively). It might have been expected that the UCA SPI, the amount of stream power coming in to a ravine from the uplands, would exert a noticeable influence on the ravine formation; but in fact, the UCA mean CTI values show a stronger influence and correlate better to the ravine area ($r = 0.45$) than the UCA SPI ($r = 0.39$). One possible explanation for the low correlation between the SPI and the ravine area is that the ravine area may be more affected by the SPI values at the intersection between the UCA and the ravine than the average SPI values of the UCA.

It is also interesting to note that the UCA mean slope has an inverse effect on ravine W/D ratio, that is, as the UCA slope increases, the W/D ratios decrease (Figure 3.4), even though the UCA slope is not correlated to the ravine area ($r = -0.26$) or the volume ($r = -0.14$). As the UCA slope increases, the ravines get deeper relative to their width, and higher-mean UCA slopes seem to have a greater influence on the W/D ratios, reducing the W/D scatter that is seen in lower UCA slope means. The UCA mean slope, in contrast, had a positive relationship with ravine relief as expected; as the UCA slope increases, the ravine relief increases.

This investigation into the terrain and geomorphic attributes of ravines and their associated UCAs is motivated by an interest in ranking the ravines according to their

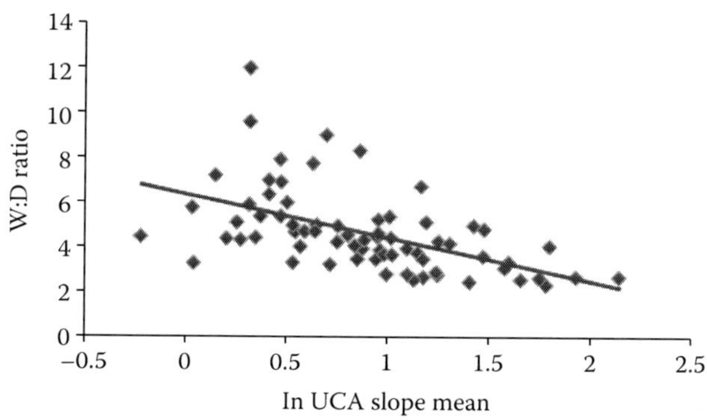

FIGURE 3.4 Scatterplot illustrating the effect of UCA slope on ravine W/D ratios ($r = 0.52$).

potential for delivering sediment to the MRB. Results of the research identified several characteristics of ravines, including area, relief, slope, and SPI, which are closely related to sediment production potential. In addition, the UCA is observed to be an important factor that affects the ravine area. These features can be used to assist with conservation planning to reduce sediment delivery from ravines. The ravines with the largest areas, highest relief, and largest UCAs have a higher priority for conservation practices than the ravines with the smallest areas, lowest relief, and smallest UCAs. Conservation practices such as conservation tillage, cover crops, or perennial crops can be installed in the UCAs in order to reduce the discharge of water to the ravines, which would cause headcutting and sediment loss from the ravine.

3.5 SUMMARY AND CONCLUSIONS

Within the context of ravine potential sediment contribution to the MNR, this study had a goal of estimating the potential contribution to sediment loads in the MRB. This goal was evaluated by collecting a complete dataset about the physical characteristics of ravines in the MRB. Seventy ravines were selected for study based on field reconnaissance. The boundaries for each ravine and its UCA were digitized. GIS techniques were used to estimate the primary and secondary attributes of each ravine and its UCA. Ravine 3-D areas range from 7745 to 1,267,816 m^2, with a mean of 171,453 m^2 and a median of 88,558 m^2. Ravine volumes range from 35,240 to 25,629,150 m^3 with a mean of 2,803,300 m^3 and a median of 1,057,230 m^3. Ravine relief values range from 7 to 68 m, with a mean of 34 m and a median of 35 m. Mean ravine slope values range from 7% to 31%, with a mean of 16% and a median of 17%. Ravine area, volume, and relief were strongly correlated with ravine slope, length, and SPI. Ravine W/D ratios were negatively correlated with the UCA mean slope. Ravine area was strongly correlated with the UCA. Another goal of the study was to evaluate whether or not ravine morphometry was affected by the characteristics of the upslope contributing areas that are associated with ravines. Ravine area was, indeed, strongly correlated with the UCA, suggesting that increasing volumes of water discharged from the UCA result in larger ravines that potentially deliver more sediment to the MRB than smaller ravines. The UCA can be used to identify ravines with the greatest potential for sediment delivery to the MRB. The UCAs for these priority ravines could then be treated with conservation practices such as conservation tillage, cover crops, or perennial grass to reduce discharge of water and sediment loading.

REFERENCES

Balco, G. 2001. Watershed Delineation Tool Script. University of Washington's Cosmogenic Isotope Lab. Available at http://depts.washington.edu/cosmolab/P_by_GIS.html (accessed January 11, 2014).

Bates, R.L., and J.A. Jackson (eds). 1984. *Dictionary of Geological Terms*, 3rd edition. Garden City, NY: Anchor Press/Doubleday.

Beven, K.J., and M.J. Kirby. 1979. A physically-based variable contributing area model of basin hydrology. *Hydrol. Sci. Bull.* 24:43–69.

CCMR. 2004. *Minnesota River Watershed Drainage Policy Reform Report*. New Ulm, MN: The Coalition for a Clean Minnesota River.

Engstrom, D.E., J.E. Almendinger, and J.A. Wolin. 2009. Historical changes in sediment and phosphorus loading to the Upper Mississippi River: Mass-balance reconstructions from the sediments of Lake Pepin. *J. Paleolimnol.* 41:563–588.

EPA. 2009. U.S. Environmental Protection Agency Website on Water Quality. Available at http://www.epa.gov/waterscience/standards/rules/303.htm (accessed January 11, 2014).

ESRI. 1994. *GRID Users Guide*. Redlands, CA: Environmental Systems Research Institute.

Galzki, J., A.S. Birr, and D.J. Mulla. 2011. Identifying critical agricultural areas with 3-meter LiDAR elevation data for precision conservation. *J. Soil Water Conserv*. 66:423–430.

Hatch, L.K., A.P. Mallawatantri, D. Wheeler, A. Gleason, D.J. Mulla, J.A. Perry, K.W. Easter, P. Brezonik, R. Smith, and L. Gerlach. 2001. Land management at the major watershed—Agroecoregion intersection. *J. Soil Water Conserv.* 56:44–51.

Kelley, D.W., S.A. Brachfield, E.A. Nater, and H.E. Wright, Jr. 2006. Sources of sediment in Lake Pepin on the Upper Mississippi River in response to Holocene climatic changes. *J. Paleolimnol*. 35:193–206.

Kelley, D.W., and E.A. Nater. 2000. Historical sediment flux from three watersheds into Lake Pepin, Minnesota, USA. *J. Environ. Qual.* 29(2):561–568.

Kober, A., P. Moore, and L. Nelson. 2008. Future of Minnesota River, One of America's Most Endangered Rivers of 2008, Still Hangs in the Balance. American Rivers Council. Available at http://www.americanrivers.org/newsroom/press-releases/future-of-minnesota-river-one-of-americas-most-endangered-rivers-of-2008-still-hangs-in-the-balance/ (accessed April 26, 2014).

Minnesota DNR. 2012. The DNR Data Deli. Available at http://deli.dnr.state.mn.us/ (accessed January 11, 2014).

Moore, I.D., P.E. Gessler, G.A. Nielsen, and G.A. Peterson. 1993. Soil attribute prediction using terrain analysis. *Soil Sci. Soc. Am. J.* 57:443–452.

Moore, I.D., R.B. Grayson, and A.R. Landson. 1991. Digital terrain modeling: A review of hydrological, geomorphological, and biological applications. *Hydrol. Process.* 5:3–30.

MPCA. 2005. Minnesota Pollution Control Agency's Minnesota River Basin TMDL Project for Turbidity Fact Sheet, Water Quality/Basins 3.33, June 2005. St. Paul, MN: MPCA.

Mulla, D.J., and A. Sekely. 2009. Historical trends affecting accumulation of sediment and phosphorus in Lake Pepin. *J. Paleolimnol*. 41(4):589–602.

Schumm, S.A. 1999. Causes and controls of channel incision. In *Incised River Channels*, eds. S.E. Darby, and A. Simon, 19–33. Chichester, England: John Wiley & Sons.

Sekely, A.C., D.J. Mulla, and D.W. Bauer. 2002. Streambank slumping and its contribution to the phosphorus and suspended sediment loads of the Blue Earth River, Minnesota. *J. Soil Water Conserv.* 57(5):243–250.

Sidorchuk, A. 2006. Stages in gully evolution and self-organized criticality. *Earth Surf. Proc. Land.* 31:1329–1344.

Simon, A., and S.E. Darby. 1999. The nature and significance of incised river channels. In *Incised River Channels*, eds. S.E. Darby, and A. Simon, 3–18. Chichester, England: John Wiley & Sons.

Tarboton, D.G. 1997. A new method for the determination of flow directions and upslope areas in grid digital elevation models. *Water Resour. Res.* 33(2):309–319.

USGS. 2012. National Elevation Dataset. Available at http://ned.usgs.gov/ (accessed January 11, 2014).

Wilson, J.P., and J.C. Gallant. 2000. Digital terrain analysis. In *Terrain Analysis: Principles and Applications*, eds. J.P. Wilson, and J.C. Gallant, 1–26. New York: John Wiley & Sons.

Yang, X., G.A. Chapman, J.M. Gray, and M.A. Young. 2007. Delineating soil landscape facets from digital elevation models using compound topographic index in a geographic information system. *Aust. J. Soil Res.* 45:569–576.

4 Siting Multiple Conservation Practices in a Tile-Drained Watershed Using LiDAR Topographic Data

Mark D. Tomer, David E. James, Jill A. Kostel, and J. Bean

CONTENTS

EXECUTIVE SUMMARY

Improving the effectiveness of conservation practices for achieving water quality goals will require conservation planning with a watershed context. Conservation planning has traditionally been field- or farm-specific and is usually aided by site-specific information that includes topographic data. Light detection and ranging (LiDAR) data, obtained from laser altimetry surveys conducted using aircraft, now provide the kind of detail needed to identify sites that may be suited for a range of conservation practices across landscapes and watersheds. Once developed, this technology could provide watershed stakeholders and landowners a new suite of tools to identify appropriate and cost-effective strategies for conservation efforts to meet water quality goals in watersheds. This chapter serves to demonstrate the use of LiDAR data for siting three different conservation practices in a 6500-ha (16,000-ac) watershed in Illinois. The chapter provides watershed topographic data and examples/instructions on how to evaluate LiDAR elevation data to identify sites in this watershed where wetlands, two-stage ditches, and grassed waterways could be recommended. The importance of on-site evaluations and landowner involvement in conservation planning is emphasized as critical aspects of applying this new technology in a real-world setting.

KEYWORDS

Light detection and ranging (LiDAR), topography

4.1 INTRODUCTION

4.1.1 Topography as a Basis for Watershed-Scale Conservation Planning

Conservation practices have been shown to improve water quality at the farm and field scale, but it has proven difficult to document the conservation's impacts on water quality at the watershed scale (Tomer and Locke 2011). To bridge the gap between scales of actual conservation implementation (farm) and desired water quality response (watershed), approaches and frameworks that facilitate conservation planning under a watershed context are needed. This chapter illustrates how topographic data for a small watershed can provide a basis to develop a "conservation planning scenario" that can address agricultural impacts on watershed water quality through a farm-scale implementation.

A "conservation planning scenario," for this discussion, is defined as a comprehensive suite of conservation practices that can be applied across a watershed to address multiple water quality contaminants and those locations that are most prone to be sources of contaminants. Topographic data, for an example watershed, are provided in the online exercises to give students the opportunity to explore the development of alternative scenarios using a range of practices and/or how modifications of criteria used to locate practices can change the planning scenario.

Topographic data are generally needed to plan conservation practices, because topography determines much about landscape/watershed hydrology at and near the land surface. Terrain analyses can be applied to topographic data to map overland flow directions and to identify pathways that are most prone to concentrated flows, where the risk of erosion is the greatest (Wilson and Gallant 2000). Also, areas susceptible to soil saturation can be mapped to indicate where crop production might be more risky due to inundation and where dissolved and sediment-bound contaminants are susceptible to water transport. These and other types of vulnerable locations require careful management to conserve soil and water resources in agricultural fields (Table 4.1). A variety of conservation practices can help protect vulnerable sites, and topographic information can help decide which practices are most beneficial. This concept, which is aimed to ensure that appropriate conservation practices are implemented where they will most effectively contribute toward solving natural resource concerns that are identified from a wider (watershed or ecosystem) perspective, is often referred to as targeted conservation (Walter et al. 2007). The use of topographic data to target conservation practices is not without precedent. These data have been shown useful to design filter strips (Dosskey et al. 2011), to locate sites to implement grassed waterways (Pike et al. 2009), and to position wetlands within a landscape (Tomer et al. 2003).

In the past, acquisition of detailed topographic survey data has been expensive and therefore has only been collected as needed on an individual-site basis. But technological advances have changed this. Airborne laser altimetry can now provide detailed topographic data across large areas. These data, most often called light detection and ranging (LiDAR) data, are becoming increasingly available. The utilities of LiDAR data are many and include mapping of floodplains (Jones et al. 2007), archeological features (Crow et al. 2007), and vegetation, including leaf area index (LAI; see Jensen et al. 2008), tree canopy heights (Hawbaker et al. 2010), and wetland distributions beneath a forest cover (Lang and McCarty 2009). It is also clear that LiDAR data offer the opportunity to map agricultural fields across watersheds,

TABLE 4.1
Examples of Environmentally Vulnerable Sites That Can Be Identified Using Topographic Data and Terrain Analysis, and Some Agricultural Conservation Practices That Can Control Environmental Losses and Limit Water Quality Impacts

Topographic Constraint That Makes a Site Vulnerable	Vulnerability	Practices to Limit Vulnerability
Slope length and steepness	Sheet-and-rill erosion	Contour farming; no-tillage; terraces; perennial strips
Concentrated surface runoff	Gully erosion	Grassed waterways; water control structures
Ponding	Crop failure; loss of dissolved contaminants	Wetland restoration; drainage management (e.g., blind inlets)
Saturation excess overland flow (i.e., convergence of surface runoff in high water table areas)	Crop failure; soil erosion; loss of dissolved constituents	Riparian buffers; filter strips

providing a data resource that has only been available for a small number of agricultural fields where ground survey data have been collected. In this chapter, we demonstrate how to analyze topographic data from a watershed-scale LiDAR survey to develop maps that locate potential sites for a variety of practices in order to develop a conservation planning scenario. For an example watershed, potential sites for nutrient-removal wetlands, two-stage ditches, and grassed waterways are mapped, and the challenges for conservation planning at the watershed scale are discussed. While the watershed is a real-world setting, this demonstration/case study should be considered hypothetical because a real-world application of these mapping techniques should always be done in collaboration with local stakeholders, especially those who own and manage land in the watershed.

4.1.2 Addressing Conservation Needs of Midwest Tile-Drained Watersheds

Agricultural watersheds in the US Midwest have been identified as major contributors to hypoxic conditions in the Gulf of Mexico due to large losses of nutrients to streams and rivers that flow to the Mississippi River (Dale et al. 2010). The combination of soil types, climate, and cropping systems that are found across much of the Midwest makes this region especially prone to nutrient loss. The Midwest has a humid to subhumid climate in which precipitation usually exceeds evapotranspiration on a seasonal basis, creating a "leaky" system that is prone to significant runoff losses that can carry sediment and phosphorus plus significant volumes of soil drainage that carry dissolved constituents, particularly nitrate–nitrogen (NO_3–N). The potential for NO_3–N losses from Midwest watersheds is exacerbated by nutrient-rich soils that are developed under prairies and wet meadows, annual row crops (especially corn) that require relatively large rates of N fertilizer to achieve potential harvest yields, and extensive areas with poor natural drainage conditions that have been remedied through artificial (or tile) drainage. Tile drains have been installed across the Midwest wherever seasonal water tables are near the surface; the area of Midwest tile drainage has been estimated at over 20×10^6 ha (50×10^6 ac) (McCorvie and Lant 1993). High water tables are prominent in areas of recent glaciation, where soils are young (little-weathered and nutrient-rich) and natural drainage patterns are not yet well developed. Lake bed and coastal plain soils of humid regions of the United States are also commonly tile-drained. The combination of soil and hydrologic conditions with row crop agriculture results in particular challenges to maintain and improve agricultural water quality (Dinnes et al. 2002). Tile drainage systems form a short-circuit pathway for losses of NO_3–N that exacerbate Gulf of Mexico hypoxia from Midwest watersheds (Dale et al. 2010).

4.1.3 Targeting Conservation Practices to Manage Tile Drainage

Conservation practices to limit nutrient loss from tile drainage must be broadly based and flexible to be applicable across the extent of tile-drained lands in the Midwest. Table 4.2 lists four types of practices that are being advocated to address nutrient losses that have design criteria either established or under development. Improved

TABLE 4.2
Conservation Practices That Can Control Nutrient Losses from Tile-Drained Croplands with Constraints on Site Suitability

Practice	Constraints on Location
Controlled drainage (water table control)	Requires flat terrain (<0.5% slope).
Woodchip bioreactor	Most suited to field drains (<100 ac in area) adjacent to drainage ditch.
Nutrient removal wetland	Low impoundment must create adequate pool area without impeding drainage of cropland.
Drainage ditch management (e.g., two-stage ditches)	Shallow ditches minimize construction costs.

crop nutrient management is a practice that could be added but is one that should improve profitability on most fields (whether tile-drained or not), and therefore, it is not a practice considered for conservation targeting in this example. Each of these practices (Table 4.2) is constrained in terms of the types of locations where they are most feasible and affordable to implement. The challenge here is to identify a consistent methodology to determine where practices will be most effective toward achieving a given nutrient load-reduction goal for a watershed at the least cost. The potential of each practice to achieve nitrate load reduction is being identified through research. This research is reasonably well progressed for wetlands (Dale et al. 2010) but less so for the other three practices, especially drainage ditch management. With any conservation practice, effectiveness for nutrient removal will be site-specific given the climatic conditions, the practice maintenance, the actual hydraulic and nutrient loads, and the variation/timing of these loads. Recognizing not only these uncertainties but also the fact that LiDAR data offer new opportunities to better manage agricultural water quality, how do we determine where to best locate each of these practices in a watershed? The answer will depend on the watershed, so we explore this through a case study as a learning exercise. In actual practice, these techniques should be applied in a planning setting with the participation of local stakeholders.

4.1.4 Case Study: Watershed—Lime Creek

The Lime Creek watershed is a 6500-ha (16,000-ac) HUC12 (Hydrologic Unit Code 12: sixth level, subwatershed) basin that is a tributary to Big Bureau Creek (BBC) in North Central Illinois, which in turn is a tributary to the Illinois and then the Mississippi Rivers (Figure 4.1). The BBC drains a 1280-km^2 (500-mi^2) watershed that is bounded to the north and east by Wisconsinan-age (12,000–22,000 years bp) terminal glacial moraines. Prior to Quaternary glaciation, the ancestral Mississippi River occupied the lower eastern valley of the BBC watershed. Glacial advances, including westward ice movement from present-day Lake Michigan, shifted the Mississippi to its present course. As glaciers receded, meltwaters carved the stream valley that drains the BBC watershed today. The BBC and Lime Creek watersheds are dominated by thick (>100 m or 300 ft) glacial moraine deposits that overlie mostly Silurian-age (approximately 420 m.y.) sedimentary bedrock. The tills are

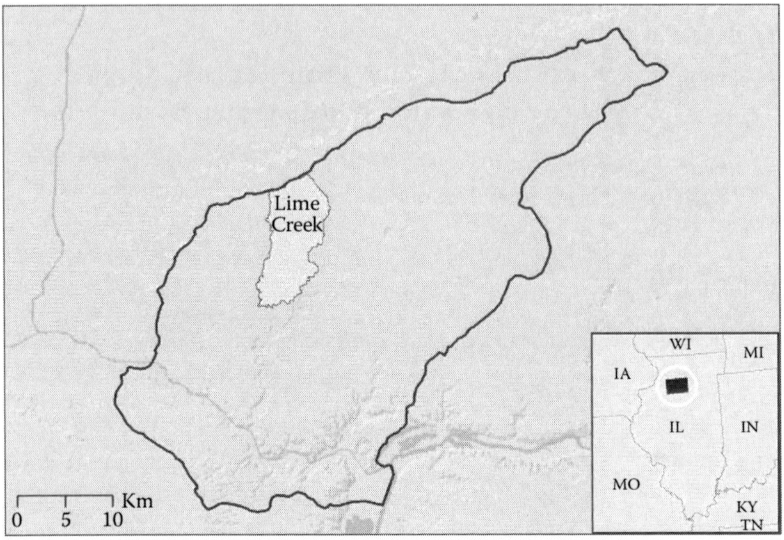

FIGURE 4.1 General map showing location of Lime Creek watershed.

often capped with a loess mantle averaging approximately 1.5 m (5 ft) thick. Lime Creek is bounded by a terminal glacial moraine to the north, with the southern two-thirds of the watershed dominated by a glacial–fluvial plain.

The BBC area was settled by European immigrants during the 1830s, who converted the native prairies and the wet meadows to agricultural use within a few decades. Improved drainage was necessary to achieve this transition, and a system of drainage ditches and subsurface (tile) drains were installed to make widespread agricultural production feasible. Surface water discharge from the Lime Creek watershed is facilitated by a network of dug drainage ditches. Today, the dominant land use is for corn and soybean production. Agricultural production results in nitrate loss to downstream surface waters, contributing to the Gulf of Mexico hypoxia.

4.2 METHODS

4.2.1 Acquisition of LiDAR Survey Data for Lime Creek

LiDAR data were obtained for the Lime Creek watershed by aircraft during December 2008 under snow-free conditions. The final return data from laser pulses were converted to a point elevation map and then processed to obtain a 1-m grid digital elevation model (DEM) of the land surface. This DEM was then processed to model hydrologic routing of flows over the land surface. The modeling was conducted using ArcMap GIS (version 9.2; see ESRI 2009). Flow routing initially comprised a "pit-filling" operation to fill depressions in the landscape and to estimate where and in what direction the filled depressions would overflow (Garbrecht and Martz 1997). A grid coverage of fill depths was generated by subtracting the original DEM from the filled DEM. This coverage of fill depths identified "potholes" or depressions that are commonly found on glaciated terrain. In addition, this coverage showed artificial impoundments upgradient of roads, because bridges and culverts that pass water beneath roads are not identifiable in a LiDAR DEM. The Lime Creek watershed has a dense network of roads that formed a number of false impoundments that had to be removed in order to represent actual flow pathways (Figure 4.2). The false impoundments most relevant to watershed-scale analysis occurred along perennial ditches and streams. For some applications, it might be necessary to address all culverts in the watershed to represent surface flows from every field as accurately as possible. Here, however, we focused on "stream" pathways and arbitrarily set a threshold of 100-ha (247-ac) contributing area to define a stream. This turned out, based on later field inspection, to be a reasonable estimate of the minimum area that generates perennial flow in ditches in the upper reaches of the watershed.

To eliminate (or "burn" through) the artificial impoundments, points were digitized where roadway centerlines crossed stream channel and ditch pathways with <100-ha contributing area (Figure 4.2). Then, a separate grid coverage was prepared in which grid cells within a given search radius of these road–stream intersection points were assigned the maximum filled depth within that radius, plus 1 m, with null values assigned to all the other cells. The search radius distance was varied by the type of roadway; a 15-m radius sufficed for most roads, but a 20-m radius was required along a state highway that crossed the watershed. This coverage of buffered road–stream intersections was then subtracted from the original DEM. This subtraction "burned" through the roads to provide an accurate flow route where road surfaces falsely impounded hydrologic flows on the original LiDAR DEM. The pit-filling operation and the hydrologic flow modeling were then repeated on the "burned" DEM. This procedure successfully kept the routing of "stream paths" within drainage ditches rather than impounding flows at road crossings (Figure 4.3). There were also several locations with small (i.e., <100 ha) contributing areas where false impoundments spilled out of (or into) the watershed. The burning operation was repeated to accurately include these drainage areas as being in (or out of) the watershed.

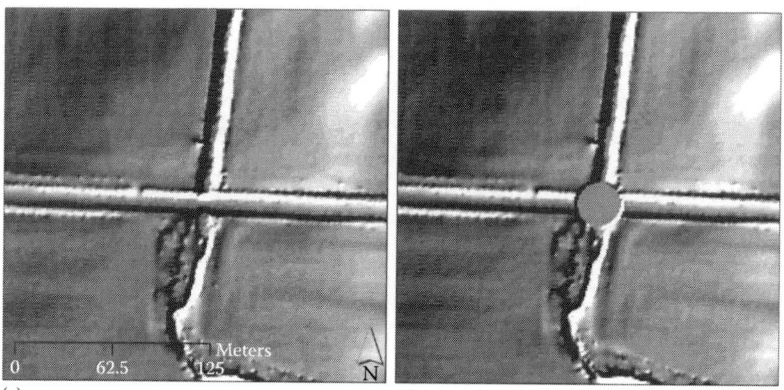

(a) (b)

FIGURE 4.2 All road–stream crossings become impoundments when hydrological processing is done using unaltered LiDAR DEMs (a). "Burning" these crossings to the impoundment depth ensures that the hydrologic modeling represents actual flow along streams (b).

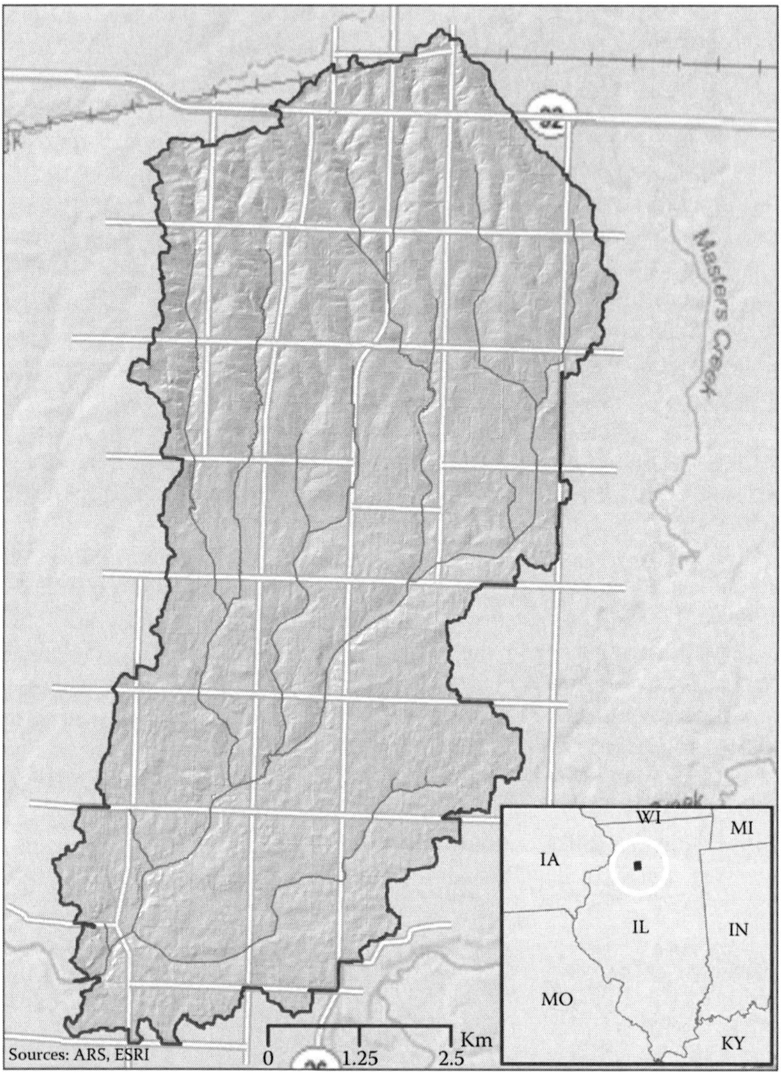

FIGURE 4.3 Image of Lime Creek LiDAR data, processed to remove falsely impounded hydrologic flows at road–stream intersections.

4.2.2 GIS Modeling to Locate Wetlands

In processing the provided grid coverage, artificial impoundments formed by road surfaces were removed. The mapping of potential wetland sites is, in contrast, a process of creating a set of artificial impoundments. These impounded sites are then tested for their suitability as nutrient removal wetlands. In this example, the initial assumption is that any candidate wetland would only be feasible if drainage beneath/along roads was not at risk of being impeded by the wetland impoundment. Further, impedance of drainage was assumed to be at risk within a "buffer" surrounding each wetland, where the surface elevation was within 1.5 m (5 ft) of the wetland pool elevation. This buffer accounts for the need to maintain subsurface drainage from the cropland adjacent to and above the wetland through tiles that are typically installed at a 1.2-m (4-ft) depth. Also, the 1.5-m buffer should amply provide for an increased pool elevation during runoff events. This buffer height criterion is regarded as conservative—perhaps unnecessarily conservative—but the results should identify sites where local landowners would most willingly accept wetlands. A wetland would be installed by building a low impoundment to flood an area of land to a shallow depth adjacent to the drainage ditch. For the example scenario (Figure 4.4), the wetland impoundment depth was the ditch depth plus 0.9 m. The maximum wetland pool depth of 0.9 m and a buffer of 1.5 m of the wetland pond

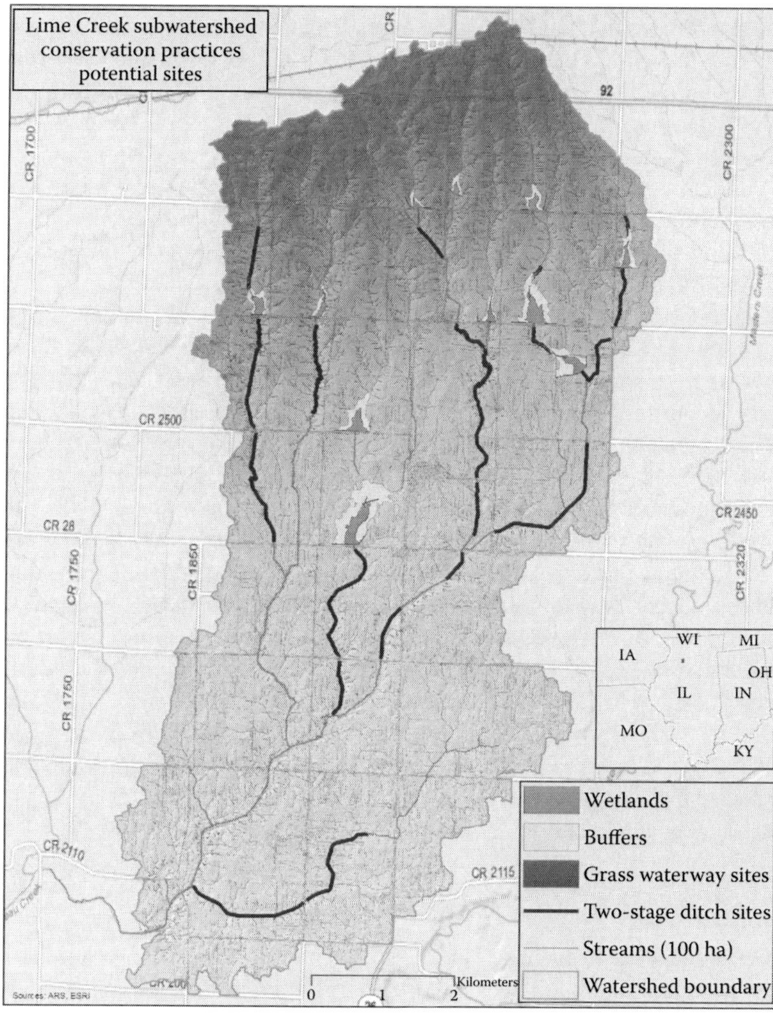

FIGURE 4.4 **(See color insert.)** Map of Lime Creek watershed illustrating a conservation planning scenario that employs three practices to identify potential artificial wetland impoundment areas for shallow flood.

elevation were borrowed from screening criteria that were used in Iowa's Conservation Reserve Enhancement Program, as discussed by Tomer et al. (2003). The LiDAR topographic data provided enough detail to estimate ditch depth, and we used the focal range (range in elevation) within 20 m of the stream, at a point 20–30 m upgradient of the burned road–stream crossings described above, as the estimate of ditch depth.

First, an impoundment was simulated that was 2.4 m (0.9 plus 1.5 m) greater in height than this focal range at these sites just above each road crossing. Locating these impoundments just above each road crossing obviously provides the best chance to avoid impeding drainage at the next upstream road crossing. If the focal minimum (i.e., estimated water elevation) at the next upstream crossing was less than this "buffer impoundment" elevation, then that location was rejected as a candidate wetland. Candidates that passed this check then had a lower impoundment simulated that was the ditch's depth (focal range) plus 0.9 m. This resulted in two impoundments: the smaller representing the potential wetland and the larger representing the wetland plus the surrounding buffer. Part or all of the entire buffer area would need to be removed from row crop production, and some land use easement from the landowner would be required. A final check was made using aerial photographic images to ensure that the wetlands and their buffers did not overlap (impede drainage from) any farmsteads. Sites passing these criteria could be moved upstream as far as possible and still avoid impeding drainage at the next upstream crossing; however, in the example (Figure 4.4), we only explored this where there was a stream confluence between road crossings. This choice was arbitrary, and students should feel free to explore a range of strategies for locating potential wetland sites.

Having identified candidate sites, their contributing watershed, wetland, and buffer areas were tabulated. A way of prioritizing among the wetland sites should be devised to assist with the implementation process. Ideally, wetlands should be located to intercept drainage from relatively large areas and should be sized at 0.5%–2.0% of the contributing area (Dale et al. 2010). With smaller areas, short hydraulic residence times may limit nitrate removal in wetlands, whereas larger wetlands may be underloaded and may not be the most efficient way to utilize limited funds that are available to incentivize wetland construction. Buffer areas should be as small as possible to minimize the purchase of easements for areas that are not converted to wetlands. A ranking scheme including these factors may be helpful if wetland sites are to be selected through a bidding system. The potential wetland sites identified by this process (Figure 4.4) were reviewed in the field and were confirmed as good candidate sites. But other sites may be suitable based on alternate and reasonable criteria. Students are encouraged to explore how new candidate sites can be identified if the selection criteria are more flexible.

4.2.3 GIS Modeling to Locate Two-Stage Ditches

Construction of a two-stage ditch involves widening a standard trapezoidal ditch to create a vegetated floodplain near the elevation of the channel's water surface under steady flow conditions. The new floodplain should have a width three to five times the width of the steady flow channel at the bottom of the existing ditch (Figure 4.5) (NRCS 2007). This requires excavation to widen the ditch, and regrading of the excavated spoil material, which becomes increasingly expensive as the ditch depth and

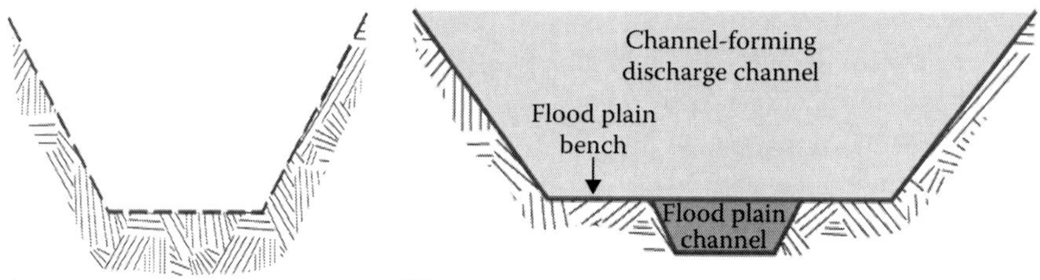

(a) (b)

FIGURE 4.5 Illustrations of trapezoidal (a) and two-stage (b) ditch. Conversion to the two-stage design requires (and should preserve) natural meandering in the ditch bottom. (From NRCS, Two-stage channel design. In *National Engineering Handbook, Part 654: Stream Restoration Design*. Chapter 10. Natural Resources Conservation Service, U.S. Dept. of Agriculture, Washington DC, 2007. Available at http://directives.sc.egov.usda.gov /OpenNonWebContent.aspx?content=17770.wba [accessed May 7, 2014].)

the channel width increase, especially if hauling spoil off-site becomes necessary. Also, more land is taken out of production with increasing ditch depth. Therefore, shallow and narrow ditches are best suited for two-stage ditches, especially when identifying locations for initial local testing and demonstration of this practice. LiDAR topographic data are capable of estimating ditch depths based on local range in elevation (focal range). Additional information is required for screening, but we expected that a map showing where ditch depths are most feasible for the practice would help screen ditches that were possible two-stage candidates. We assigned focal range values (within a 20-m window) to "stream" (>100-ha contributing area) cells and classified those values between 1 and 2.5 m as being potentially suited for a two-stage ditch construction. Conversion of any trapezoidal ditch to the two-stage configuration requires that some natural channel meandering has occurred in the bottom of the ditch prior to converting the ditch. This naturally results from transport of fine sediment under baseflow conditions (NRCS 2007) but usually takes several years following any ditch maintenance that reinforced a straightened trapezoidal channel. The sites shown as suitable for two-stage ditches in Figure 4.4 have been adjusted to only include those ditch lengths that had developed some meandering based on field review. In that review, we also found that focal ranges less than 1 m have generally occurred where the >100-ha flow path was along a wide grassed waterway, which probably covered a subsurface tile main in most cases.

4.2.4 GIS Modeling to Locate Grassed Waterways and Riparian Filter Strips

Two-stage ditches and nutrient removal wetlands are practices aimed to treat tile drainage, which would presumably be a conservation priority in the Lime Creek watershed. But the maintenance and longevity of these practices will depend on effective erosion control to minimize sediment delivery and accumulation. This emphasizes the need for comprehensive conservation planning in watersheds, including multiple practices that address multiple resource concerns. Grassed waterways are a practice aimed to control concentrated flow erosion along ephemeral waterways in upland areas. Recently, Pike et al. (2009) demonstrated that terrain analysis and neural network analysis could be used to convert key terrain attributes toward mapping locations where concentrated flow erosion would be most likely. Their results were expressed in a scalar form with values varying between 0 and 1, reflecting the probability of concentrated flow erosion (p):

$$p = \frac{1}{\left(1+\exp^{-(-3.63+1.11\mathrm{LS}+0.21\mathrm{WET}-12.1\mathrm{PLAN})}\right)} \tag{4.1}$$

in which LS is the length–slope (sometimes the erodibility index):

$$\mathrm{LS} = 1.4\left[\frac{A_{sc}}{22.13}\right]^{0.4}\left[\frac{\sin\beta}{0.0896}\right]^{1.3} \tag{4.2}$$

WET is the topographic wetness index:

$$\mathrm{WET} = \ln\left[\frac{A_{sc}}{\tan\beta}\right] \tag{4.3}$$

and PLAN is plan curvature, which is the rate of change in the aspect of an elevation contour (i.e., the second derivative of a quadratic expression describing the shape of the elevation contour; see Wilson and Gallant 2000). Terms A_{sc} and β are the specific contributing area (contributing area divided by cell width) and slope (in degrees), respectively. It is beyond the scope of this chapter to provide detailed information on terrain analysis. Readers interested in a more complete discussion of the topic are directed to Chapter 1 and the background readings listed in the references. There are terrain analysis tools in ARC/GIS with commands for slope, flow accumulation, and curvature that facilitate key steps in these calculations. The exercises in Chapter 1 explore these commands in detail.

While the wetland and the two-stage ditch siting used the original 1-m DEM scale, here, we resampled the data to a 5-m grid to more closely match the scale that is used in the Pike et al. (2009) study. A threshold value for p of 0.7 is highlighted in Figure 4.4. Grassed waterways could be recommended where there are clusters of grid cells with p values above this (or an alternate) threshold. Where a number of concentrated flow pathways are indicated near a stream, this might be a priority site for installation of a riparian filter strip (see Dosskey et al. 2011).

4.3 RESULTS AND DISCUSSION

4.3.1 Implications for Water Quality Management in Lime Creek

The distribution of combined practices shown (Figure 4.4) illustrates a concept that can be explored as conservation planning scenarios are developed in this exercise and discussed. Consider the following statement: "Distinct landscapes within a watershed can be distinguished as being best suited to a distinct set of conservation practices." In the Lime Creek example, the uppermost reaches of the watershed are occupied by a terminal glacial moraine, with slopes that may be prone to erosion and require protection against concentrated flow erosion. Moving south, there are positions along streams at the base of the moraine where low impoundments can be placed to create small wetlands that do not impede tile drainage in their contributing area. If wetlands were to be established further south, the central part of the watershed is dominated by near-level terrain, and there appear to be a few appropriate wetland sites that would not impede significant areas of tile drainage. Toward the mouth of the watershed, Lime Creek becomes somewhat incised again, and it is possible that wetlands could be installed there, although they would not meet the criteria described above. In the upper-central parts of the watershed, there are drainage ditches with low bank heights where they should be feasible to implement two-stage ditches, more so than in the lower watershed where the ditches are wider and deeper. In most of the lower third of the watershed, none of the three practices that we evaluated above are suitable. In these areas, flat terrain may be suited to controlled drainage, and sites that are suitable for installation of bioreactors could be prioritized in this area. Students are encouraged to develop maps depicting other conservation planning scenarios for this watershed, including any practice that might be appropriate, to further explore the idea of "putting the right conservation practice in the right place."

4.4 CONCLUSION—ISSUES IMPACTING TRANSPORTABILITY OF LESSONS LEARNED

The Lime Creek watershed is an ideal watershed to explore the concepts considered here because of its small size and the simplicity of its landforms. Lime Creek is an archetypical watershed for the glaciated Midwest, including a terminal moraine that drains down to a glacial–fluvial plain. In particular, not all watersheds originating as Wisconsinan-age glacial deposits will have LiDAR data that are as simple to hydrologically process as Lime Creek. More hummocky terrain, deeper ditches, and multiple glacial features including complex moraines, kames, outwash channels, etc., are often present, which complicate the interpretation of these young landscapes for conservation opportunities. Students may want to explore these concepts in watersheds that are formed on older geologic surfaces. However, as landscape regions vary, so does the climate, the management systems, and the water quality issues that will be needed to be addressed. This means that the set of conservation practices that are most appropriate may be completely different. Different practices of course will mean different criteria that need to be developed in identifying conservation planning scenarios for each watershed.

It is important to recognize that this exercise was essentially conducted without any context of social setting. Keep this in mind: the landowners who work the lands in any watershed are those who implement conservation practices. They will do so either voluntarily or not at all. Therefore, conservation planning in watersheds should be done in close partnership with the landowners who will implement and maintain the practices that are being planned. Such partnerships, to be effective, will need to be based on trust. Successful watershed planning will require that we address management challenges and opportunities on technical and social levels.

4.5 CASE STUDY

4.5.1 Practice Exercises to Develop a Watershed Planning Scenario

A LiDAR-based 1-m grid DEM is included with this chapter to allow students to develop conservation planning scenarios on their own. This DEM has been processed as described above (see Section 4.2.1; Figures 4.2 and 4.3). Approaches to targeting the different practices could follow the criteria suggested in the following or could be developed independently. To become familiar with the process of developing a conservation planning scenario, one can exactly follow the procedures to arrive at a scenario that is similar to the one depicted in Figure 4.4. Alternatively, experienced students can read through the sections and then develop independent scenarios. Students should feel free to work at a variety of scales, examining the entire watershed, subbasins, and even individual fields to consider how terrain analysis might influence conservation planning from a variety of perspectives, and to include a wider range of practices that address surface runoff as well as tile drainage. Note that the conservation planning scenario depicted (Figure 4.4) resulted from a process that included a field review of candidate sites and ditches, which led to the slight alteration of initial screening results. This point is important for two reasons. First, students should not expect to exactly replicate the scenario mapped in Figure 4.4; and second, conservation planning results done through any desktop exercise like this need to be verified through a field review, preferably one that includes input from local landowners.

4.5.2 Suggestions and Questions for Practice and Discussion

If there was a need to identify potential wetland sites near the bottom of the Lime Creek watershed that added up to 1% of the watershed area (65 ha or 160 ac), could these sites be found, and how would selection criteria need to be altered to do so? What are the risks involved with relaxing wetland site selection criteria, and how could they be overcome?

What other conservation practices can you identify that could be implemented in this watershed, or parts of the watershed, and sited using terrain attributes? Begin by considering the practices listed in Table 4.2 and developing your own criteria for targeting these practices in Lime Creek. What other practices can you include in a conservation planning scenario for this watershed? How do you go about developing criteria for targeting these practices? Zoom in on a part of the watershed and ask the same questions again. Does narrowing your spatial scale alter your answer? How would you use other information like soil survey to refine planning scenarios in this watershed?

This exercise raises questions that are active areas for research and are brought up here to stimulate discussion. Assume that croplands in the watershed are losing 25 kg of NO_3–N/ha/year and 250 mm of discharge by tile lines (average concentration of

10 mg NO_3–N/L). How would you determine whether a conservation scenario could reduce that average concentration by 45%? What information would you need to be able to estimate the effectiveness of a conservation planning scenario? Start by discussing a scenario that only includes wetlands before including multiple practices. What are some challenges for addressing this question using a watershed-scale simulation model?

REFERENCES

Crow, P., S. Benham, B.J. Devereux, and G.S. Amable. 2007. Woodland vegetation and its implications for archaeological survey using LiDAR. *Forestry* 80(3):241–252.

Dale, V., C. Kling, J. Meyer, J. Sanders, H. Stallworth, T. Armitage, D. Wangsness, T. Bianchi, A. Blumberg, W. Boynton, D. Conley, W. Crumpton, M. David, D. Gilbert, R. Howarth, R. Lowrance, K. Mankin, J. Opaluch, H. Paerl, K. Reckhow, A. Sharpley, T. Simpson, C. Snyder, and D. Wright. 2010. *Hypoxia in the Northern Gulf of Mexico*. Springer, New York.

Dinnes, D.L., D.L. Karlen, D.B. Jaynes, T.C. Kaspar, J.L. Hatfield, T.S. Colvin, and C.A. Cambardella. 2002. Nitrogen management strategies to reduce nitrate leaching in tile-drained Midwestern soils. *Agronomy Journal* 94(1):153–171.

Dosskey, M.G., M.J. Helmers, and D.E. Eisenhauer. 2011. A design aid for sizing filter strips using buffer area ratio. *Journal of Soil and Water Conservation* 66(1):29–39.

ESRI. 2009. *ArcMap 9.2*. ESRI, Redlands, CA.

Garbrecht, J., and L.W. Martz. 1997. The assignment of drainage direction over flat surfaces in raster digital elevation models. *Journal of Hydrology* 193:204–213.

Hawbaker, T.J., T. Gobakken, A. Lesak, E. Tromborg, K. Contrucci, and V. Radeloff. 2010. Light detection and ranging-based measures of mixed hardwood forest structure. *Forest Science* 56(3):313–326.

Jensen, J.L.R., K.S. Humes, L.A. Vierling, and A.T. Hudak. 2008. Discrete return LiDAR-based prediction of leaf area index in two conifer forests. *Remote Sensing of Environment* 112(10):3947–3957.

Jones, A.F., P.A. Brewer, E. Johnstone, and M.G. Macklin. 2007. High-resolution interpretive geomorphological mapping of river valley environments using airborne LiDAR data. *Earth Surface Processes and Landforms* 32(10):1574–1592.

Lang, M.W., and G.W. McCarty. 2009. LiDAR intensity for improved detection of inundation below the forest canopy. *Wetlands* 29(4):1166–1178.

McCorvie, M.R., and C.L. Lant. 1993. Drainage district formation and the loss of Midwestern wetlands. *Agricultural History* 67(4):13–39.

NRCS. 2007. Two-stage channel design. In *National Engineering Handbook, Part 654: Stream Restoration Design*, Chapter 10, J.M. Bernard, J. Fripp, and K. Robinson (eds.) Natural Resources Conservation Service, U.S. Dept. of Agriculture, Washington, DC. Available at http://directives.sc.egov.usda.gov/OpenNonWebContent.aspx?content=17770.wba (accessed May 7, 2014).

Pike, A.C., T.G. Mueller, A. Schörgendorfer, S.A. Shearer, and A.D. Karathanasis. 2009. Erosion index derived from terrain attributes using logistic regression and neural networks. *Agronomy Journal* 101:1068–1079.

Tomer, M.D., D.E. James, and T.M. Isenhart. 2003. Optimizing the placement of riparian practices in a watershed using terrain analysis. *Journal of Soil and Water Conservation* 58(4):198–206.

Tomer, M.D., and M.A. Locke. 2011. The challenge of documenting water quality benefits of conservation practices: A review of USDA-ARS's Conservation Effects Assessment Project watershed studies. *Water Science & Technology* 64(1):300–310.

Walter, T., M. Dosskey, M. Khanna, J. Miller, M. Tomer, and J. Weins. 2007. The science of targeting within landscapes and watersheds to improve conservation effectiveness. In *Managing Agricultural Landscapes for Environmental Quality: Strengthening the Science Base*, eds. M. Schnepf and C. Cox, 63–89. Soil & Water Conservation Society, Ankeny, IA.

Wilson, J.P., and J.C. Gallant. 2000. *Terrain Analysis: Principles and Applications*. John Wiley & Sons, New York.

5 Erosion Modeling in 2D with the Revised Universal Soil Loss Equation–Version 2

A Tool for Conservation Planning

Seth M. Dabney, Daniel C. Yoder, and Dalmo A.N. Vieira

CONTENTS

EXECUTIVE SUMMARY

The current version of the Revised Universal Soil Loss Equation—Version 2 (RUSLE2) does not estimate concentrated flow erosion, which can be substantial in agricultural fields. The primary objective of this chapter is to describe our ongoing efforts using geographical information system (GIS) tools and high-resolution topographic elevation data to develop a distributed version of RUSLE2 that can be linked with a process-based channel erosion model to account for concentrated flow erosion. This is being accomplished by modifying the RUSLE2 so that runoff can be estimated and local slope length can be determined based on accumulated runoff. This chapter provides (1) a detailed description of these proposed methodologies; (2) results from a case study in a research watershed located near Treynor, Iowa, demonstrating how these techniques can be used to assist conservation planning decisions; and (3) an example dataset and step-by-step procedure that will help practitioners apply RUSLE2 in 2D.

KEYWORDS

Erosion, light detection and ranging (LiDAR), geographical information system (GIS), Revised Universal Soil Loss Equation—Version 2 (RUSLE2), runoff

5.1 INTRODUCTION

The Revised Universal Soil Loss Equation–Version 2 (RUSLE2; ARS 2008; Renard et al. 2011) is the most recent in the family of universal soil loss equation (USLE) models that compute sheet-and-rill erosion in complex, one-dimensional (1D) hillslopes. RUSLE2 computes erosion along a 1D flow path that extends from the top of the hill, where runoff begins, to a location where runoff meets a concentrated flow channel, where sheet and rill computations are no longer applicable.

The RUSLE2 is an advancement of the erosion prediction technology of the RUSLE (Renard et al. 1997), as it implements sediment transport methods that permit the determination of sediment deposition that occurs in areas of reduced slope steepness frequently found in concave areas that may occur typically at hillslope bottoms, resulting in more accurate determination of the amount of sediment that is actually lost from agricultural fields.

Inherited from the application of the RUSLE and the USLE before that, the determination of erosion estimates for conservation planning using the RUSLE2 requires the selection by an expert of hillslope profiles that are representative of the erosion in the entire field. The recent availability of high-resolution topographic information has created the opportunity for the automatic determination of realistic runoff flow paths. Historically, automatic identification of the locations of concentrated flow in a geographical information system (GIS) has been less than satisfactory because of the poor resolution of readily available topographic data sources. The lack of resolution precluded the identification of low-gradient sections where substantial deposition may occur and did not allow the proper definition of small flow channels. New high-resolution topographic data, such as those available through the use of Light Detection and Ranging (LiDAR) and similar technologies, make it possible to overcome these limitations.

The primary objective of this chapter is to describe the approach being taken by the RUSLE2 development team to extend RUSLE2 as a two-dimensional (2D) application in a GIS environment that provides high-resolution, spatially distributed estimations of soil erosion. The new approach implements terrain analysis algorithms to define overland flow paths that are a realistic representation of runoff accumulation and convergence created by field topography. By computing runoff and erosion over each grid cell, the new model accounts for the spatial variability in the generation and accumulation of runoff over the terrain and eliminates uncertainties that are associated with the selection of representative 1D profiles.

This chapter provides a detailed description of the proposed methodologies and shows erosion prediction results for a research watershed that is located near Treynor, Iowa. It also describes some of the problems encountered when applying the new model and proposes solutions that are currently being researched and implemented into a new generation of erosion prediction tools. A practical GIS exercise demonstrates how terrain analysis can be used to determine overland flow paths that help define the location and topographical characteristics of representative hillslope profiles for erosion estimation with RUSLE2.

5.2 GRID-BASED RUSLE2

Tools are being developed to extend the existing RUSLE2's computational engine to calculate runoff and soil erosion over the entire agricultural fields using digital elevation models (DEMs) that are generated from high-resolution LiDAR elevation data. The tools, being developed in cooperation with Agren, Inc. (Agren 2012), reuse RUSLE2's methods for 1D hillslope profiles to compute erosion, sediment transport, and deposition for each cell on the 2D computational grid that matches the resolution of the DEM describing the terrain.

The method involves determining overland flow paths by defining a flow direction for each grid cell using terrain elevation data. The flow directions are also used to determine the position of "channels" that represent pathways of concentrated runoff, which indicate locations where hillslopes end and where ephemeral gullies may form. Figure 5.1a depicts an area where a channel was identified, shown in the figure as a sequence of raster cells marked by a black outline. For each channel cell, flow directions are used to identify the corresponding drainage areas (shown in different shades of gray) to define the linkage between cells. These interconnected grid cells become 2D "profiles" that represent realistic flow pathways from hilltops to locations of concentrated flows. In the RUSLE2 computation, each 2D cell is equivalent to a 1D slope segment, as shown in Figure 5.1b. This approach allows the immediate application of the RUSLE2 computational engine in a 2D-gridded terrain description.

The method is conceived in three independent phases: phase 1 encompasses most of the user interaction through a graphical interface and consists of identifying the simulation area, generating soil and management layers, retrieving a DEM in the required resolution, and then determining the drainage networks and the locations of the concentrated flow channels that output RUSLE2 hillslope profiles. In phase 2, a C++ program analyzes GIS layers that are created in phase 1 to set up the RUSLE2 model for the 2D simulation area. It also executes the simulations and exports simulation results for postprocessing. In phase 3, results are converted into user-friendly formats such as maps, graphs, and summary tables according to user requirements and are optionally linked to an ephemeral gully or a channel erosion model. The user interface for this application and phase 3 programming remains rudimentary at this time.

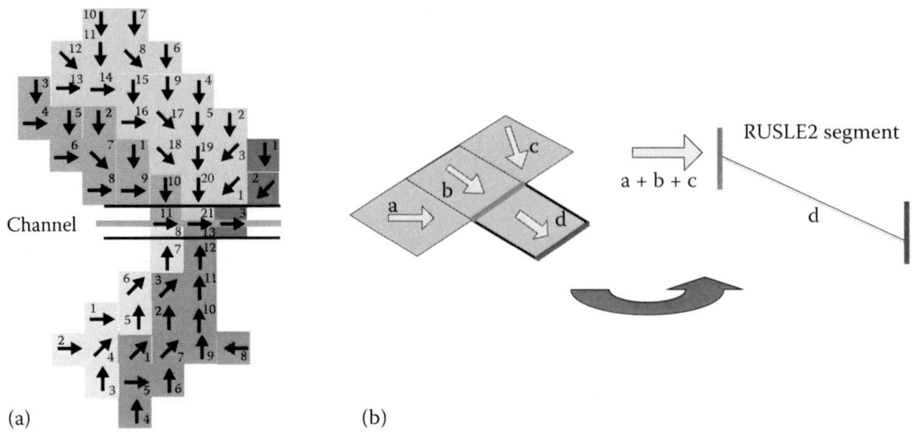

FIGURE 5.1 (a) Starting with each channel cell (black outline), a computer algorithm analyzes the flow direction map to determine the connectivity and computational sequence of the cells that compose a 2D "profile." The figure shows profiles on each side of the channel, identified in different shades of gray. (b) RUSLE2 utilizes the linkage between cells to determine the runoff and sediment loads entering each cell. It also calculates an effective slope length for each cell based on the ratio of runoff leaving the cell to that generated within the cell, enabling the appropriate accounting for topographic, soil, and management effects on local erosion estimates.

5.2.1 Overland Flow Paths

First, terrain elevation data processing is used to determine flow directions for all DEM cells. Flow directions are based on the Deterministic 8 (D8) method (O'Callaghan and Mark 1984), which assigns flow to the neighbor cell with the steepest downward slope. The raster map of flow directions is then used to determine a network of concentrated flow channels based on the accumulated contributing area. Because the locations of the channels determine the lower end of each hillslope, the channel map is used to define the areas that are draining into each channel cell. These become the "simulation units" for the distributed version of the RUSLE2 model, as hydrology and erosion processes can be computed independently for each of these units.

5.2.2 Channel Networks

In this study, two alternative methods were used to define the location of concentrated flow channels. In the first method, ArcGIS 10.1 was used to process the DEM so that all enclosed depressions or pits were removed. Pits (local minima) create difficulties for algorithms that simulate the flow of water across the landscape. The pits were removed by raising the elevation of raster cells within each pit until they matched the elevation of the lowest boundary pixel that permitted flow out of the pit. ArcGIS implements the method of Jenson and Domingue (1988). Flow directions for each cell were determined using the D8 method, which assigns a single flow direction for each cell, in the direction of the steepest downward slope. The location of areas of concentrated flow was determined by computing the accumulated drainage area for each cell using the D8 flow directions. All cells with an accumulated drainage area larger than the threshold value were considered to contain a channel. A second method, developed by Agren, Inc., includes initial DEM smoothing and flow directions that are determined by the D8 method using a DEM with limited pit filling. Because some pits remain, the flow connectivity is incomplete, and the resulting channel network is usually interrupted. An additional algorithm based on a neural network application developed for computer vision is applied to improve the connectivity of channels. Optionally, an algorithm for edge detection is used to enforce flow blockages, such as terraces in agricultural fields, by modifying the flow directions of cells near identified obstructions.

5.2.3 2D "Profiles"

Once the raster maps containing flow directions and channels are determined, these are further processed by algorithms that define the 2D "profiles" on which erosion calculations will be performed. Each cell crossed by a channel defines a drainage area outlet corresponding to the end of an overland flow path (Figure 5.1a). Therefore, for each channel cell, the flow direction map is analyzed to determine which of the neighboring cells drain to that channel cell. The process is recursive and somewhat complex: if a cell drains into the cell that is being inspected, the focus is shifted to that second cell. This process is repeated in checking cells uphill until no inflow is detected for a cell. The no-inflow cell identifies the beginning of a flow path. The cell is marked and numbered. A reverse process is then started, following flow paths downhill, defining the connectivity among the several cells that compose the area draining to the original channel cell. The result is a 2D RUSLE2 profile that comprises an ordered collection of raster cells. The channel cell itself is divided by the channel and potentially contributes to the slope length of the left and right bank overland flow paths. The process is repeated for each channel cell in the network.

5.2.4 Erosion Computations

The computational module of phase 2 accesses RUSLE2's computational engine through its application programming interface (API), distributed as a dynamic-link library. A C++ program uses data layers prepared in phase 1 (flow directions, channel network cells, slope steepness, soil map, agricultural management map) as input and user-defined parameters such as RUSLE2 simulation options and requirements for data output. The program creates the 2D "profiles" as described above and then manages the execution of the simulations through RUSLE2 API functions. RUSLE2 internally manages the transfer of runoff and sediment among the cells according to the cell interconnectivity that is prescribed. The RUSLE2 profiles can be computed independently, in any order, which creates the potential for optimization through parallel computations in multiprocessor or multicore computers. RUSLE2 outputs of distributed soil erosion, sediment delivery to channels, and sediment deposition in channels and sediment basins are retrieved and saved. Ultimately, channel (ephemeral gully) erosion will be estimated by linkage of RUSLE2 results with a channel model.

5.2.5 Enhancements to RUSLE2

A couple of enhancements to the RUSLE2 engine were necessary to implement the new 2D approach. RUSLE2 was modified so that the slope length at a hillslope cell could be determined based on accumulated runoff. The slope length is based on the ratio of runoff leaving a cell to that of runoff generated within a cell. This approach is more powerful than determining the slope length from a contributing area alone since it also accounts for spatial variability in runoff generation related to soil type and management effects.

RUSLE2 was enhanced to generate a representative series of runoff events based on existing monthly climate data (Dabney et al. 2011b). The new methods allow the prediction of daily runoff and sediment amounts that are suitable for linkage with a physically based ephemeral gully model. These changes were incorporated into the version of RUSLE2 that is available at the ARS website (USDA-ARS 2012).

5.3 MATERIALS AND METHODS

5.3.1 Model Application to Watershed "W-11" in Iowa

The 2D version of RUSLE2 was applied and compared with observations on Watershed 11 (Rachman et al. 2008) of the USDA-ARS Deep Loess Research Station located near Treynor, Iowa (Karlen et al. 2009). The predominant soil was Monona silt loam (fine–silty, mixed, superactive, mesic Typic Hapludolls). This 6.6 watershed was selected because of the extensive research archive that exists for it. Specifically, four hillslope watersheds labeled 1 to 4 (Figure 5.2a) were used in the original RUSLE1.04 documentation (Renard et al. 1997) to illustrate the proper selection of hillslope profiles, which should extend from ridge tops to areas of concentrated flow.

Daily runoff and sediment yield were monitored at the watershed outlet from 1975 to 2002. Throughout the period of record, a grassed waterway was located in the lower portion of the watershed, and the field was tilled and planted with contour-rows. Beginning in 1991, contour grass hedges began to be established (Figure 5.2d). Hedges were approximately 1 m wide and were spaced 15.4 m apart to accommodate 16 crop rows. From 1975 to 1996, the field was farmed with contour-planted, conventional-tilled (CT) corn (*Zea mays*, L.). From 1997 to 1999, the field was planted with no-till (NT) soybean (*Glycine max*, L.), and an NT corn–soybean rotation was used from 2000 to 2002 (Rachman et al. 2008). The average corn yield for the period 1987 to 1996 was 7.6 Mg ha^{-1} (Eghball et al. 2000).

A DEM was created at a 3-m resolution from the 0.31-m contour lines shown in Figure 5.2a using the ArcGIS function TopoToGrid, based on the package ANUDEM (Hutchinson 1989). The method used pit filling and enforced flow along the digitized gullies from Figure 5.2a. Channels were assumed to start where the accumulated drainage area reached four criteria: 300, 600, 900, or 4000 m^2. Among these alternatives, the 600 and 900 m^2 resulted in similar flow networks, and the 600-m^2 value was used in subsequent analyses. A 0.5-m DEM, which was used to compute water depths with an overland flow model for a large runoff event (Figure 5.2c), was created by interpolation from a triangulated irregular network generated from the RTK

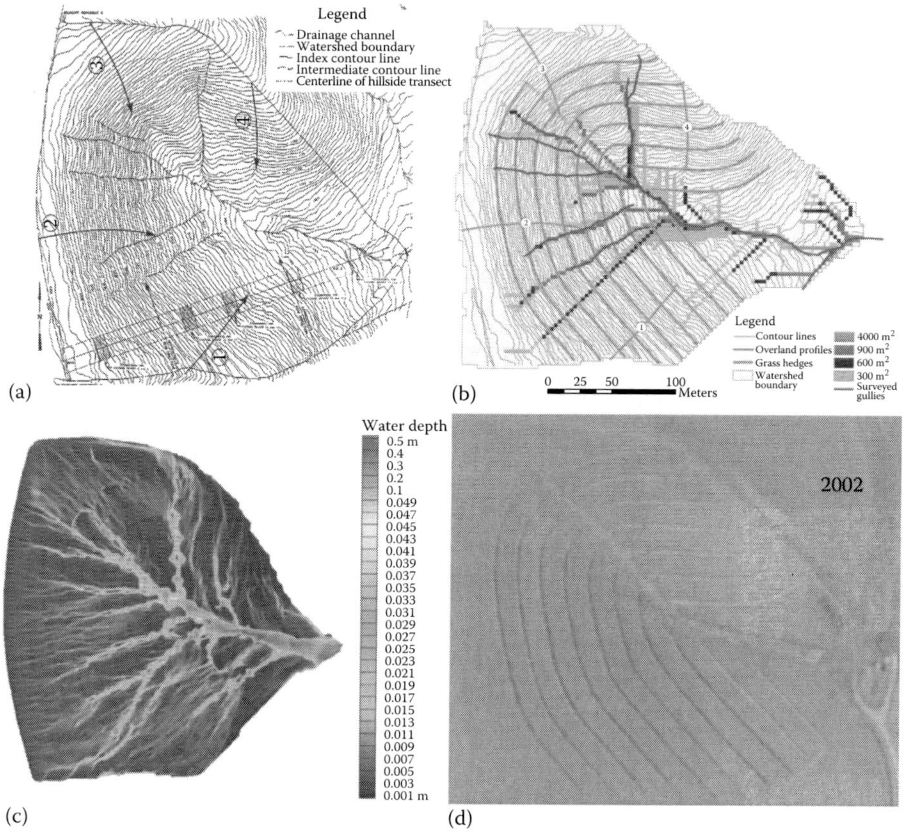

FIGURE 5.2 (See color insert.) (a) Topographic map (0.31-m contour interval) of Watershed 11 at Treynor, Iowa, illustrating appropriate RUSLE hillslope flow paths (From Renard, K.G. et al., *Predicting Soil Erosion by Water: A Guide to Conservation Planning with the Revised Universal Soil Loss Equation* [*RUSLE*]. Agric. Handbook 703, U.S. Department of Agriculture–Agricultural Research Service: Washington, DC, Figure 4-5B, 1997); (b) concentrated flow paths created from a 3-m DEM for alternative minimum contributing areas, indicating that 600 m^2 is the best approximation of the location of concentrated flow areas that end RUSLE hillslope flow paths; (c) water flow depths after 1 h of steady runoff of 50 mm h^{-1} based on a 0.5-m DEM, anisotropic roughness, and a diffusive wave solution (From Vieira, D.A.N., and S.M. Dabney, *Hydrological Processes*, 26:2225–2234, 2012); and (d) aerial photo showing the locations of grass hedges and waterways in 2002.

survey points collected in 1999. Slope steepness was computed using the ArcGIS, which uses a moving 3 × 3 kernel (Horn 1981).

RUSLE2 simulations were conducted using a 3-m DEM for three scenarios for growing spring-plowed CT corn yielding 7.6 Mg ha^{-1}: (1) no additional conservation practices; (2) with grass hedges; and (3) NT soybean yielding 3.0 Mg ha^{-1} rotated with NT corn yielding 7.6 Mg ha^{-1} with grass hedges. For cells with grass hedges, the entire cell was treated as a hedge to approximate the effect in the 1D hillslope version of RUSLE2, where backwater effects are treated by creating an extra depositional length based on local steepness and the hydraulic resistance of the vegetation (USDA-ARS 2013). A limitation of the current study is that no channel model was employed to compute either sediment deposition or ephemeral gully erosion in the channels. Rather, sediment delivered to the channels by RUSLE2 is reported as soil loss in this report. RUSLE2 runs were also performed for profile 2 (Figure 5.2a and b) for three management scenarios: CT corn, CT corn with grass hedges, and NT corn–soybean rotation. The standard database climate records for Pottawattamie County, IA were used in all RUSLE2 simulations.

5.4 RESULTS

Observed rainfall depths during the three observation periods were similar to the 30-year county average in the official National Resources Conservation Service database (Table 5.1). The RUSLE2-predicted runoff was about twice the observed runoff during the first two observation periods, possibly because the RUSLE2 simulation did not include a representation of the grassed waterway, which slows down runoff and increases infiltration. During the 1997–2002 NT period, the differences between the observed runoff and the predicted runoff were smaller; but the RUSLE2 predicted less runoff relative to earlier periods, while observed runoff, like rainfall, was slightly higher. The RUSLE2 runoff estimates were only slightly reduced by the inclusion of grass hedges.

The RUSLE2 hillslope sediment yields greatly exceeded sediment yield at the watershed outlet. This large difference occurred because RUSLE2 estimates represent sediment delivery to the concentrated flow channels that are delineated in the watershed and do not consider sediment deposition in the channels. RUSLE2 can estimate deposition in channels, but this feature was not implemented in the current simulations. The higher erosion in profile 2 (Figure 5.2a and b) than in the distributed watershed average reflects its higher than average slope steepness and higher local erosion rates. Grass hedges reduced sediment delivery by approximately 50%

TABLE 5.1
Long-Term County Average Rainfall, Observed Average Annual Rainfall, Observed and Predicted Average Annual Runoff, and Observed and Predicted Sediment Yield for Watershed 11 Near Treynor, Iowa

	Pottawattamie County	CT (1975–1991)	CT Hedge (1991–1997)	NT Hedge (1997–2002)
Annual rainfall (mm)	801	811	766	835
Runoff observed (mm)		50	48	78
Runoff predicted (mm)		110	107	88
Sediment yield (Mg ha^{-1} year^{-1})				
Observed		14.6	5.6	11.3
Profile 2		83.9	39.9	2.2
2D simulation		46.5	22.3	0.9

in both observations and predictions. However, conversion to NT reduced erosion much more in the RUSLE2 simulations than was observed in the watershed. It seems likely that ephemeral gully erosion caused this discrepancy (Tomer et al. 2007).

Erosion patterns that are predicted by the 2D simulation reflected slope steepness and flow accumulation effects. Overall treatment effects were similar in the distributed estimates to those estimated using the 1D version of RUSLE2 applied to profile 2. Sheet and rill erosion were lower downslope of areas where accumulated runoff was intercepted by channels. Figure 5.3a and b shows computed soil loss for the CT management using flow directions and channel networks that are created using the D8/flow accumulation method (Figure 5.3a) and the method developed by Agren, Inc. (Figure 5.3b). Both networks are similar, but the latter avoided the creation of long straight channels that do not appear realistic. Neither procedure reproduced the complex patterns rendered by the overland flow simulation using a higher-resolution DEM (Figure 5.2c), but both were adequate to allow reasonable estimation of distributed patterns of sheet and rill erosion with the 2D version of RUSLE2.

Where sediment loads were high with CT management, grass hedges became depositional zones (Figure 5.3c), whereas they were merely areas of very low soil loss when sediment loads were low with NT management (Figure 5.3d). The discrepancy between the low predicted soil loss with NT management and the considerable sediment

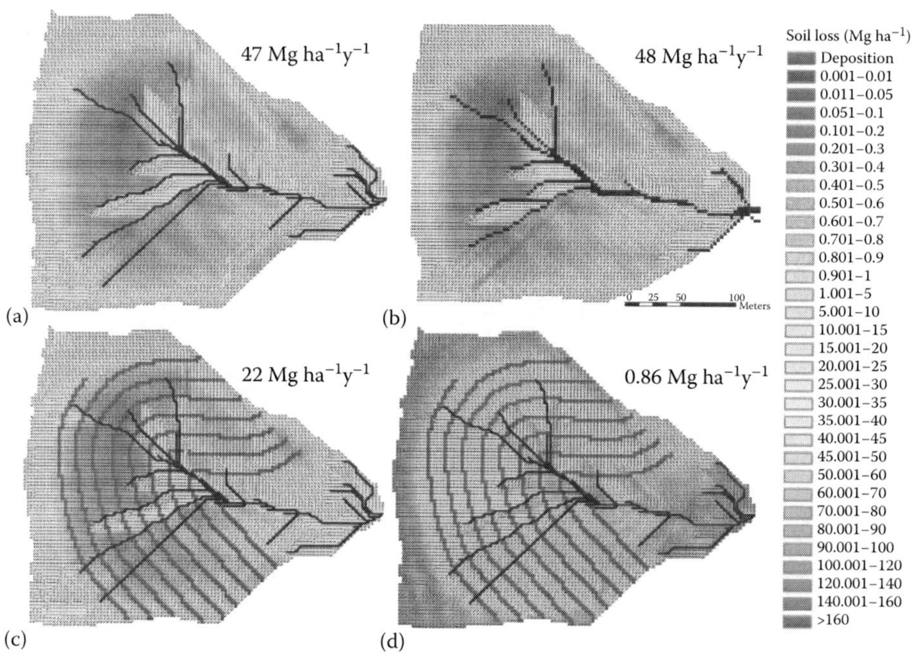

FIGURE 5.3 **(See color insert.)** Flow vectors and patterns of annual RUSLE2 soil loss on a 3-m grid and the field average soil loss for corn yielding 7 Mg ha^{-1} with spring-plow tillage (a, b), with grass hedges (c), or no-till corn with grass hedges (d). Black lines show concentrated flow channels based on 600-m^2 minimum contributing area using a D8 flow-routing algorithm (a, c, d) or using the channel network generation algorithm being developed by Agren, Inc. (b).

delivery that is observed during the period 1997–2002 (Table 5.1) again emphasizes the importance of adding an ephemeral gully estimate to the sheet and rill erosion estimate that is currently provided by RUSLE2. The channel network definition combined with the RUSLE2 estimates of runoff and sediment loading to the channels for a sequence of storm events provides the basis for driving a channel erosion model.

5.5 PROBLEMS THAT REMAIN TO BE SOLVED

The 2D RUSLE2 simulations exposed several problems that remain to be solved: (1) developing flow vector maps that respect linear barriers such as terraces; (2) automating conditional or "selective" pit filling; (3) improving calculation of slope steepness downslope of topographic breaks; and (4) reducing runtime of the distributed RUSLE2 erosion computations.

The first two of these problems are related and are illustrated in Figure 5.4, which depicts a ~100-ha area in Boone County, Iowa. LiDAR data were aggregated to a 3-m DEM, and a flow network was determined with complete pit filling using ArcGIS procedures. The difference in elevation between the original and the filled DEMs was calculated and is shown in blue shades in Figure 5.4a to illustrate the depth of

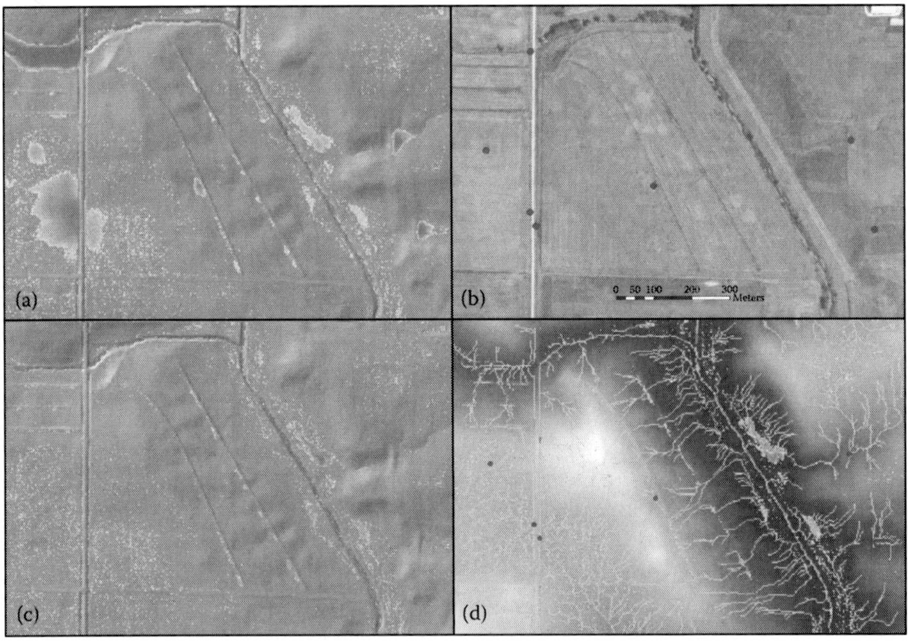

FIGURE 5.4 (See color insert.) Shaded relief map of fields in Boone County, Iowa, (41.9315′N, 93.7623′W) showing depths of filled pits (in shades of blue) when flow vectors were determined with a standard D8 algorithm with full pit filling for a 3-m DEM derived from LiDAR (a); color orthophoto showing locations of seven user-selected pixels that were declared "NoData" (b); filled pits resulting from processing the DEM with the seven "NoData" pixels (c); and an elevation map showing a channel network based on 675-m² minimum contributing area using a D8 flow-routing algorithm (d).

pit filling needed so that all cells would flow to an outlet. Several areas of deep (anomalous) pit filling are noted, including the following: (1) a stream channel was filled to overflow a road embankment in the northwest part of the scene; (2) two debris basins were filled in the eastern part of the scene; and (3) an area behind the upper terrace was filled in the center of the scene. Each of these areas actually has a physical outlet: a road culvert or a tile inlet. Other areas of the scene show large areas of filled pits that represent relatively flat areas that may be pot holes or may drain into road ditches through outlets that are not resolved in the 3-m DEM. Through inspection of Figure 5.4a, seven pixels were selected, and their elevation values were set to "NoData" (Figure 5.4b), essentially turning them into scene edges and thus declaring those points as outlets for routed flow. The pit filling was repeated for the modified DEM, and the resulting difference map (Figure 5.4c) shows depths of filled pits. Elevation and flow channels defined with a minimum contributing area of 675 m^2 are shown in Figure 5.4d.

Several problems may be noted with the channel network. Many channels still cross the visible terraces despite the fact that there is relatively little pit filling. This is a problem related to the aggregation of LiDAR data to a 3-m resolution, although it may also reflect real breaches in the terraces. At 3-m resolution, terraces may, in places, be only one pixel wide. Pixels will seldom fall directly on the terrace so that even though the terraces are clearly visible in the shaded relief map, the elevation of terrace pixels may be lowered by averaging with the surrounding cells. When diagonal terraces are only a single pixel wide, the D8 algorithm can easily find a flow path that crosses the terraces at touching corners (Figure 5.5). This problem would not be solved by increasing the elevation of the terrace cells. One solution would be to buffer the terraces so that they are at least two pixels wide at every point. Another alternative is to define or recognize a linear feature that would act as a barrier that a modified D8 search algorithm would respect so that it would not find an outlet that would cause flow to cross the defined line. A refinement of this approach would consider the height of the barrier with respect to runoff amount as flow directions may depend on runoff depth (Vieira and Dabney 2012).

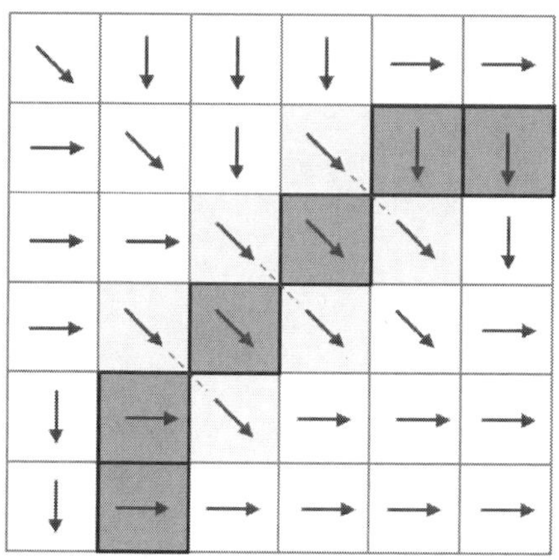

FIGURE 5.5 Illustration of how flow directions determined by the D8 flow-routing algorithm can cross single-pixel terraces (dark gray), passing through light gray-shaded cells.

The seven "NoData" pixels avoided filling some large pits where real outlets of runoff were postulated to exist. A method to automate this "selective pit filling" is needed. Pit-filling algorithms could be developed in which the search for an outlet would be abandoned if the depth, area, or volume of the depression exceeded specified thresholds and the lowest pixel in the depression would be declared an outlet. Such a method would fill all small pits to ensure flow connectivity but not large depressions such as those behind terraces or representing natural potholes. A still more sophisticated tool would search for and allow definition of linkages between the declared outlets and the channel network.

Dabney et al. (2011a) showed that using a 3 × 3 kernel method to calculate slope steepness resulted in anomalously steep values downslope of abrupt topographic breaks such as terraces. This problem is illustrated for a 1D schematic at point "D" in Figure 5.6, in which a more accurate representation of the steepness of the tilled area downslope of a terrace would be estimated using only points "D" and "E," ignoring point "C."

The problem is generalized to the 2D situation, and a possible solution is illustrated in Figure 5.7. A linear feature, similar to that suggested to prevent flow vectors from crossing barriers, could be used to limit slope steepness searches to points on the downhill side of the linear feature. "Ghost" values for points "D," "G," and "H" could be interpolated before the value for point "E" was determined from the 3 × 3 kernel. This approach has been used successfully for the computation of slope steepness near barriers in the modeling of erosion caused by tillage (Vieira and Dabney 2009).

Incorrectly determining slope steepness downslope of topographic breaks may not greatly affect overall erosion rates, but errors can be compounded if erosion estimates are used as feedback to predict patterns of landscape evolution (Dabney et al. 2011a). Adding feedback to erosion predictions is essential if long-term effects on soil properties are to be assessed. Ephemeral gully erosion estimation must be included

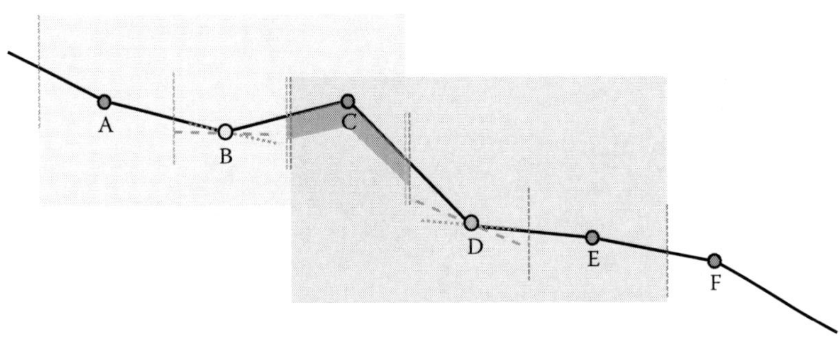

FIGURE 5.6 Illustration of problems with determining correct slope steepness in the vicinity of sharp breaks in topography such as those caused by the existence of a terrace. If a channel is defined upslope of the terrace, the channel elevation can be used to solve the problem there, but downslope of the terrace, slope steepness is generally estimated to be anomalously high by a 3 × 3 moving kernel. For cell D, estimating slope from points C and E gives the slope steepness represented by the dashed line; a more accurate representation is shown as the dotted line, which is the steepness between cells D and E.

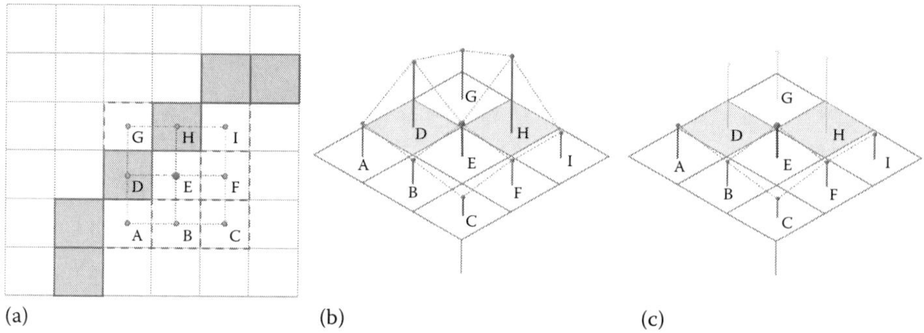

FIGURE 5.7 The slope steepness of cell "E" near a feature such as a terrace, shown in (a) as gray cells, is overestimated if a standard 3 × 3 kernel method uses elevation points that lie on the top of the wall (D and H) or on the opposite side (G), as illustrated in (b). A better method that recognized wall cells would exclude points D, G, and H and utilize only data points on the same side of the wall. As shown in (c), excluded cells would be substituted by interpolated values computed using only valid cells. For example, the value cell H in (c) could be substituted by linear interpolation using cells B and E.

if such simulations are to be realistic, since the channel system in the topography reflects cumulative and progressive incision over time. Tillage erosion is another important cause of soil redistribution that affects topography and spatial patterns of soil organic carbon loss and accumulation (Van Oost et al. 2007).

A final limitation of the distributed version of RUSLE2 is excessive runtime. The simulations illustrated in Figure 5.3 each took approximately 30 min to complete. The calculation of soil biomass, soil residue cover, soil roughness, and similar properties for every day of a simulation is one of the most time-consuming steps in RUSLE2 computations, so reusing this information for the limited number of soil and management combinations found in a typical site simulation reduces the runtime of a grid-based simulation. To improve computational efficiency, RUSLE2 has also been recoded to allow reuse of common information in grid-based calculations (Figure 5.1a). To use this "cell-based" approach, information about the linkage between the cells such as that used by the shell program described above is passed to the RUSLE2 through its API, and RUSLE2 calculates the runoff, erosion, and sediment transport for the entire domain. Memory management currently limits the success of this approach, and development work continues.

5.6 CONCLUSIONS

Distributed erosion models have the potential for identifying critical areas in landscapes. Critical areas can be transformed into management zones for the application of precision conservation. Tools are needed for a proper and efficient representation of linear roughness and flow barrier elements and slope steepness breaks to create realistic representations of flow paths and then to define appropriate locations of concentrated flow areas that end RUSLE2 hillslopes. These flow channels include the following: terrace channels that do not erode due to modest slope gradients; grassed waterways that do not erode due to vegetative cover; and ephemeral gullies that gradually burn their pattern into the landscape topography as they periodically erode and are filled by subsequent tillage operations. A model that can predict erosion, sediment transport, and deposition in upland channels must be linked to the distributed runoff and sheet and rill erosion estimates that are created by RUSLE2 in order to improve the correspondence between predicted and observed sediment delivery from field-sized watersheds.

REFERENCES

Agren, Inc. 2012. Estimating soil loss with RUSLE2 & LiDAR. Available at http://www.agren-inc.com/projects.php?proj=15 (accessed January 10, 2014).

Dabney, S.M., D.A.N. Vieira, and D.C. Yoder. 2011a. Effects of topographic feedback on erosion and deposition prediction. Paper #11091. *International Symposium on Erosion and Landscape Evolution CD-Rom Proceedings* (September 18–21, 2011, Hilton Anchorage, Anchorage, Alaska), St. Joseph, MI, ASABE.

Dabney, S.M., D.C. Yoder, D.A.N. Vieira, and R.L. Bingner. 2011b. Enhancing RUSLE to include runoff-driven phenomena. *Hydrological Processes* 25:1373–1390.

Eghball, B., J.E. Gilley, L.A. Kramer, and T.B. Moorman. 2000. Narrow grass hedge effects on phosphorus and nitrogen in runoff following manure and fertilizer application. *Journal of Soil and Water Conservation* 55:172–176.

Horn, B.K.P. 1981. Hill shading and the reflectance map. *Proceedings of the IEEE* 69:14–47.

Hutchinson, M.F. 1989. A new procedure for gridding elevation and stream line data with automatic removal of spurious pits. *Journal of Hydrology* 106:211–232.

Jenson, S.K., and J.O. Domingue. 1988. Extracting topographic structure from digital elevation data for geographic information system analysis. *Photogrammetric Engineering and Remote Sensing* 54:1593–1600.

Karlen, D.L., D.L. Dinnes, M.D. Tomer, D.W. Meek, C.A. Cambardella, and T.B. Moorman. 2009. Is no-tillage enough? A field-scale watershed assessment of conservation effects. *Electronic Journal of Integrative Biosciences* 7(2):1–24.

O'Callaghan, J.F., and D.M. Mark. 1984. The extraction of drainage networks from digital elevation data. *Computer Vision, Graphics and Image Processing* 28:328–344.

Rachman, A., S.H. Anderson, E.E. Alberts, A.L. Thompson, and C.J. Gantzer. 2008. Predicting runoff and sediment yield from a stiff-stemmed grass hedge system for a small watershed. *Transactions of the ASABE* 51:425–432.

Renard, K.G., G.R. Foster, G.A. Weesies, D.K. McCool, and D.C. Yoder. 1997. *Predicting Soil Erosion by Water: A Guide to Conservation Planning with the Revised Universal Soil Loss Equation (RUSLE).* Agric. Handbook 703, Washington, DC: U.S. Department of Agriculture–Agricultural Research Service.

Renard, K.G., D.C. Yoder, D.T. Lightle, and S.M. Dabney. 2011. Universal soil loss equation and revised universal soil loss equation. In Morgan, R.P.C. and M.A. Nearing (eds.), *Handbook of Erosion Modeling*, pp. 137–167. Oxford, England: Blackwell Publishing Ltd.

Tomer, M.D., T.B. Moorman, J.L. Kovar, D.E. James, and M.R. Burkhart. 2007. Spatial patterns of sediment and phosphorus in a riparian buffer in Western Iowa. *Journal of Soil and Water Conservation* 62:329–338.

USDA-ARS. 2012. Revised Universal Soil Loss Equation 2—Changes in 2010. Available at http://www.ars.usda.gov/Research/docs.htm?docid=20222 (accessed January 10, 2014).

USDA-ARS. 2013. Science documentation. Revised Universal Soil Loss Equation, Version 2 (RUSLE2). Available at http://www.ars.usda.gov/sp2UserFiles/Place/64080510/RUSLE/RUSLE2_Science_Doc.pdf (accessed June 06, 2014).

Van Oost, K., T.A. Quine, G. Govers, S. De Gryze, J. Six, J.W. Harden, J.C. Ritchie, G.W. McCarty, G. Heckrath, C. Kosmas, J.V. Giraldez, J.R. Marques da Silva, and R. Merckx. 2007. The impact of agricultural soil erosion on the global carbon cycle. *Science* 318:626–629.

Vieira, D.A.N., and S.M. Dabney. 2009. Modeling landscape evolution due to tillage: Model development. *Transactions of the ASABE* 52:1505–1522.

Vieira, D.A.N., and S.M. Dabney. 2012. Two-dimensional flow patterns near contour grass hedges. *Hydrological Processes* 26:2225–2234.

6 Application of Geographical Information System and Terrain Analysis for Designing Filter Strips

Michael G. Dosskey, Surendran Neelakantan, Tom Mueller, Todd Kellerman, and Eduardo A. Rienzi

CONTENTS

EXECUTIVE SUMMARY

Filter strips are installed along the margins of agricultural fields to trap sediment and other pollutants conveyed in overland runoff. A geographical information system (GIS) procedure is presented for designing a filter strip that will achieve a constant, user-selected level of trapping efficiency along a field margin. The design model is based on relationships between the buffer area ratio and trapping efficiency, which account for pollutant type and site conditions. The design model has been coupled with GIS and terrain analysis in a computer program called AgBufferBuilder to run with ArcGIS for quickly producing filter strip designs that can vary in width along a field margin to compensate for concentrated runoff flows. The design procedure is demonstrated on a sample agricultural field. The computer program, user's guide, and demonstration datasets can be downloaded from the website http://www2.ca.uky.edu/BufferBuilder.

KEYWORDS

Best management practices, conservation planning tools, geospatial analysis, nonpoint source pollution, precision conservation, vegetative buffers, water quality

6.1 INTRODUCTION

Filter strips are installed along the margins of agricultural fields to reduce the load of sediment and other pollutants transported from fields in overland runoff to waterways (USDA 2013). The filter strips function mainly by promoting infiltration of runoff water and deposition of sediments. They perform best where field runoff is uniformly dispersed across the entire filter strip. Where overland runoff is not uniform, however, the performance is limited. To counter this problem, additional measures may be taken such as land shaping and level spreaders to disperse runoff flow through a filter strip. Alternatively, the filter strip could be made larger where more runoff flows and smaller where less runoff flows. This latter approach bolsters the effectiveness of the filter strip without the added cost of additional measures for dispersing concentrated flows.

A performance-based, quantitative model for designing filter strips has been developed (Dosskey et al. 2011), which matches filter size to spatial patterns of runoff flow. In general, the model sizes short segments of the filter strip in proportion to the size of the upslope contributing area that drains to each segment, i.e., the buffer area ratio. Each segment of a filter strip is determined independently. Automation of the design model for use with the GIS software ArcGIS (ESRI, Redlands, California) enables rapid design of many segments around large agricultural fields. The computer program, AgBufferBuilder version 1.0, produces a design that will yield an approximately constant, user-selected level of pollutant-trapping efficiency (i.e., percent of runoff pollutant load that is retained in the filter strip) around an entire field. The program can also estimate the performance of alternative user-defined configurations. The tandem of design and assessment modules is used to develop an idealized design that can then be modified into more practical configurations and can be compared for size and performance. The computer program, user's guide, and practice datasets are downloadable from the website http://www2.ca.uky.edu/BufferBuilder. In this chapter, the operations of the AgBufferBuilder 1.0 program are described, and the program is demonstrated on a sample agricultural field.

6.2 MATERIALS AND METHODS

6.2.1 Design Model

The core design model (Dosskey et al. 2011) guides the user to a buffer area ratio that will achieve a user-selected level of pollutant-trapping efficiency under the existing field conditions (i.e., slope, soil texture, and tillage and soil cover) and pollutant type during a design rainfall event (61 mm in 1 h). Multiplying that buffer area ratio by the size of the runoff contributing area to a given segment of field margin determines the size of the filter strip to be installed along that segment. This process is repeated for each segment of field margin where the filter strip is to be installed.

Mathematical relationships between the pollutant-trapping efficiency and the buffer area ratio were developed by simulation modeling using a process-based vegetative filter strip model (VFSMOD; see Muñoz-Carpena and Parsons 2005). For a complete description of VFSMOD and its development, refer to the website http://abe.ufl.edu/carpena/vfsmod/. For a detailed description of the core model and its development from VFSMOD simulations, refer to Dosskey et al. (2011).

The use of computer technology can speed up the design process for many small segments. In a GIS, AgBufferBuilder 1.0 uses a digital elevation model (DEM) to divide the entire field into grid cells where each field margin segment is the size of one grid cell. The DEM is analyzed to determine the number of upgradient grid cells comprising the contributing area to each field margin cell and their average slope. The user inputs two field site conditions: texture class of the surface soil and the tillage and soil cover condition for the basis of the design. The user also inputs design criteria: pollutant type and the level of trapping efficiency for which to design. Then, executing the design program determines which grid cells must contain a filter strip to achieve the desired trapping efficiency, provides a table of statistics including filter size and performance level, and displays a map of the resulting design around the entire field margin.

Modifications to that design can also be made and assessed for comparative size and performance. In the GIS, the user can draw an alternative filter strip configuration. Then, executing the program determines the reconfigured buffer area ratio and the associated trapping efficiency for each field margin segment and subsequently computes the average trapping efficiency of all segments that are weighted by the size of their respective contributing areas.

6.2.2 Computer Hardware, Software, and Data

The AgBufferBuilder 1.0 program runs in coordination with ArcGIS version 10.0 with SP5 and version 10.1 with SP2 (ESRI). It is assumed that the user has the ArcGIS program loaded onto a computer that has a sufficient computing capability, network connectivity, and peripheral hardware for acquiring digital datasets, conducting the GIS operations, and displaying the results. It is further assumed that the user has training and experience working with ArcGIS and these system components.

Using the program requires a digital aerial orthophotograph and a DEM of the general area of interest. Standardized and reliable sources of DEM and aerial imagery include the USGS National Map (http://nationalmap.gov) and the USDA–NRCS Geospatial Data Gateway (http://datagateway.nrcs.usda.gov). The recommended procedures for obtaining DEM and imagery from the USGS are provided in Appendix 6.1 on the CRC Press website at http://www.crcpress.com/prod/isbn/9781439867228 or from the AgBufferBuilder website (http://www2.ca.uky.edu/BufferBuilder). Make sure that the DEM and the orthophotograph are in a proper projection and have the desired resolution. Keep in mind that finer-resolution DEM data will produce results with greater precision but will require longer computation times.

The AgBufferBuilder 1.0 program has two components: (1) Clip & Project, Editing Tools that prepare the imagery and DEM files for subsequent use in the (2) Buffer Builder Main tool. The Clip & Project, Editing Tools component is used to define the field area and its border and, additionally, the filter polygons if conducting an assessment using a package of basic ArcGIS functions. The Buffer Builder Main tool then executes the model to design a new filter strip and/or to assess the size and performance of one that is defined by the user.

6.2.3 Steps in the Design Procedure

The design procedure will produce a map showing locations to place the filter strip around an agricultural field to yield an approximately constant, user-selected trapping efficiency along the field margin for sediment or sediment-bound pollutants. The output includes a map and summary statistics including the total area of the designed filter strip and its estimated level of performance under the design storm conditions.

The major steps in the design procedure include the following.

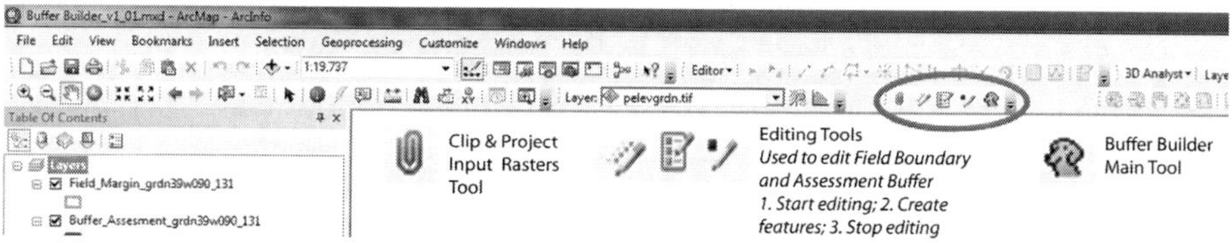

FIGURE 6.1 The ArcMap tool bar after opening the BufferBuilder_v1_01.mxd ArcMap project file. The AgBufferBuilder 1.0 toolbar icons are highlighted.

6.2.3.1 User Downloads the AgBufferBuilder Program

From the website http://www2.ca.uky.edu/BufferBuilder, download the program zip file for the version of ArcGIS that will be used. Extract the .zip file to C:/. Open the folder named "Buffer_Builder." Then, open the file named "BufferBuilder_v1_01.mxd," which will automatically open an ArcMap project in ArcGIS. The ArcGIS toolbar will contain five new icons, including the Clip & Project Input Rasters tool, three editing tools (Start Editing, Create Features, and Stop Editing), and the Buffer Builder Main tool (Figure 6.1).

6.2.3.2 User Prepares the Orthophotograph and the DEM Data Files

Clip each dataset to the same general field area, and set the projection to be used in the design program. The Clip operation will create shape files with generic names to save the clipped data layers and will give the user the option to rename them. It is recommended that the user also run the ArcGIS Fill tool on the DEM to remove sink errors and improve subsequent drainage mapping.

6.2.3.3 User Creates the Field Border File

On the orthophotograph, manually digitize the outside margin of the field area, i.e., the field border. It must form a closed polygon. The outside margin is the furthest that a functional filter strip could realistically extend away from the adjacent agricultural field. Typically, it will be located along a property line or a roadside ditch, or along the upper bank of a stream or a lakeshore that you want to protect with the filter strip. In some locations, the field border may coincide with the existing cultivated field margin. In other cases, it may lie at a considerable distance from the existing cultivated field margin with herbaceous and/or forest cover in between. If the forest or shrub cover obscures the precise location of stream banks, the National Hydrography Dataset 1:24,000 scale stream map can be downloaded (http://nhd.usgs.gov/data.html) and overlaid on the orthophotograph to aid in locating an outside margin for the filter strip. Later, when the Buffer Builder Main tool is executed, the field border polygon automatically will be clipped and projected on the DEM grid to create a field border raster file.

6.2.3.4 User Inputs Site Conditions and Design Criteria

Open the Buffer Builder Main tool. Select from a drop-down menu one of two pollutant types for the design (sediment or sediment-bound chemical). Also, from the drop-down menu, the user selects a soil texture class number (0, 1, or 2), which describes the texture class name of the surface soil that dominates the field (Table 6.1). From the third drop-down menu, the user selects one of two soil protection conditions for the field area (no-till or chisel after corn leaving good residue cover, or plow tillage after corn or chisel after soybeans). Soil protection condition is equivalent to the Universal Soil Loss Equation (USLE) cover and management factor (*C* factor) (Wischmeier and Smith 1978), and the model uses *C* factor values of 0.50 and 0.15, which correspond to these condition names at the seedbed stage, respectively. Examples of different soil cover and management conditions and their associated *C* factor values are shown in Table 6.2; more extensive tables are found in the work of Wischmeier and Smith (1978). For rotation systems, it is recommended that the soil protection condition be based on the part that has the higher *C* factor value (i.e., would produce the higher sediment load). Finally, the user inputs the level of trapping efficiency (in percent) to be achieved by this design. Trapping efficiency is the percentage of current runoff load from the field area that would be retained in the filter strip. The user can either type in a numerical value or use the slider bar.

TABLE 6.1
Soil Texture Class Names and Their Acronyms and the Corresponding Design Model Number

Soil Texture Class Name	Acronym	Class Number
Clay	C	0
Silty clay	SiC	0
Clay loam	CL	0
Silty clay loam	SiCL	0
Silt	Si	0
Sandy clay	SC	1
Silt loam	SiL	1
Loam	L	1
Very fine sandy loam	VFSL	1
Fine sandy loam	FSL	2
Sandy clay loam	SCL	2
Loamy very fine sand	LVFS	2
Sandy loam	SL	2
Loamy fine sand	LFS	2
Very fine sand	VFS	2
Coarse sandy loam	CSL	2
Fine sand	FS	2
Sand	S	2
Loamy coarse sand	LCS	2
Coarse sand	CS	2

Note: Class name modifiers such as gravelly, cobbly, channery, etc., are ignored for design model purposes.

TABLE 6.2
Example Values of the USLE *C* Factor for Different Soil Cover and Management Conditions

Cover Type and Management	Other Characteristics	USLE *C* Factor
Undisturbed forest land	45%–70% canopy cover and 75%–85% duff cover	0.002 to 0.004
Idle land: tall grass, weeds, short brush	75% canopy cover and 60% litter cover	0.03 to 0.07
Idle land: tall grass, weeds, short brush	25% canopy cover and 20% litter cover	0.17 to 0.20
Chisel or no-tillage after corn	Seedbed stage and 50% residue cover	0.13 to 0.17
Disk plow tillage after corn	Seedbed stage and 10% residue cover	0.45 to 0.52
Chisel tillage after soybeans	Seedbed stage	0.40 to 0.58
Moldboard plow tillage after soybeans	Seedbed stage	0.72 to 0.83

Source: Values were extracted from among the more extensive tables contained in Wischmeier, W.H., and D.D. Smith, *Predicting Rainfall Erosion Losses—A Guide to Conservation Planning*. Agric. Handbook No. 537, USDA, Washington, DC. Available at http://topsoil.nserl.purdue.edu/usle/AH_537.pdf (accessed May 1, 2014), 1978.

Upon completion of parameterization, clicking "OK" executes the Buffer Builder Main tool, which automatically conducts the following steps.

6.2.3.5 AgBufferBuilder Calculates Terrain Attributes for Field Border Grid Cells

The number of field area grid cells that contribute runoff flow to each field border grid cell is determined by running the ArcGIS flow accumulation function on the DEM layer. The average slope (in percent) of each contributing area is determined. Then, a stream power index (SPI) for each grid cell is calculated where SPI = ln (contributing area × average slope).

6.2.3.6 AgBufferBuilder Determines Buffer Area Requirement for Each Field Border Grid Cell

The appropriate mathematical relationship between pollutant trapping efficiency and the buffer area ratio for each grid cell in the field border layer is determined based on the pollutant type and the site conditions in its contributing area. It is then used to calculate the buffer area ratio corresponding to the input trapping efficiency, which, in turn, is multiplied by the contributing area of that cell to calculate its buffer area requirement in units of grid cell rounded to the nearest whole number (i.e., the number of upgradient grid cells that need to be filled with a filter strip). If the input trapping efficiency is greater than the maximum capability for those site conditions, then the buffer area requirement is equal to the entire contributing area.

6.2.3.7 AgBufferBuilder Locates Filter Strip in Grid Cells

For each grid cell in the field border that has a buffer area requirement ≥ 0.5, the field border grid cell is filled first, and then the cells in its contributing area that are closest to it and are upgradient from it are filled sequentially based on flow direction until the total number of cells in its buffer area requirement is filled. A higher value for SPI is used to break ties. This algorithm balances the need for the filter to contact greater runoff load with the practical need to encourage compact field margin designs.

6.2.3.8 AgBufferBuilder Creates a Map and Table of Summary Statistics

An output map is displayed that consists of four output layers: orthophotograph image, field border raster, topographic contours, and the designed buffer raster. The designed buffer raster shows the location of grid cells that must be filled with a filter strip to achieve the desired trapping efficiency. An output file table named "Buffer_analysis.dbf" is created and placed in the "Tool" folder within the "Buffer_Builder" folder. To view it, the user must add it to the map and then open it. The table contains the summary statistics for the buffer design project (Table 6.3), including the size of the field (AOIAREA), the size of the designed filter strip (BUFFERAREA), and the whole-field trapping efficiency of the designed filter area (CAWATE1). (Note that the values shown for CAWATE2–4 in this table are for research purposes only.) The value for CAWATE1 likely will differ from the input value for trapping efficiency due to rounding errors associated with the raster analysis and design structure, and to site conditions in some contributing areas under which the input trapping efficiency cannot be achieved. The CAWATE1 value is the best estimate of whole-field trapping efficiency that would be achieved by a filter placed in the locations indicated by the designed buffer raster.

Finally, the user must save the map project to a desired location before starting another analysis.

6.2.4 Steps in the Assessment Procedure

The assessment procedure produces an estimate of total area and trapping efficiency of a user-defined filter strip. To conduct an assessment, the user must add these steps to the design procedure (Section 6.2.3).

6.2.4.1 User Creates a Filter Strip Polygon File

At the bottom of the "Clip & Project Input Rasters" dialog box, place a check mark in the box labeled "Create Buffer Assessment" to create an empty shape file in which to edit filter strip polygons. On the orthophotograph, manually digitize where the buffer is, or would be, located. The polygon(s) must be closed polygons with their outer margins snapped to the outer field border. Later, the Buffer Builder Main tool will convert the polygon(s) to raster.

6.2.4.2 User Inputs a Command to Run an Assessment

Click on the box near the bottom of the Buffer Builder Main tool page labeled "Use Buffer Assessment" to place a check mark in it.

Then, clicking "OK" executes the Buffer Builder Main tool, which automatically conducts the assessment procedure on the user-defined filter polygon file. The assessment procedure conducts the following steps.

TABLE 6.3
Names, Units, and Definitions of Parameters Appearing in the Output Table "Buffer_Analysis"

Parameter	Definition
OID	ArcGIS row designation in the output table. A design or an assessment will show only one row (OID = 0); both a design and an assessment will show two rows (OID = 0 and 1).
TIMESTAMP	Date and time of the analysis; year, month, day, hour, minute, second.
IMAGERY	Filename containing the aerial orthophotograph of the field area as specified in the "Buffer Builder Main" dialog box.
FLDMARGIN	Filename containing the field margin vector as specified in the "Buffer Builder Main" dialog box under "Field Margin Polygon."
POLLUTYPE	Type of pollutant to analyze for as selected in the "Buffer Builder Main" dialog box.
SOILTEX	Soil texture category as selected in the "Buffer Builder Main" dialog box.
CFACTOR	USLE *C* factor as selected in the "Buffer Builder Main" dialog box.
INTRAPEFF	Level of pollutant trapping efficiency to design a filter for, in percent, as specified in the "Buffer Builder Main" dialog box.
ELCOUNTOUR	Filename containing the contour lines of the field area. Contour interval is 1 m.
AOI	Filename containing the raster of the assessed filter polygon for an assessment procedure, and the DEM raster clipped to the field area for a design procedure.
AOISR	Coordinate system (i.e., projection) of the map project files.
AOIEXTENT	Coordinates of the maximum extent of the AOI in cardinal directions.
CELLSIZE	Raster grid dimensions = length of one side of a grid cell in meters. (Note that in version 1.0, this column always shows "9" regardless of the value specified on the Clip & Project Input Rasters dialog box. It does not affect the analysis. Simply ignore this value and refer to the value placed in the dialog box.)
AOIAREA	Total area of grid cells in the AOI in acres.
BUFFERAREA	Total area of grid cells containing a filter strip in acres.
CAWATE1	Contributing area-weighted average trapping efficiency in percent = $\left(\sum_{n}^{1} \text{contributing\_area} \times \%\text{trapping\_efficiency}\right) \div \text{AOIAREA}$ where n is the number of field border grid cells through which runoff exits the AOI.
CAWATE2	(For research purposes).
CAWATE3	(For research purposes).
CAWATE4	(For research purposes).
ANALYSIS	Type of analysis (design or assessment) entered in the "Buffer Builder Main" dialog box.
COMMENTS	User-supplied identifiers and notes entered in the "Buffer Builder Main" dialog box.

6.2.4.3 AgBufferBuilder Calculates Terrain Attributes for Field Border Grid Cells

The number of field area grid cells that contribute runoff flow to each field border grid cell is determined by running the ArcGIS flow accumulation function on the DEM layer. The average slope (in percent) of each contributing area is determined.

6.2.4.4 AgBufferBuilder Determines the Buffer Area Ratio Associated with Each Field Border Grid Cell

The number of grid cells in the contributing area to each field margin cell that lies within buffer polygons is determined, and the buffer area ratio (the number of cells within the buffer polygon ÷ the total number of cells in the contributing area) for each contributing area is calculated.

6.2.4.5 AgBufferBuilder Calculates the Trapping Efficiency Associated with Each Field Border Grid Cell

The appropriate mathematical relationship between the pollutant trapping efficiency and the buffer area ratio for each grid cell in the field border layer is determined based on pollutant type and site conditions in its contributing area. That equation is then used to calculate the trapping efficiency corresponding to the buffer area ratio that exists for each field border grid cell. Then, the contributing area-weighted trapping efficiency is calculated among all field border cells through which runoff exits the field area.

6.2.4.6 AgBufferBuilder Creates a Map and Table of Summary Statistics

An output map is produced consisting of four (or five) output layers: orthophotograph image, field border raster, topographic contours, and the existing buffer raster (and the designed buffer raster if the design procedure is also conducted). The output file table "Buffer_analysis.dbf," located in the "Tool" folder within the "Buffer Builder" folder, can be viewed by adding it to the map and opening it. The table contains summary statistics for the buffer assessment project (Table 6.3), including the size of the field (AOIAREA), the total size of the assessed filter strip polygons (BUFFERAREA), and the field-scale trapping efficiency of the filter strip polygons (CAWATE1). (Note that values that are shown for CAWATE2–4 are for research purposes only.)

6.3 DEMONSTRATION OF THE PROCEDURES

6.3.1 Demonstration Site and Datasets

The procedures are demonstrated on a 35.7-ac crop field in the Midwest United States. The reader can conduct the following analysis and then determine if it was properly conducted by comparing his/her results to those provided here. Before starting, it is recommended that the User's Guide be downloaded from the CRC Press website at http://www.crcpress.com/prod/isbn/9781439867228. It contains detailed tips for ensuring that the program is set up correctly and working properly on a computer. It also identifies a couple of idiosyncrasies in the output that could create some confusion. Appendix 6.5 of the User's Guide on the CRC Press website contains detailed step-by-step instructions and hints for running the program.

The demonstration site is in rolling topography where undulations and swales in crop fields are common. The field grows corn and soybeans using no-till practices. The surface soil is a predominantly silty clay loam. A grassed waterway has been installed in this field to prevent gully erosion. For this demonstration, the goal was to design a filter strip that would trap approximately 75% of the sediment leaving this field during a large spring thunderstorm (61 mm in 1 h) that occurs soon after spring planting.

All of the digital data layers that are needed for conducting a design procedure and an assessment procedure on this field are contained in the zip file "Demonstration Project–Field B," which can be downloaded from the website http://www2.ca.uky.edu/BufferBuilder. For running the design procedure, the zip file contains a prepared aerial photo, a DEM (5-m grid size), and field border files. For running an assessment procedure, a prepared filter polygon file is also provided. Finally, the zip file contains a map of the designed filter for this field and the output table containing statistics for both the design filter and the assessed filter using the prepared data files and input parameters provided here.

6.3.2 Design and Assessment

The procedures in Sections 6.2.3 and 6.2.4, as detailed in Appendix 6.5 of the User's Guide on the CRC Press website, were followed to produce a model-designed filter strip around the field border layer and to assess a user-defined filter strip along a portion of the field border. Both analyses (design and assessment) were run simultaneously. To do so, the default location/file names for the DEM and the photograph files in the "Clip & Project Input Rasters" dialog box (Figure 6.2) were replaced with those of the extracted Demonstration Project-Field B files. Clicking "OK" brings up the

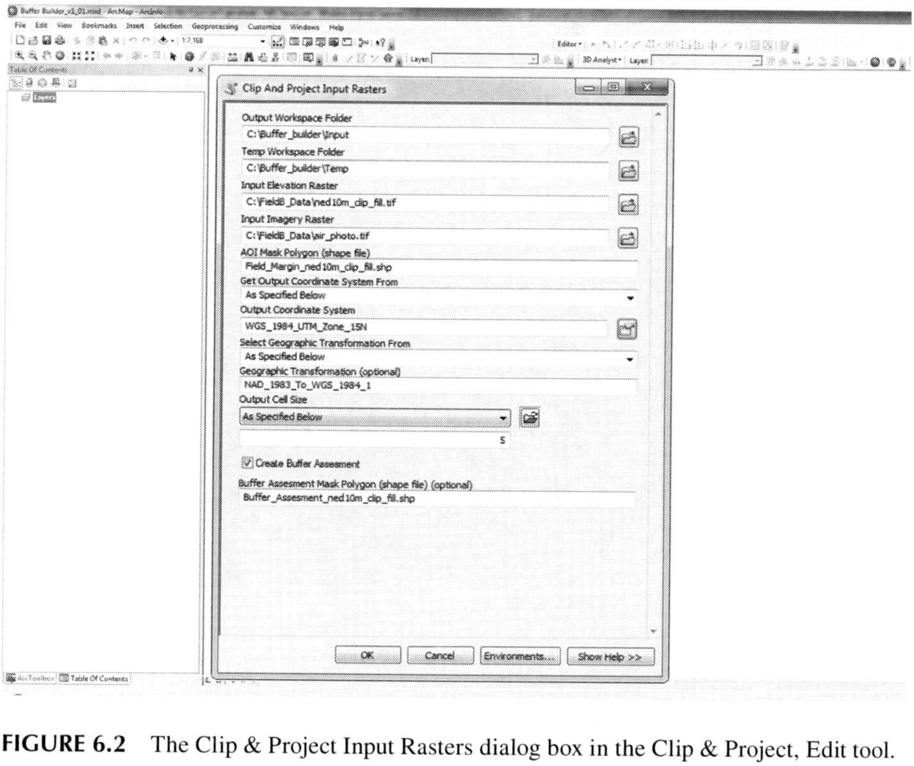

FIGURE 6.2 The Clip & Project Input Rasters dialog box in the Clip & Project, Edit tool.

aerial photo map (Figure 6.3). In a normal project, the user would next edit the "Field Margin" and "Buffer Assessment" layers. However, for this demonstration, the edited files have already been prepared. To use the prepared layers, the default location/filenames were removed, and the location/filenames of the extracted Demonstration Project-Field B files were added. Clicking the "puppy" icon displays the Builder Main tool dialog box showing these new file location/names (Figure 6.4). In this dialog box, the site conditions and the design criteria specified in Section 6.3.1 were also entered.

Upon executing the Buffer Builder Main tool, the screen will display an output map (Figure 6.5) showing both the designed filter "Buffer_75%" and the assessed filter polygon "existingbuffer." The designed filter strip is located almost entirely on the right side of the field where the grassed waterway ends. This is where both terrain analysis of the DEM and empirical placement of the grassed waterway indicate nearly all of the runoff from this field flows and where the model indicates that a filter strip would be most effective for trapping sediment. Adding the output table "Buffer_Analysis" (Figure 6.6) shows the designed filter to be 2.38 ac in size with an estimated trapping efficiency of 70% for sediment. This value for trapping efficiency is less than the input value of 75% due to the effects of mathematical rounding and site conditions, as explained in Section 6.2.3.8. The assessed filter polygon in this demonstration (Figure 6.5) was drawn to be two grid cells 10 m wide along the right side of the field. The assessed filter polygon is 1.41 ac in size with an estimated trapping efficiency of 18% for sediment (Figure 6.6). The reader can compare his/her results with those shown in Figures 6.5 and 6.6 by simply adding these layers from the extracted Demonstration Project-Field B files to the current ArcMap project and toggling between the two results.

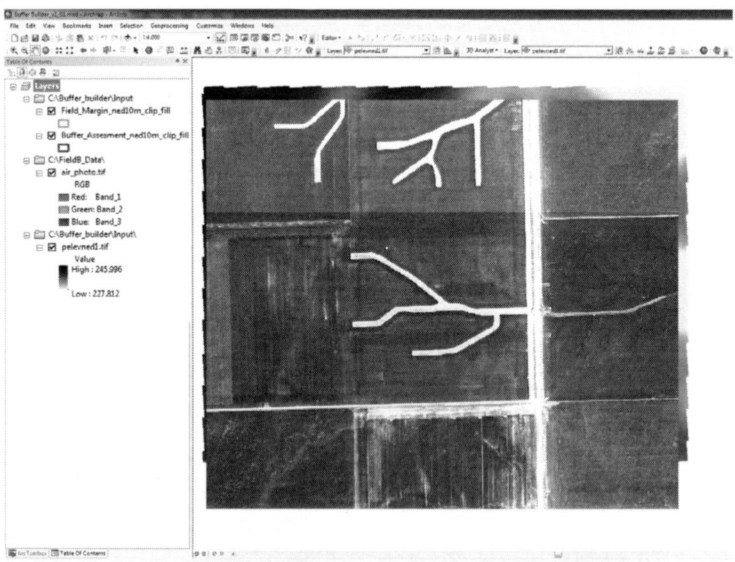

FIGURE 6.3 The aerial photo layer is displayed after clicking "OK" in the Clip & Project Input Rasters dialog box. This layer is used to edit the "Field_Margin..." and "Buffer_Assesment..." shape files with field margin and filter polygons, respectively.

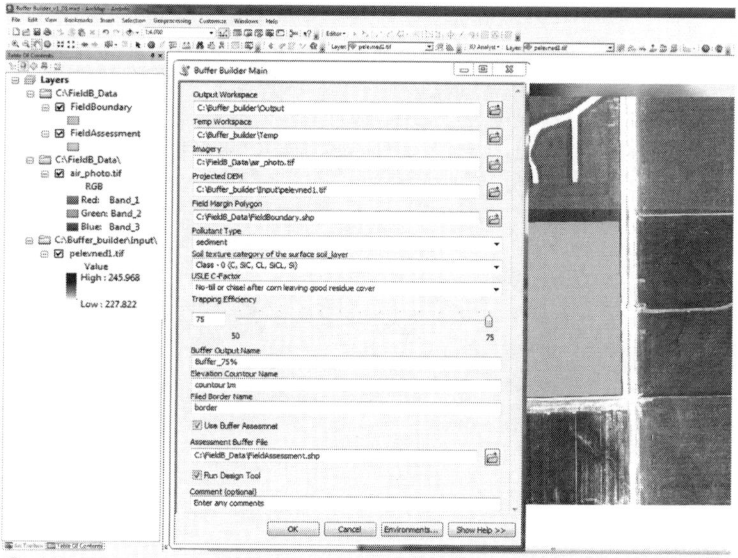

FIGURE 6.4 The Buffer Builder Main tool dialog box.

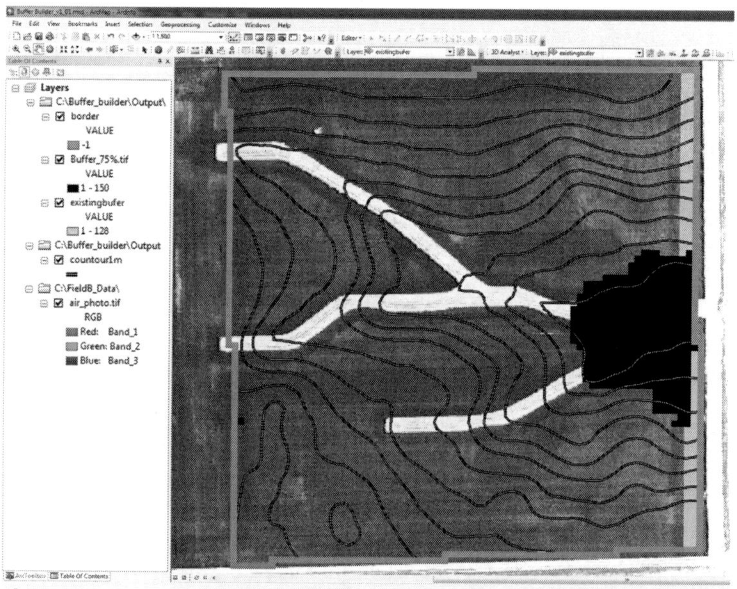

FIGURE 6.5 **(See color insert.)** The display after executing the Buffer Builder Main tool to run both the design and the assessment procedures. The map shows five output layers: designed filter areas, assessed filter polygons, field border raster, and topographic contours on the photo of the field.

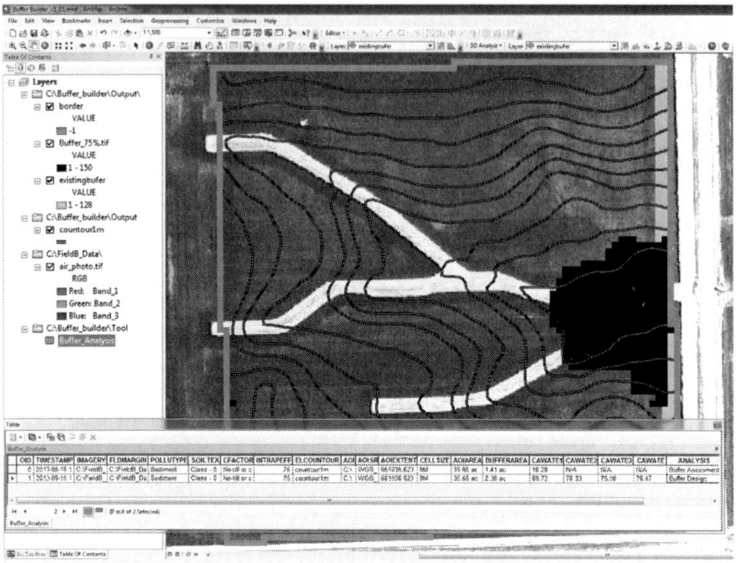

FIGURE 6.6 The display after adding the output table "Buffer_Analysis" to the output map. Only some of the columns are shown here. The complete set of parameter columns are listed in Table 6.3.

6.3.3 How to Use the Design and Assessment Results

The configuration of the designed filter in this example, based solely on water quality considerations, may not be easily adoptable by the landowner for other reasons, such as its overall size and difficulty to farm around. The designed filter map can be used as a guide for drawing alternative sizes and configurations, such as the constant-width strip, that are more compatible with the landowner's needs. The AgBufferBuilder program can be used to assess those alternative designs and to compare levels of performance. The output statistics can be used to calculate and compare installation costs and incentives.

6.4 LIMITATIONS

The key advancement of AgBufferBuilder 1.0 is the use of the DEM for identifying spatial patterns of field runoff at a sufficiently fine scale of resolution to efficiently treat agricultural runoff with a filter strip. The user must keep in mind, however, that DEM datasets may contain significant inaccuracies. For example, there may be influential microtopography on the ground that is too subtle to be detected by the original topographic data from which the DEM was developed. There may also have been land shaping and drainage modifications made by farmers after the original data were collected. These issues may cause runoff to flow to locations that are not indicated by the DEM. The D8 flow direction algorithm used by ArcGIS (Jenson and Domingue 1988) may produce some inaccuracies as well (Tarboton 1997).

The application of filter strips and the use of AgBufferBuilder 1.0 are not appropriate where flow concentrates to an extent that significant gullying occurs. Practices like grassed waterways or vegetative barriers should be installed to stabilize gully-prone pathways. BufferBuilder can best be used to design filter areas either at the outflow of these stabilized pathways (such as in the Demonstration Project presented in this chapter) or at their upgradient by drawing the field margin around the perimeter of a waterway.

Because of these limitations and uncertainties, design maps and assessments produced by this program should be used cautiously.

6.5 SUMMARY

A method is described for designing and/or assessing the performance of filter strips that utilizes GIS and terrain analysis. The procedure identifies locations where a filter strip is required in order to achieve an approximately constant level of pollutant trapping efficiency around an agricultural field. The procedure can create a design that provides a user-specified level of trapping efficiency for sediment and sediment-bound pollutants. The use of this procedure enables a precise fit between filter size and runoff load where the runoff from an agricultural field is nonuniform. Coupling the design model with geospatial technologies greatly increases the speed of designing performance-based filter strips, especially for large or multiple fields.

ACKNOWLEDGMENTS

The funding for the development of AgBufferBuilder 1.0 was provided by the U.S. Department of Agriculture, Forest Service, National Agroforestry Center, Lincoln, Nebraska, and Southern Research Station, Asheville, North Carolina, to the University of Kentucky (grant no. 10-JV-11330152-119) and from Kentucky state water quality grants (SB-271) administered by the University of Kentucky, College of Agriculture, Food and Environment, Lexington, Kentucky.

REFERENCES

Dosskey, M.G., M.J. Helmers, and D.E. Eisenhauer. 2011. A design aid for sizing filter strips using buffer area ratio. *J. Soil Water Cons.*, 66:29–39.

Jenson, S.K., and J.O. Domingue. 1988. Extracting topographic structure from digital elevation data for geographic information system analysis. *Photogramm. Eng. Remote Sens.*, 54:1593–1600.

Muñoz-Carpena, R., and J.E. Parsons. 2005. *VFSMOD-W: Vegetative Filter Strip Hydrology and Sediment Transport Modeling System, v. 2.6.* University of Florida, Homestead, FL. Available at http://abe.ufl.edu/carpena/vfsmod/index.shtml (accessed May 1, 2014).

Tarboton, D.G. 1997. A new method for the determination of flow directions and upslope areas in grid digital elevation models. *Water Resour. Res.*, 33:309–319.

U.S. Department of Agriculture (USDA). 2013. *Filter Strip Code 393.* National Conservation Practice Standards, Natural Resources Conservation Service, Washington, DC. Available at http://www.nrcs.usda.gov/Internet/FSE_DOCUMENTS/stelprdb1241319.pdf (accessed May 1, 2014).

Wischmeier, W.H., and D.D. Smith. 1978. *Predicting Rainfall Erosion Losses—A Guide to Conservation Planning.* Agric. Handbook No. 537, USDA, Washington, DC. Available at http://topsoil.nserl.purdue.edu/usle/AH_537.pdf (accessed May 1, 2014).

7 Modeling Landscape-Scale Nitrogen Management for Conservation

*James C. Ascough II and Jorge A. Delgado**

CONTENTS

EXECUTIVE SUMMARY

There are concerns about how agriculture will adapt to a changing climate and other environmental challenges during the twenty-first century in order to provide food security for the ever-increasing global population. Designing and implementing best nitrogen management practices will be critical in global food security efforts because of the positive correlation between nitrogen inputs and crop yield and between nitrogen inputs and nitrogen losses to the environment. Computer models are important tools that can help nutrient managers implement conservation practices on the ground to improve nitrogen use

* Contact author: jorge.delgado@ars.usda.gov.

efficiency and reduce losses of nitrogen from agricultural systems. One such tool is the Nitrogen Losses and Environmental Assessment Package (NLEAP) with Geographic Information System (GIS) capabilities, Version 4.2 (NLEAP GIS 4.2), as well as the next generation NLEAP GIS 5.0, which allows geospatial analysis of multiple fields simultaneously. To use NLEAP GIS 4.2, users need to download location-specific Natural Resources Conservation Service (NRCS) SSURGO soil and weather databases from an Internet server and develop the nitrogen management scenario that they want to test. If desired, the more advanced NLEAP GIS 5.0 model can be used to conduct geospatial simulations and display nitrogen losses to the environment using embedded GIS-integrated NASA World Wind™ technology and other tools for spatial statistical analysis. This advanced technology allows a collection of components that interactively display 3D geographic information within Java applications. This chapter includes a hands-on exercise that presents the NLEAP GIS 5.0 prototype with instructions on how to perform quick evaluations across large areas and identify the effects of best management practices. The user will also learn how to conduct a Nitrogen Trading Tool analysis and determine the potential benefits of implementing management practices and the quantity of nitrogen savings that could potentially be traded in future air or water quality markets.

KEYWORDS

Nitrogen, Nitrogen Losses and Environmental Assessment Package with Geographic Information System (NLEAP GIS) capabilities, Nitrogen Trading Tool (NTT), nitrogen use efficiency, soil and water conservation, water quality

7.1 INTRODUCTION

Environmental concerns and issues are being raised as to whether food security for the increasing global population will be achievable during the twenty-first century, given the collective pressure on natural resources (Verdin et al. 2005; Montgomery 2007; Bryan et al. 2009) and especially the anticipated effects of climate change (Delgado et al. 2011; Lal et al. 2011). Recent research has illustrated the importance of soil and water conservation, which can help us mitigate and adapt to climate change and other environmental concerns and challenges (Montgomery 2007; Johnson et al. 2010; Delgado et al. 2011; Lal et al. 2011; Eagle et al. 2012).

Nitrogen management will be critical in helping to achieve food security on a global scale because of the positive correlation between nitrogen inputs and both crop yield and environmental nitrogen losses to the environment. Unfortunately, only about 35% of the total cropland in the United States is under nitrogen management that meets all of the three criteria (rate, timing, and method criteria) for best management practices (BMPs; Ribaudo et al. 2011a). Designing and implementing the best nitrogen management practices will be essential in helping to achieve food security and the soil and water conservation needed for global sustainability (Figure 7.1). Conservation practices and computer models can be fundamental tools

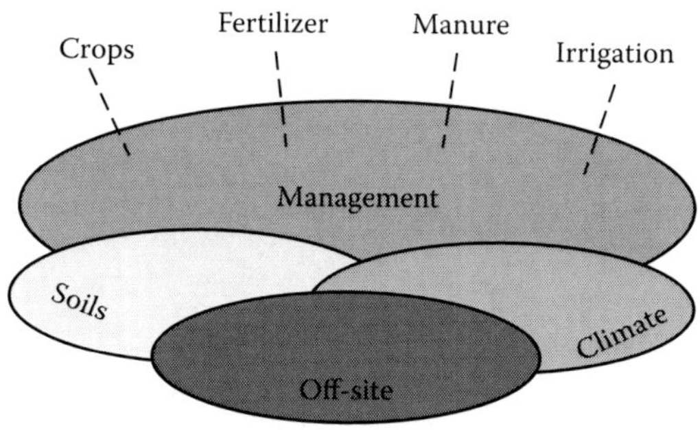

FIGURE 7.1 Management is key for the reduction of nitrogen losses (e.g., nitrate leaching). (From Shaffer, M.J. and J.A. Delgado, *J. Soil Water Conserv.*, 57:327–335, 2002.)

for improving nitrogen use efficiency and reducing the losses of reactive nitrogen to the environment. In the remainder of this introduction, we will discuss nitrogen use efficiency, nitrogen models, and the Nitrogen Losses and Environmental Assessment Package (NLEAP) with Geographic Information System (GIS) version 4.2 (NLEAP-GIS 4.2) and NLEAP GIS 5.0 modeling tools.

7.1.1 Nitrogen and Nitrogen Use Efficiency

Nitrogen is a very dynamic and mobile element, and considerable work has been done to study nitrogen losses including ammonia (NH_3) emissions, nitrous oxide (N_2O) emissions (IPCC 2007), and nitrate (NO_3) leaching losses. Environmental losses of reactive nitrogen can negatively impact surface water quality (e.g., hypoxic zones) and groundwater (Figure 7.2). There are many economic impacts as well; for example, one study found that the cost of removing nitrate that originated from agriculture from the US drinking water is approximately $1.7 billion per year (Ribaudo et al. 2011a). While the environmental damage from excessive application of fertilizer is clear, loss of nitrogen from agricultural systems also represents a significant economic loss, at least to crop farmers. Improved nitrogen management and use efficiency converts expensive applied fertilizers into higher harvested yields and nitrogen uptake.

7.1.2 Nitrogen Tools Can Provide Opportunities to Assess Field Scenarios

In this chapter, the term "nitrogen tools" will refer to computer software programs that can be used to assess the effects of management practices on nitrogen use efficiency and/or nitrogen losses to the environment. Several authors have reviewed the literature on nitrogen modeling and have discussed how nitrogen tools integrate biogeochemistry, physics, agronomy, soils, crops, computer technology, and other areas of science to simulate the nitrogen cycle in order to conduct assessments of the effects of management practices on the losses of nitrogen from agricultural systems

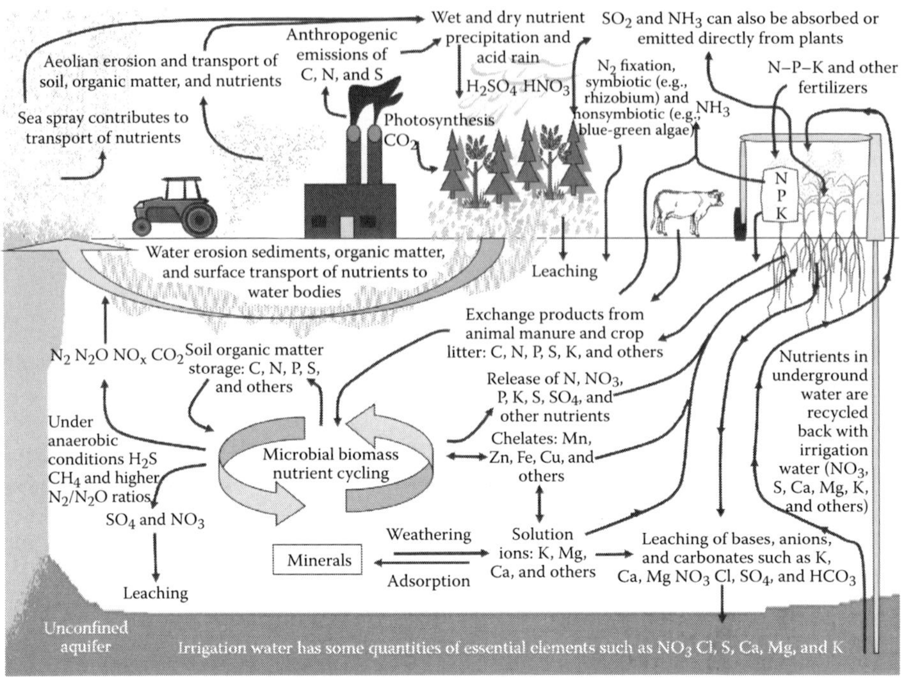

FIGURE 7.2 Nutrient cycles (e.g., nitrogen) for elements essential for crop production, and the fate and transport of these nutrients in the environment. (From Delgado, J.A. and R.F. Follett, *J. Soil Water Conserv.*, 57:455–464, 2002.)

(e.g., Shaffer et al. 2001; Cabrera et al. 2008; Ahuja and Ma 2011). Examples of modeling approaches, and how to apply them for evaluating nitrogen management under commercial field scenarios, have been discussed by Delgado (2001), Shaffer and Delgado (2001), and Ahuja and Ma (2011), among others. A detailed review of how some of these models simulate nitrogen processes and nitrogen transformations has also been conducted by Ma and Shaffer (2001) and Cabrera et al. (2008). Other models that can be used to evaluate the potential benefits of conservation practices and management decisions in terms of reduced losses of reactive nitrogen include the Crop Estimation through Resource and Environmental Synthesis (CERES; Ritchie et al. 1985), Erosion/Productivity Impact Calculator (EPIC; Williams et al. 1983), Leaching Estimation and Chemistry Model (LEACHM; Wagenet and Hutson 1989), and the Root Zone Water Quality Model2 (Ahuja et al. 2000).

Another well-known nitrogen model is the NLEAP-GIS 4.2 (Delgado et al. 2010a; Shaffer et al. 2010). NLEAP-GIS 4.2 was released in 2010 and is a revised, more advanced, and flexible version of the older DOS version released about two decades ago (Shaffer et al. 1991; Delgado et al. 1998a,b). Delgado (2001), Shaffer and Delgado (2001), and Delgado and Shaffer (2008) have presented examples of how to use this model to assess nitrogen management in order to better understand how conservation practices under commercial agricultural systems can contribute to increased nitrogen use efficiency. NLEAP simulation of residual soil nitrate and soil water content was comparable to other models such as EPIC, CERES, and LEACHM (Khakural and Robert 1993; Beckie et al. 1994). Khakural and Robert (1993) also evaluated the nitrate leaching assessment capability of NLEAP and found that it performed similarly to the LEACHM model. Delgado et al. (2008) reported that NLEAP has been tested at several national and international sites and has been found to be capable of evaluating the effects of management practices on nitrogen losses and residual soil nitrate in the soil profile. Additional information on the advantages and disadvantages of NLEAP can be found in Shaffer and Delgado (2001), Delgado and Shaffer (2008), Shaffer et al. (2010), and Delgado et al. (2010a).

The NLEAP GIS 5.0 prototype, a successor to NLEAP-GIS 4.2, was developed and released to the general public in early 2013. This next-generation version of the model allows for improved GIS presentation and analysis of spatial data. The NLEAP GIS 5.0 also has an enhanced version of the Nitrogen Trading Tool (NTT) that was available in NLEAP-GIS 4.2 (Delgado et al. 2008, 2010c) that can be used to assess how conservation practices can provide savings in nitrogen losses that could potentially be traded in emissions markets.

The NLEAP GIS 5.0 prototype is written in the Java™ programming language and has GIS capabilities as provided by embedded NASA World Wind™ technology. The data generated and used by NLEAP-GIS 4.2, as well as supporting ESRI GIS shape files, can be imported into the new NLEAP GIS 5.0 for better GIS analysis and visualization. This chapter will present some examples and discussion of data and analysis that has been generated with NLEAP-GIS 4.2. Additionally, this chapter will present hands-on exercises (see CRC Press website, http://www.crcpress.com/prod/isbn/9781439867228) with the NLEAP GIS 5.0 prototype, which can be used to quickly assess the effects of conservation practices on nitrogen use efficiency as well as the potential to trade savings in nitrogen losses resulting from improved practices.

There are opportunities to use computer modeling tools to assess how conservation practices can contribute to increased nitrogen use efficiencies and reduced losses of nitrogen (Delgado et al. 2008; Ribaudo et al. 2011a,b; Saleh et al. 2011). Nitrogen models have been used for over two decades to increase our understanding of field plot research and potential upscale application to watersheds (Williams et al. 1983; Ritchie et al. 1985; Shaffer and Larson 1987; DeCoursey et al. 1989; Wagenet and Hutson 1989; Follett et al. 1991). More recently, simulation models have been applied to assess the effects of BMPs under commercial farm operations and the potential to increase nitrogen use efficiency and reduce nitrogen losses (Delgado et al. 2000, 2001; Delgado 2001). The development of new markets, such as water quality and emissions trading markets, provides new opportunities for farmers to implement BMPs and reduce nitrogen losses, generating savings in nitrogen loss that can potentially be traded (Delgado et al. 2008; Lal et al. 2009; Ribaudo and Gottlieb 2011; Ribaudo et al. 2011a,b; Hansen et al. 2012). These prospective new markets are bringing to the forefront of nutrient management the possibility of using models for water quality trading programs. There is also potential to use the Internet to apply models on a tactical basis not only as nitrogen management tools but also as potential NTTs for new noncommodity markets (Delgado et al. 2008, 2010c; Lal et al. 2009; Saleh et al. 2011; Ribaudo et al. 2011a,b). Ribaudo and Gottlieb (2011) reported that models such as the NTT (Delgado et al. 2008) could potentially be used as tools to provide information and that these tools could be accepted by trading markets if the models are perceived by the markets and peers as reducing uncertainty and providing accurate information.

There is potential to use nitrogen models to assess the effect of management decisions on risky landscape-cropping system combinations to determine how BMPs can contribute to reduced nitrogen losses to the environment or even to the identification of hot spots across a field. Shaffer and Delgado (2001) and Delgado and Shaffer (2008) reported that the type of model required for a given field situation will depend on, among other factors, the availability of databases, the collection of input data from field scenarios if needed, the development of management scenarios for assessing conservation practices, the installation and operation of the model to be used (as well as its calibration and validation for the region that is being assessed), and the knowledge of how to interpret model simulation results. Nitrogen models have also been used in combination with field studies to identify hot spot areas of a field where nitrogen losses are higher and to quantify how conservation practices such as management zones, precision farming, and remote sensing can reduce these losses (Delgado and Bausch 2005; Delgado et al. 2005). Additional information on how to apply a nitrogen model to a field situation can be found in Shaffer and Delgado (2001), Delgado and Shaffer (2008), and Ahuja and Ma (2011).

7.1.3 Basics of NLEAP-GIS 4.2

NLEAP-GIS Version 4.2 (Delgado et al. 2010a) provides more flexibility to conduct spatial and long-term analyses simultaneously across a large number of fields (Delgado et al. 2010a; Shaffer et al. 2010). Version 4.2 has a new graphical user interface that can be used to create and update database files to be used with multiple GIS software packages. The NLEAP-GIS 4.2 allows for rapid connection to Internet databases, including current Natural Resources Conservation Service (NRCS) soil (National Soil Information System [NASIS] and Soil Survey Geographic Database [SSURGO]) and climate databases that can be downloaded and used with the model. Detailed instructions on how to use the model can be found in the NLEAP-GIS 4.2 user manual (Delgado et al. 2010a,c), which contains examples and exercises.

The basic algorithms of the model describing how nitrogen dynamics and pathways for losses are simulated are described in Shaffer et al. (2010). These basic algorithms apply to both versions of the model: NLEAP-GIS 4.2 and NLEAP GIS 5.0. Shaffer et al. (2010) presented these basic algorithms, which are used for the nitrogen dynamics and transformations conducted by both models. The basic algorithms for the NTT concept used in both models were described in Delgado et al. (2008, 2010c). A user-friendly interface developed with Visual Basic® programming that runs within the Microsoft Excel® environment allows the user to navigate within the program and to communicate with Internet web pages and GIS software. Potential applications of the NLEAP-GIS 4.2 software are described in the book edited by Delgado and Follett (2010), especially the chapter from Shaffer et al. (2010). Additionally, Delgado et al. (2010a,c) describe examples of applications of the model.

NLEAP-GIS 4.2 and NLEAP GIS 5.0 use soil, weather, and management databases. The user-friendly interface with both of these models allows the user to create these databases by downloading web-based NRCS data, or the user can develop site-specific databases. Examples of management scenarios (e.g., scenarios with different crops, fertilizer rates, and conservation practices such as cover crops) are provided, or the user can create their own management scenarios. The management database has detailed information about agricultural operations during the growing season such as planting, harvesting, cultivation, irrigation, fertilization, or other activities conducted during the year, such as tillage after harvest. Each specific management scenario has detailed information accompanying the event, such as the type of fertilizer used (e.g., urea, ammonium nitrate), the amount and method of application (e.g., incorporated, surface applied), and the date of fertilizer application.

The NRCS soil databases have a collection of soil types at the region where the model can be applied and contain detailed soil property information for the soil profile such as bulk density, soil organic matter content, and soil texture. The weather databases contain daily weather (e.g., precipitation, temperature, etc.) at the site; NRCS climate data from nearby weather stations can be downloaded from an Internet server. The interface is designed so that the soil and weather databases across the United States can be downloaded and quickly set up to run the model.

In general, NLEAP-GIS 4.2 studies (Delgado et al. 2010a,b; Hansen et al. 2012; Shumway et al. 2012) show that even with the model's current limitations, it is a useful tool for conducting spatial and temporal analysis of nitrogen management practices. Details on how to create a GIS-based analysis from scratch are provided in step-by-step examples presented in the NLEAP-GIS 4.2 user manual. In summary, users need to download geospatial NRCS soil databases, download NRCS weather databases for the climate station nearest to where the analysis will be conducted, and develop the nitrogen management scenario that they want to test. NLEAP-GIS 4.2 input databases can be seamlessly exported to the new and more advanced NLEAP GIS 5.0 prototype, which has been enhanced with the incorporation of various geospatial visualization and analysis features, i.e., the NLEAP GIS 5.0 supports GIS-based simulations and can display nitrogen losses to the environment using GIS-integrated NASA World Wind™ technology.

7.1.4 NLEAP-GIS 4.2 Simulation Capabilities and Limitations

It is very important that users of nitrogen models understand their capabilities and limitations in simulating the nitrogen cycle for a given site and management scenario. There are a large number of these types of models in the world, and several authors have discussed nitrogen model capabilities and limitations (e.g., Ma and Shaffer 2001; McGechan et al. 2001; Shaffer et al. 2001; Cabrera et al. 2008; Ahuja and Ma 2011). It is important that the user is familiar with the capabilities and limitations of a particular model and how it will be used at a given site, what databases are available, how complex the model is, and whether calibration will be required at a given site. Additionally, the selection of a model by a specific user may depend on model complexity, user-friendliness, and the input parameter information that is needed for a given project.

NLEAP-GIS 4.2 is a simple model that, when provided with sufficient and good-quality input, has been shown to simulate and evaluate agricultural systems accurately. Across the literature, NLEAP has been found to be a strong model that can accurately simulate residual soil nitrate and nitrate leaching. Delgado et al. (2008) reported that for over 200 US and international sites studied by different researchers, the correlation between residual nitrate in the soil profile predicted by NLEAP-GIS 4.2 and observed residual nitrate in the soil profile was highly significant ($r^2 = 0.92$). Additional information on how to calibrate and validate the model for commercial field evaluations is available in Delgado (2001), Delgado et al. (2000, 2001, 2010a), Delgado and Bausch (2005), and Shaffer and Delgado (2001). The NLEAP-GIS 4.2 user manual also has detailed examples showing how to create a GIS field scenario from scratch or using GIS examples that are included with the NLEAP-GIS 4.2 software package.

One limitation of the NLEAP-GIS 4.2 is that the model does not have a predictive crop growth module to simulate yields. The expected yield at the site or the observed site yields from previous years must be entered for model applications, e.g., evaluation of the impact of conservation practice on nitrogen use efficiency at a given site. Another limitation is that the model does not simulate any negative impacts from having weed infestations or disease problems. Additionally, NLEAP-GIS 4.2 cannot assess nitrogen uptake for weeds nor can it simulate the growth of two crops simultaneously (i.e., intercropping) at a given site. In other words, the model assumes that there is a weed and disease control program where yields are not affected and the average yield is representative of the expected yield (the user is required to enter an expected total yield that is representative of the field conditions or a yield that was measured at the site). The use of site-specific average or measured yields has been demonstrated as sufficient for model calibration and validation across the United States and internationally (Shaffer and Delgado 2001; Delgado et al. 2008).

Nitrogen uptake is simulated by NLEAP-GIS 4.2 using the yield and a nitrogen uptake index that multiplies the nitrogen uptake per unit of yield by the total expected yield at a given site. The model uses a sigmoid curve growth pattern to simulate crop growth and nitrogen uptake. The user may edit the nitrogen uptake index or add new crop varieties for a given site or situation with specific properties that describe cases of sufficient nitrogen uptake (or cases where there are nitrogen deficiencies). Similarly, if the model is being applied across a large area, the user needs to know if there are significant regional changes in elevation that may result in differences in climate (e.g., sharp deviations in temperature or precipitation). If so, the user should divide the region by zones and use climate data that are specific for each zone, or use the climate data collected close to the site where the evaluation is being conducted. Additionally, there could be hundreds of fields in a simulation, and each could have different irrigation management, nitrogen management, and soil series; however, the weather (e.g., precipitation, temperature) across these hundreds of fields is assumed to be uniform. The user could divide the fields by zones if there are different weather patterns and use a more site-specific (and uniform) weather database for each zone.

NLEAP-GIS 4.2 conducts simulations on a daily-event time interval so the model is not able to capture events at a given site occurring at smaller time intervals (e.g., sub-daily or hourly events such as rapid infiltration events, rapid weather/storm events, or rapid ammonia volatilization events). The model can perform long-term simulations of several fields for 25–50 or more years but only at daily intervals. The NLEAP-GIS 4.2 model can also quickly conduct long-term spatial analyses across many fields with multiple management practices and across many soil types in a region, conducting a mass and water balance for each polygon. However, a spatial limitation of the model is that there is no water and nitrogen mass transfer at a surface or subsurface scale from polygon to polygon. Even with mass transfer limitation, NLEAP-GIS 4.2 has been able to identify hot spots for nitrogen losses across different regions and accurately simulate predicted versus observed values at plot, commercial field, and grid scales (Wylie et al. 1995; Hall et al. 2001; Delgado and Bausch 2005; Lavado et al. 2010). All the capabilities and limitations of NLEAP-GIS 4.2 listed above also apply to NLEAP GIS 5.0. When experimental data are collected at different regions and sites, NLEAP-GIS 4.2 has been able to accurately assess the effects of management practices on nitrogen and water dynamics (Delgado et al. 2008, 2010c).

7.2 CASE STUDIES

7.2.1 General Methods and Procedures

The first step in using NLEAP is to identify the region where the model is going to be used and then to download a site-specific NRCS SSURGO soil database. Additionally, an NRCS weather database from a weather station close to the site should also be downloaded. The user will then need to develop a management database that is representative of the management practices at the site.

After setting up the soil, weather, and management databases, NLEAP can be used to do a rapid assessment that can help identify the magnitude of nitrogen losses from hot spots, such as soils with a sandier texture (Hydrology A), which have a higher nitrate leaching potential compared to finer-textured clay soils (Hydrology D) (see Sections 7.2.2 and 7.2.3). Another quick assessment approach for a given site-specific field is to perform a model evaluation using the spatial GIS soil properties for the soils and to export the simulated model outputs to a GIS software program for subsequent analysis of nitrate leaching across a field that is sampled with a grid for precision farming studies (see Section 7.2.4). A more advanced GIS assessment can be conducted across a larger region and a large number of fields, soil types, and a set of different management scenarios for different fields to quickly identify (1) risky landscape crop and soil combinations, (2) the potential to use BMPs to reduce nitrogen losses, and (3) the potential to trade nitrogen (see Sections 7.2.5 and 7.2.6 [Delgado et al. 2008, 2010a,c], and the hands-on exercise for this chapter).

7.2.2 Assessment under Rainfed Systems

Shumway et al. (2012) used NLEAP to quantify the best nitrogen management practices to reduce the potential flux of nitrogen from the Arkansas Delta. Two years of model calibration and validation showed that even when employing good irrigation management practices, a year with higher-than-average precipitation can contribute to higher nitrogen losses that are driven by wetter, above-average precipitation.

For example, their assessment found that when 280 mm of irrigation water was applied per year, during a drier year that had 1031 mm of precipitation, the nitrate leaching losses were half of the leaching losses assessed for a year with greater precipitation. In a wetter year that had a lower irrigation amount of 200 mm but a higher precipitation of 1840 mm per year, the assessed nitrate leaching losses were twice those estimated for the drier year.

A long-term temporal analysis showed higher nitrate leaching losses for the wetter years and significantly higher losses in the sandier, coarser areas of the field (Figure 7.3). Long-term simulation modeling results also show that BMPs can be implemented to significantly reduce leaching losses from coarser-textured soil areas of a field during wetter years (Shumway et al. 2012). If a leguminous crop with BMPs (such as application of nitrogen at the right time with the right method and rate) is added, the leaching losses from a corn–corn rotation system will be approximately 20 kg N ha^{-1}, and the finer-textured soil areas will be less than 10 kg N ha^{-1}. These evaluations show that using crop rotations and/or good nitrogen management practices with the right timing, rate, and method can minimize leaching losses and other nitrogen losses across space and time.

7.2.3 Assessment of Manure Systems

Another study using the NLEAP model that was conducted to assess the effects of manure systems in the Midwest showed that high rates of manure application can significantly increase the potential for nitrate leaching, denitrification, and N_2O emissions (Delgado et al. 2008). NLEAP simulation results for a high rate of manure

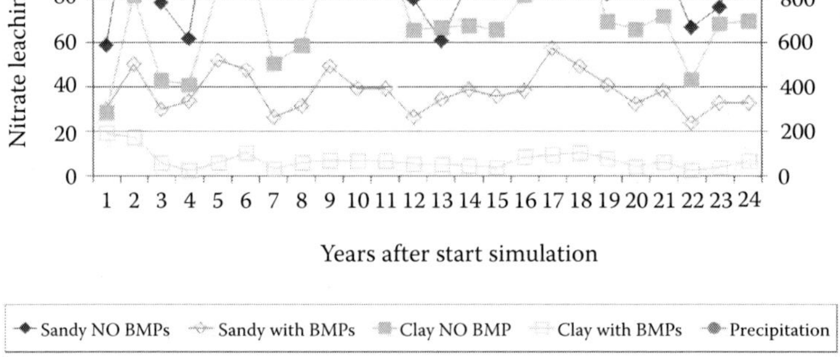

FIGURE 7.3 Long-term evaluations using the NLEAP in coarser, sandy soils, and finer, clay soils with and without BMPs for the best rate, time, and method of application. Nitrate leaching for the corn-corn rotation is higher during wetter years, but the BMPs could significantly minimize these losses.

application on a Haskin sandy loam (Ha) showed annual nitrogen loss rates 131, 66, and 8 kg N ha^{-1} for nitrate leaching, denitrification, and N_2O emission pathways, respectively. Nitrate leaching was lower in a Fulton loam (66 kg N ha^{-1}), while losses due to denitrification and N_2O emissions of 112 and 10 kg N ha, respectively, were much higher than those estimated for the coarser-textured Haskin soil. A reduced rate of manure applied with a timely application in spring minimized leaching, denitrification, and N_2O emission losses to 22, 31, and 5 kg N ha^{-1}, respectively, for the coarser-textured soil areas. Nitrogen losses in the finer-textured soil areas were also significantly reduced to 13, 38, and 4 kg N ha^{-1}, respectively. The simulation conducted by the NLEAP model suggests that lower rates of manure significantly reduce the potential for leaching, especially from the sandier coarse-textured areas, which have higher permeability.

7.2.4 Synchronizing Fertilizer Application with Crop Uptake

NLEAP has been used to evaluate the potential effects of better synchronization of nitrogen fertilizer applications with crop uptake demands to assess the potential for reduction of nitrogen losses. The model was also able to simulate the spatial variability of residual soil nitrate across a commercial field. Using precision conservation techniques such as remote sensing to coordinate nitrogen fertilizer application with crop nitrogen demands, Bausch and Delgado (2003, 2005) found that remote sensing techniques can be used to apply nitrogen in close synchronization with nitrogen uptake demands. This resulted in nitrogen fertilizer rates that were one-half of traditional practice application rates without reducing crop yields. Delgado and Bausch (2005) used NLEAP to evaluate commercial farm operations and found that there was a direct correlation between the excess nitrogen applied and the nitrogen losses from the system. In addition, nitrate leaching losses were cut by approximately 50% using remote sensing techniques to apply nitrogen in fertigation events in synchronization with crop demands. Delgado and Bausch (2005) also found that leaching losses were significantly correlated with sand content in the soil profile and that the coarser-textured soil areas (which had lower yields but received the same amount of fertilizer) had much higher nitrate leaching losses (Figure 7.4).

7.2.5 Use of Cover Crops to Mine and Recover Nitrate

NLEAP has also been used to assess the benefits of cover crops (Delgado 1998; Delgado et al. 1998b, 2001). There are significant benefits from these crops since they help increase the nitrogen use efficiency of the system and can even mine nitrate from groundwater to help restore groundwater quality (Delgado 1998, 2001; Delgado et al. 2001). Dabney et al. (2010) conducted an analysis with the NLEAP-GIS 4.2 model and reported that nitrate losses from cropping systems without cover crops in south-central Colorado averaged 70 kg N ha^{-1}, but with cover crops, the nitrate leaching losses were reduced to 45 kg N ha^{-1} (with summer cover crops, the leaching losses were even lower; less than 30 kg N ha^{-1}). For crops like potatoes, summer cover crops can significantly contribute to the reduction of nitrogen losses from the system and to the cycling of nitrogen to the next potato crop, thus increasing

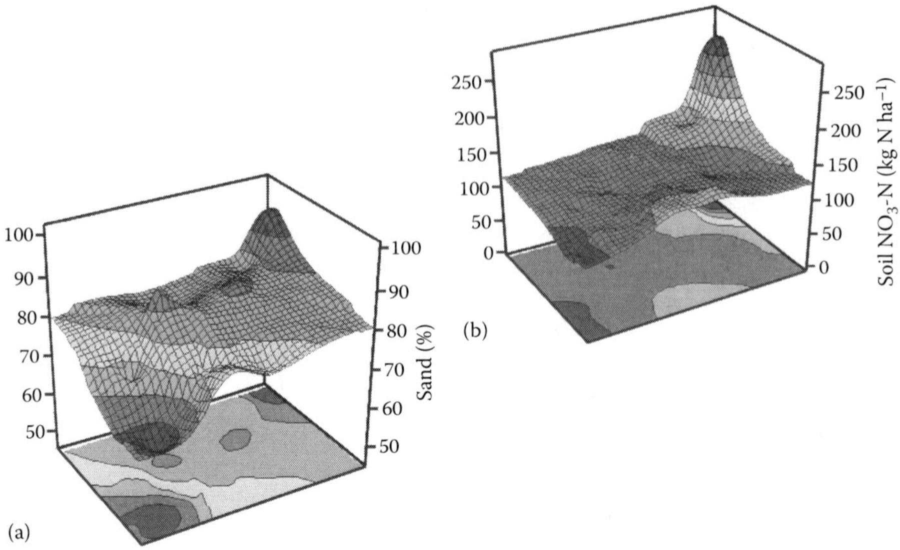

FIGURE 7.4 **(See color insert.)** Spatial distribution of sand content (a) and residual soil nitrate (NO_3-N) (b) in the top 1.5 m of soil across different productivity zones. (From Delgado, J.A. and W. Bausch, *J. Soil Water Conserv.*, 60:379–387, 2005.)

the nitrogen use efficiency for the potato systems and reducing the nitrate leaching losses from the potato–cover crop system (Delgado et al. 2007; Dabney et al. 2010). The hands-on practice of this chapter will present an exercise using summer cover crops.

7.2.6 Potential of Using NTT

The concept of trading nitrogen environmental credits is not new, and researchers have proposed that we can use environmental payments/conservation practices in water quality programs to reduce the transport of nitrate to large water bodies (Hey 2002; Greenhalch and Sauer 2003; Hey et al. 2005; Ribaudo et al. 2005; Glebe 2006). For example, Hey (2002) and Hey et al. (2005) proposed the application of ecological engineering concepts to strategically establish or develop wetlands to remove nitrate from water flowing to the Gulf of Mexico (improving water quality), and to consider programs that can provide environmental payments to farmers for the remediation of waters containing this nitrate.

The first use of a model to develop an NTT concept was developed and published by a joint USDA-ARS and NRCS effort (Delgado et al. 2008, 2010c). The NTT concept was developed and based on NLEAP model simulations, which can be used to quickly conduct a spatial analysis to assess the potential to trade nitrogen that was prevented from being lost to the environment due to improvement in management and conservation practices. This concept was later expanded to develop a Nutrient Tracking Tool that can assess not only nitrogen but also other nutrients (e.g., phosphorus) and soil erosion (Saleh et al. 2011), which are not assessed with NLEAP-GIS 4.2 or NLEAP GIS 5.0. For details on the development and advantages/disadvantages of the NTT, the reader is referred to Delgado et al. (2008, 2010c) and Shumway et al. (2012). The NTT has been shown to be capable of assessing nitrogen losses and evaluating nitrogen management practices across several regions and areas of the United States (Delgado et al. 2008, 2010a,c; Hansen et al. 2012; Shumway et al. 2012).

There are several water quality programs in the United States that can potentially utilize nitrogen modeling tools for assessing the potential of various management practices to reduce the losses of nitrogen from agricultural fields (Lal et al. 2009; Ribaudo and Gottlieb 2011). Tools such as the NTT can potentially be used to reduce the degree of uncertainty and to assess the benefits of nitrogen management practices (Delgado et al. 2008; Gross et al. 2008; Ribaudo et al. 2011a,b). For example, the NTT can be used to do an assessment of a given nitrogen management practice and to quickly compare it to a baseline. If the new practice reduces nitrogen losses, there will be an assessment and quantification of the nitrogen that was saved (i.e., the quantity of nitrogen that was prevented from being lost) and can potentially be traded. If the new practice increases the nitrogen losses over the baseline, then there is no potential to trade nitrogen, and the NTT will show a negative number for nitrogen loss (Delgado et al. 2008, 2010c). The hands-on exercise in this chapter will present an exercise with a new NTT embedded within NLEAP GIS 5.0.

7.3 DISCUSSION AND CONCLUSION

Several studies about nitrogen losses to the environment report on the impacts to groundwater, large water bodies, the atmosphere, and terrestrial areas, as well as for the need to increase nitrogen use efficiency to reduce losses of reactive nitrogen (Rabalais et al. 2002a,b; Bricker et al. 2007; IPCC 2007; Galloway et al. 2008; Rupert 2008; De Paz et al. 2009; Dubrovsky et al. 2010). A recent book on nitrogen management (Delgado and Follett 2010) clearly shows that nitrogen management principles can be applied nationally and internationally to reduce losses of reactive nitrogen to the environment, and that modeling tools such as NLEAP-GIS 4.2 and NLEAP GIS 5.0 can help us to (1) better understand the soil–plant–hydrologic cycle to identify hot spots across the landscape and (2) implement management practices in agroecosystems across the world to increase nitrogen use efficiency.

The included hands-on exercise covers the NLEAP GIS 5.0 nitrogen modeling framework, an improved prototype of the NLEAP-GIS 4.2 model (Figure 7.5). The user will be shown how to perform rapid evaluations across large areas and identify the effects of nitrogen BMPs. The user will also be instructed on how to perform an NTT analysis to determine the benefits of implementing nitrogen BMPs across agricultural landscapes.

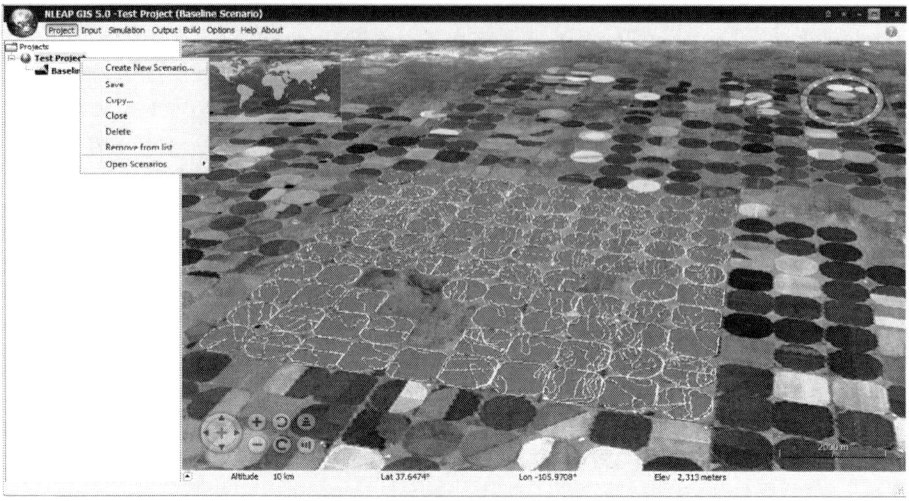

FIGURE 7.5 **(See color insert.)** NLEAP GIS 5.0 Project example showing that the image can be zoomed out so that the entire landscape is visible. Evaluation with color ramping allows identification of the hot spots in the landscape with high risks of nitrogen losses.

DISCLAIMER

Mention of trade names is necessary to report factually on available data; however, the USDA neither guarantees nor warrants the standard of the product, and the use of a given name by the USDA does not imply the approval of that product to the exclusion of others that may be suitable.

REFERENCES

Ahuja, L.R., and L. Ma, eds. 2011. *Methods of Introducing System Models into Agricultural Research*. Advances in Agricultural Systems Modeling, American Society of Agronomy, Madison, WI.

Ahuja, L.R., K.W. Rojas, J.D. Hanson, M.J. Shaffer, and L. Ma, eds. 2000. *Root Zone Water Quality Model—Modelling Management Effects on Water Quality and Crop Production*. Water Resources Publication, Englewood, CO.

Bausch, W.C., and J.A. Delgado. 2003. Ground base sensing of plant nitrogen status in irrigated corn to improve nitrogen management. In *Digital Imaging and Spectral Techniques: Applications to Precision Agriculture and Crop Physiology*, ed. T. VanToai, pp. 151–163. ASA Spec. Publ. 66, American Society of Agronomy, Madison, WI.

Bausch, W.C., and J.A. Delgado. 2005. Impact of residual soil nitrate on in-season nitrogen applications to irrigated corn based on remotely sensed assessment of crop nitrogen status. *Prec. Agric.*, 6:509–519.

Beckie, H.J., A.P. Moulin, C.A. Campbell, and S.A. Brandt. 1994. Testing the effectiveness of four simulation models for estimating nitrates and water in two soils. *Can. J. Soil Sci.*, 75(1):135–143.

Bricker, S., B. Longstaff, W. Dennison, A. Jones, K. Boicourt, C. Wicks, and J. Woerner. 2007. *Effects of Nutrient Enrichment in the Nation's Estuaries: A Decade of Change*. NOAA Coastal Ocean Program Decision Analysis Series No. 26, National Centers for Coastal Ocean Science, Silver Spring, MD.

Bryan, E., T.T. Deressa, G.A. Gbetibouo, and C. Ringler. 2009. Adaptation to climate change in Ethiopia and South Africa: Options and constraints. *Environ. Sci. Policy*, 12:413–426.

Cabrera, M., J. Molina, and M. Vigil. 2008. Modeling the nitrogen cycle. In *Nitrogen in Agricultural Systems*, eds. J.S. Schepers, and W.R. Raun, pp. 695–730. Agronomy Monograph No. 49, Agronomy Society of America, Madison, WI.

Dabney, S.M., J.A. Delgado, J.J. Meisinger, H.H. Schomberg, M.A. Leibig, T. Kaspar, J. Mitchell, and W. Reeves. 2010. Using cover crops and cropping systems for nitrogen management. In *Advances in Nitrogen Management for Water Quality*, eds. J.A. Delgado, and R.F. Follett, pp. 231–282. Soil and Water Conservation Society, Ankeny, IA.

De Paz, J.M., J.A. Delgado, C. Ramos, M.J. Shaffer, and K.K. Barbarick. 2009. Use of a new Nitrogen Index—GIS assessment for evaluation of nitrate leaching across a Mediterranean region. *J. Hydrol.*, 365:183–194.

DeCoursey, D.G., K.W. Rojas, and L.R. Ahuja. 1989. Potentials for non-point source groundwater contamination analyzed using RZWQM. Paper No. SW 89562, *1989 Int. Winter Mtg. of the Am. Soc. of Agric. Eng.*, New Orleans, LA, December 12–15, Am. Soc. Agric. Eng., St. Joseph, MI.

Delgado, J.A. 1998. Sequential NLEAP simulations to examine effect of early and late planted winter cover crops on nitrogen dynamics. *J. Soil Water Conserv.*, 53:241–244.

Delgado, J.A. 2001. Use of simulations for evaluation of best management practices on irrigated cropping systems. In *Modeling Carbon and Nitrogen Dynamics for Soil Management*, eds. M.J. Shaffer, L. Ma, and S. Hansen, pp. 355–381. CRC Press, Boca Raton, FL.

Delgado, J.A., and W. Bausch. 2005. Potential use of precision conservation techniques to reduce nitrate leaching in irrigated crops. *J. Soil Water Conserv.*, 60:379–387.

Delgado, J.A., and R.F. Follett. 2002. Carbon and nutrient cycles. *J. Soil Water Conserv.*, 57:455–464.

Delgado, J.A., and R.F. Follett. 2010. *Advances in Nitrogen Management for Water Quality*. Soil and Water Conservation Society, Ankeny, IA.

Delgado, J.A., and M.J. Shaffer. 2008. Nitrogen management modeling techniques: Assessing cropping systems/landscape combinations. In *Nitrogen in the Environment: Sources, Problems and Management*, eds. R.F. Follett, and J.L. Hatfield, pp. 539–570. Elsevier Science, New York.

Delgado, J.A., R.F. Follett, J.L. Sharkoff, M.K. Brodahl, and M.J. Shaffer. 1998a. NLEAP facts about nitrogen management. *J. Soil Water Conserv.*, 53:332–337.

Delgado, J.A., M.J. Shaffer, and M.K. Brodahl. 1998b. New NLEAP for shallow- and deep-rooted crop rotations. *J. Soil Water Conserv.*, 53:338–340.

Delgado, J.A., R.F. Follett, and M.J. Shaffer. 2000. Simulation of NO_3-N dynamics for cropping systems with different rooting depths. *Soil Sci. Soc. Am. J.*, 64:1050–1054.

Delgado, J.A., R.R. Riggenbach, R.T. Sparks, M.A. Dillon, L.M. Kawanabe, and R.J. Ristau. 2001. Evaluation of nitrate–nitrogen transport in a potato–barley rotation. *Soil Sci. Soc. Am. J.*, 65:878–883.

Delgado, J.A., R. Khosla, W.C. Bausch, D.G. Westfall, and D.J. Inman. 2005. Nitrogen fertilizer management based on site-specific management zones reduce potential for nitrate leaching. *J. Soil Water Conserv.*, 60:402–410.

Delgado, J.A., M.A. Dillon, R.T. Sparks, and S.Y.C. Essah. 2007. A decade of advances in cover crops: Cover crops with limited irrigation can increase yields, crop quality, and nutrient and water use efficiencies while protecting the environment. *J. Soil Water Conserv.*, 62:110A–117A.

Delgado, J.A., M.J. Shaffer, H. Lal, S.P. McKinney, C.M. Gross, and H. Cover. 2008. Assessment of nitrogen losses to the environment with a Nitrogen Trading Tool (NTT). *Comput. Electron. Agric.*, 63:193–206.

Delgado, J.A., P.M. Gagliardi, D. Neer, and M.J. Shaffer. 2010a. *Nitrogen Loss and Environmental Assessment Package with GIS Capabilities (NLEAP GIS 4.2) User Guide*. USDA Agricultural Research Service, Soil Plant Nutrient Research Unit, Fort Collins, CO. Available at http://www.ars.usda.gov/SP2UserFiles/ad_hoc/54020700 NitrogenTools/NLEAP_GIS_4_2_Manual_Nov_29_2010.pdf (accessed April 27, 2014).

Delgado, J.A., S.J. Del Grosso, and S.M. Ogle. 2010b. ^{15}N Isotopic crop residue exchange studies suggest that IPCC methodologies to assess N_2O-N emissions should be reevaluated. *Nutr. Cycl. Agroecosyst.*, 86:383–390.

Delgado, J.A., C.M. Gross, H. Lal, H. Cover, P. Gagliardi, S.P. McKinney, E. Hesketh, and M.J. Shaffer. 2010c. A new GIS nitrogen trading tool concept for conservation and reduction of reactive nitrogen losses to the environment. *Adv. Agron.*, 105:117–171.

Delgado, J.A., P.M. Groffman, M.A. Nearing, T. Goddard, D. Reicosky, R. Lal, N.R. Kitchen, C.W. Rice, D. Towery, and P. Salon. 2011. Conservation practices to mitigate and adapt to climate change. *J. Soil Water Conserv.*, 66:118A–129A.

Dubrovsky, N.M., K.R. Burow, G.M. Clark, J.A.M. Gronberg, P.A. Hamilton, K.J. Hitt, D.K. Mueller, M.D. Munn, B.T. Nolan, L.J. Puckett, M.G. Rupert, T.M. Short, N.E. Spahr, L.A. Sprague, and W.G. Wilber. 2010. *The Quality of Our Nation's Waters—Nutrients in the Nation's Streams and Groundwater, 1992–2004*. Circular-1350, U.S. Geological Survey, Reston, VA. Available at http://water.usgs.gov/nawqa/nutrients/pubs/circ1350 (accessed April 27, 2014).

Eagle, A., L. Olander, L.R. Henry, K. Haugen-Kozyra, N. Millar, and G.P. Robertson. 2012. *Greenhouse Gas Mitigation Potential of Agricultural Land Management in the United States: A Synthesis of the Literature*, Third Edition. Report NI R 10-04, Nicholas Institute for Environmental Policy Solutions, Duke University, Durham, NC. Available at http://nicholasinstitute.duke.edu/sites/default/files/publications/ni_r_10-04_3rd_edition.pdf (accessed April 27, 2014).

Follett, R.F., D.R. Keeney, and R.M. Cruse, eds. 1991. *Managing Nitrogen for Groundwater Quality and Farm Profitability*. Soil Science Society of America, Madison, WI.

Galloway, J.N., A.R. Townsend, J.W. Erisman, M. Bekunda, Z. Cai, J.R. Freney, L.A. Martinelli, S.P. Seitzinger, and M.A. Sutton. 2008. Transformation of the nitrogen cycle: Recent trends, questions, and potential solutions. *Science*, 320(5878):889–892.

Glebe, T.W. 2006. The environmental impact of European farming: How legitimate are agri-environmental payments? *Rev. Agric. Econ.*, 29:87–102.

Greenhalch, S., and A. Sauer. 2003. *Awakening the Dead Zone: An Investment for Agriculture, Water Quality, and Climate Change*. World Resources Institute, Washington, DC.

Gross, C.M., J.A. Delgado, S.P. McKinney, H. Lal, H. Cover, and M.J. Shaffer. 2008. Nitrogen Trading Tool (NTT) to facilitate water quality credit trading. *J. Soil Water Conserv.*, 63:44A–45A.

Hall, M.D., M.J. Shaffer, R.M. Waskom, and J.A. Delgado. 2001. Regional nitrate leaching variability: What makes a difference in Northeastern Colorado? *J. Am. Water Resour. Assoc.*, 37:139–150.

Hansen, L., J.A. Delgado, M. Ribaudo, and W. Crumpton. 2012. Minimizing cost of reducing agricultural nitrogen loadings: Choosing between on- and off-field conservation practices. *Environ. Econ.*, 3:98–113.

Hey, D.L. 2002. Nitrogen farming: Harvesting a different crop. *Restor. Ecol.*, 10:1–10.

Hey, D.L., L.S. Urban, and J.A. Kostel. 2005. Nutrient farming: The business of environmental management. *Ecol. Eng.*, 24:279–287.

Intergovernmental Panel on Climate Change (IPCC). 2007. Summary for policymakers. In *Climate Change 2007: The Physical Science Basis. Contribution of Working Group I to the Fourth Assessment Report of the Intergovernmental Panel on Climate Change*, eds. S. Solomon, D. Qin, M. Manning, Z. Chen, M. Marquis, K.B. Averyt, M. Tignor, and H.L. Miller. Cambridge University Press, Cambridge, and New York. Available at http://www.ipcc.ch/publications_and_data/publications_ipcc_fourth_assessment_report_wg1_report_the_physical_science_basis.htm (accessed April 27, 2014).

Johnson, J.M.F., D.L. Karlen, and S.S. Andrews. 2010. Conservation considerations for sustainable bioenergy feedstock production: If, what, where, and how much? *J. Soil Water Conserv.*, 65:88A–91A.

Khakural, B.R., and P.C. Robert. 1993. Soil nitrate leaching potential indices: Using a simulation model as a screening system. *J. Environ. Qual.*, 22:839–845.

Lal, H., J.A. Delgado, C.M. Gross, E. Hesketh, S.P. McKinney, H. Cover, and M. Shaffer. 2009. Market-based approaches and tools for improving water and air quality. *Environ. Sci. Policy*, 12:1028–1039.

Lal, R., J.A. Delgado, P.M. Groffman, N. Millar, C. Dell, and A. Rotz. 2011. Management to mitigate and adapt to climate change. *J. Soil Water Conserv.*, 66:276–285.

Lavado, R.S., J.M. de Paz, J.A. Delgado, and H. Rimski-Korsakov. 2010. Evaluation of best nitrogen management practices across regions of Argentina and Spain. In *Advances in Nitrogen Management for Water Quality*, eds. J.A. Delgado, and R.F. Follett, pp. 314–343. Soil and Water Conservation Society, Ankeny, IA.

Ma, L., and M.J. Shaffer. 2001. A review of carbon and nitrogen processes in nine U.S. soil nitrogen dynamics models. In *Modeling Carbon and Nitrogen Dynamics for Soil Management*, eds. M.J. Shaffer, L. Ma, and S. Hansen, pp. 55–102. CRC Press, Boca Raton, FL.

McGechan, M.R., D.R. Lewis, L. Wu, and L.P. McTaggart. 2001. Modeling the effects of manure and fertilizer management options on soil carbon and nitrogen processes. In *Modeling Carbon and Nitrogen Dynamics for Soil Management*, eds. M.J. Shaffer, L. Ma, and S. Hansen, pp. 427–458. CRC Press, Boca Raton, FL.

Montgomery, D.R. 2007. Soil erosion and agricultural sustainability. *Proc. Natl. Acad. Sci.*, 104:13268–13272.

Rabalais, N.N., R.E. Turner, and W.J. Wiseman, Jr. 2002a. Gulf of Mexico Hypoxia, a.k.a. "The Dead Zone." *Ann. Rev. Ecol. Evol. Syst.*, 33:235–263.

Rabalais, N.N., R.E. Turner, and D. Scavia. 2002b. Beyond science into policy: Gulf of Mexico hypoxia and the Mississippi River. *Bioscience*, 52(2):129–142.

Ribaudo, M., and J. Gottlieb. 2011. Point–nonpoint trading—Can it work? *J. Am. Water Resour. Assoc.*, 47:5–14.

Ribaudo, M.O., R. Heimlich, and M. Peters. 2005. Nitrogen sources and Gulf hypoxia: Potential for environmental credit trading. *Ecol. Econ.*, 52(2):159–168.

Ribaudo, M., J. Delgado, L. Hansen, M. Livingston, R. Mosheim, and J. Williamson. 2011a. *Nitrogen in Agricultural Systems: Implications for Conservation Policy*. Economic Research Report (ERR-127), Economic Research Service, Washington, DC.

Ribaudo, M., J. Delgado, and M. Livingston. 2011b. Preliminary assessment of the potential for nitrous oxide offsets in a cap and trade program. *Agric. Resour. Econ. Rev.*, 40:266–281.

Ritchie, J.T., D.C. Godwin, and S. Otter-Nacke. 1985. *CERES Wheat: A Simulation Model of Wheat Growth and Development*. Texas A&M University Press, College Station, TX.

Rupert, M.G. 2008. Decadal-scale changes of nitrate in ground water of the United States, 1988–2004. *J. Environ. Qual.*, 37:S240–S248.

Saleh, A., O. Gallego, E. Osei, H. Lal, C. Gross, S. McKinney, and H. Cover. 2011. Nutrient Tracking Tool—A user-friendly tool for calculating nutrient reductions for water quality trading. *J. Soil Water Conserv.*, 66:400–410.

Shaffer, M.J., and J.A. Delgado. 2001. Field techniques for modeling nitrogen management. In *Nitrogen in the Environment*, eds. R.F. Follett, and J.L. Hatfield, pp. 391–411. CRC Press, New York.

Shaffer, M.J., and J.A. Delgado. 2002. Essentials of a national nitrate leaching index assessment tool. *J. Soil Water Conserv.*, 57:327–335.

Shaffer, M.J., and W.E. Larson. 1987. *NTRM, a Soil–Crop Simulation Model for Nitrogen, Tillage, and Crop Residue Management.* USDA-ARS Conserv. Res. Rep. 34-1. USDA-ARS, Washington, DC.

Shaffer, M.J., A.D. Halvorson, and F.J. Pierce. 1991. Nitrate Leaching and Economic Analysis Package (NLEAP): Model description and application. In *Managing Nitrogen for Groundwater Quality and Farm Profitability*, eds. R.F. Follett, D.R. Keeney, and R.M. Cruse, pp. 285–322. Soil Science Society of America, Madison, WI.

Shaffer, M.J., L. Ma, and S. Hansen, eds. 2001. *Modeling Carbon and Nitrogen Dynamics for Soil Management.* CRC Press, Boca Raton, FL.

Shaffer, M.J., J.A. Delgado, C. Gross, R.F. Follett, and P. Gagliardi. 2010. Simulation processes for the Nitrogen Loss and Environmental Assessment Package (NLEAP). In *Advances in Nitrogen Management for Water Quality*, eds. J.A. Delgado, and R.F. Follett, pp. 361–372. Soil and Water Conservation Society, Ankeny, IA.

Shumway, C., J.A. Delgado, T. Bunch, L. Hansen, and M. Ribaudo. 2012. Best nitrogen management practices can reduce the potential flux of nitrogen out of the Arkansas Delta. *Soil Sci.*, 177:198–209.

Verdin, J., C. Funk, G. Senay, and R. Choularton. 2005. Climate science and famine early warning. *Phil. Trans. R. Soc.*, 360:2155–2168.

Wagenet, R.J., and J.L. Hutson. 1989. *LEACHM: Leaching Estimation and Chemistry Model: A Process-Based Model of Water and Solute Movement, Transformations, Plant Uptake, and Chemical Reactions in the Unsaturated Zone.* Dept. of Agronomy, Cornell University, Ithaca, NY.

Williams, J.R., P.T. Dyke, and C.A. Jones. 1983. EPIC—A model for assessing the effects of erosion on soil productivity. In *Proceedings of the Third International Conference on State of the Art in Ecological Modeling*, ed. W.K. Lauenrroth, pp. 555–572. Colorado State University, Elsevier, New York.

Wylie, B.K., M.J. Shaffer, and M.D. Hall. 1995. Regional assessment of NLEAP NO_3-N leaching indices. *Water Resour. Bull.*, 31:399–408.

8 Use of Advanced Information Technologies for Water Conservation on Salt-Affected Soils

Dennis L. Corwin

CONTENTS

EXECUTIVE SUMMARY

Water scarcity is an identifying feature of arid and semiarid regions of the world, causing water conservation to be a constant consideration in these areas. Due to the unpredictability and scarcity of natural precipitation in arid and semiarid regions, irrigation is essential for maintaining crop productivity. In general, irrigation and soil salinization go hand in hand, particularly in the arid zones of the world. Water conservation on arid and semiarid soils must be done with constant and careful consideration of the distribution of salinity across the landscape and through the soil profile. Salinity is of concern because it causes a significant decrease in crop productivity due to osmotic and toxic ion effects on plant growth. However, soil salinity can be managed through leaching and the application of various soil amendments. The field-scale management of soil salinity is best handled with knowledge of its spatial and temporal distribution. Ideally, water conservation on irrigated agricultural lands is best achieved by applying irrigation water where, when, and in the amounts needed to adequately leach salts and to meet the crop's water needs. This is not easily done since water content and salinity are highly variable both spatially and temporally across a field and through the root zone. The goal of site-specific irrigation, however, is to account for within-field variation of water content and salinity. Field-scale salinity measurement and mapping protocols have been developed by Corwin and his colleagues at the U.S. Salinity Laboratory in Riverside, California. These protocols utilize advanced information technologies (i.e., geophysical techniques measuring apparent soil electrical conductivity [EC_a], geographic information system [GIS], geostatistics, spatial statistical analysis, and spatial statistical sampling designs) to map the spatiotemporal distribution of soil salinity for management applications. These protocols and technologies also have the potential to map soil water content and texture in most instances. The goal of this chapter is to provide an overview of the approach for delineating site-specific irrigation management units (SSIMUs) from the field-scale characterization of soil salinity, water content, and textural distributions using advanced information technologies. Guidelines, special considerations, protocols, and strengths and limitations are presented. Maps of SSIMUs provide irrigation management information to ameliorate crop yield reduction on salt-affected soils with minimal irrigation water requirements. Land resource managers, water conservation specialists, farmers, extension specialists, and Natural Resource Conservation Service field staff are the beneficiaries of field-scale maps of soil salinity, water content, texture, and SSIMUs, which can be used for crop selection, irrigation and salinity management, and remediation. These tools are important to provide adequate water for crop production while protecting soils from excessive salinization that will degrade soil quality and impair future productivity.

KEYWORDS

Electrical conductivity, sampling, soil salinity, spatial variability

8.1 INTRODUCTION

Due to an ever-growing population with its increasing demand on finite water supplies, the world faces an unprecedented crisis in water resources management, with profound implications for global food security, protection of human health, and maintenance of aquatic ecosystems. Jury and Vaux (2007) provide an insightful look into the emerging global water crisis, identifying a definitive and imminent need for water conservation throughout the world, particularly for water-scarce and water-stressed countries.

The increase in population and urbanization has resulted in severe water shortages throughout the world, which are exacerbated by changes in climate patterns. For instance, the United States has been experiencing an increase in moderate to severe levels of drought particularly in the Southwest, but other areas of the United States are not exempt. This has caused reductions in irrigation water allocations to farmers in the San Joaquin Valley and heightened water conservation measures in urban areas. It is estimated that in the near future four-fifths of the states in the USA are expected to face localized or statewide water shortages. In the mid-1990s, 80 countries that constitute 40% of the world's population were suffering from serious water shortages (CSD 1997). The World Water Council (2000) forecasts that by 2020, the world will be 17% short of the water supply needed to feed the world's population. The United Nations FAO (2011) indicates that by 2025, nearly 2 billion people will be living in countries or regions with absolute water scarcity, and two-thirds of the world population could be living under stressed water conditions. These statistics just scratch the surface of the compelling data pointing to the need for global water conservation.

In late 2011, the world population passed 7 billion; the United Nations forecasts that the world population will reach 9.3 billion by 2050. There are grave concerns that, at that time, 40% of the global population (i.e., 3.6 billion people) may suffer from food shortage, economic deprivation, and poor health because of water stress. Even though water covers 70% of the earth's surface, less than 3% of the world's water is fresh—the remainder is undrinkable seawater.

8.1.1 Importance of Irrigated Agriculture

Agriculture accounts for more than 70% of the freshwater drawn from lakes, rivers, and underground sources (CSD 1997). In 2000, roughly 57% of the world's freshwater withdrawal and 70% of its consumption took place in Asia, where the world's major irrigated lands are found (Shiklomanov and Roda 2003). From 1970 to 2000, the area of land under irrigation increased from less than 200 million ha to over 270 million ha (United Nations FAO 2001). From a global perspective, irrigated agriculture makes an essential contribution to the food needs of the world, with only 15% of the world's farmland irrigated, and yet it produces 40% of the total food and fiber (World Water Council 2000). Without a doubt, irrigated agriculture is essential to meet the world's ever-increasing demand for food.

Ironically, some of the most agriculturally productive areas of the world occur in water-scarce regions, such as the arid Southwestern United States (e.g., California's

San Joaquin and Imperial–Coachella Valleys) and other arid regions of the world, including the Middle East; the Hai He, Huang He, and Yangtze basins in China; and along the Nile River in Egypt and Sudan. In most cases, these areas owe their successful crop productivity to mild year-round climates and irrigation to supplement inadequate rainfall. Furthermore, global climate change model predictions indicate decreased precipitation for drier regions of the world, with annual average precipitation decreases likely to occur in most of the Mediterranean, Northern Africa, Northern Sahara, Central America, the American Southwest, the Southern Andes, and Southwestern Australia (IPCC 2007).

8.1.2 Need for Water Conservation in Irrigated Agriculture

There is no doubt that the Green Revolution, which dramatically changed the world's crop productivity over the past half century, was successful in large part due to worldwide advances in the use of irrigation. Continued reliance on irrigated agriculture is necessary to meet the growing food demands of an ever-increasing global population, but the high consumptive water use of irrigated agriculture places heavy demands on finite water resources. Agriculture is the largest consumer of freshwater worldwide (UNESCO 2009). Irrigated agriculture cannot possibly meet future food production expectations if it cannot conserve water while concomitantly increasing productivity.

The prospect of feeding a projected additional 2.3 billion people over the next 40 years poses more challenges than encountered in the past 40 years. Global resource experts predict that in the short term, there will be adequate global food supplies, but the distribution of those supplies to malnourished people will be the primary problem. However, in the long term, the obstacles become more formidable though not insurmountable. Although total yields continue to rise on a global basis, there is a disturbing decline in yield growth, with some major crops such as wheat and maize reaching a "yield plateau" (World Resources Institute 1998). Feeding the ever-increasing world population will require a sustainable agriculture that can keep pace with population growth and can balance crop yield with resource utilization, particularly water resources. The concept of sustainable agriculture is predicated on a delicate balance of maximizing crop productivity and maintaining economic stability while minimizing the utilization of finite natural resources and the detrimental environmental impacts of associated agrichemical pollutants. In irrigated agricultural regions throughout the world, which are often located in water-vulnerable arid and semiarid climates, water will be the crucial resource that must be conserved to maintain productivity.

8.1.3 Interrelationship of Salinity and Irrigated Agriculture

The accumulation of soil salinity is a consequence of a variety of processes. In arid and semiarid areas, for example, where precipitation is less than evaporation, salts can accumulate at the soil surface when the depth to the water table is less than 1 to 1.5 m, depending on the soil texture. The accumulation of salts at the soil surface is the consequence of the upward flow of water and the subsequent transport of salts due to capillary rise driven by the evaporative process. However, by far, the most common cause for the accumulation of salts is evapotranspiration (ET) by plants, which results in an increase in salt concentration with depth through the root zone and the accumulation of salts below the root zone. The level of salt accumulation within and below the root zone due to ET depends upon the fraction of irrigation and/or precipitation that flows beyond the root zone, referred to as the leaching fraction (LF). As the LF increases, the total salts within the root zone decrease due to their removal from the root zone by leaching.

The accumulation of salts in the root zone goes hand in hand with irrigated agriculture. Irrigation management of arid and semiarid agricultural soils is more often than not a matter of salinity management through irrigation. Irrigation management in arid and semiarid regions must concomitantly manage salinity to be viable.

Vast areas of irrigated land are known to be threatened by salt accumulation. According to CSD (1997), salinization of approximately 20% of the world's irrigated land results from poor water management practices, with an additional 1.5 million ha affected annually, significantly reducing crop production. Rhoades and Loveday (1990) estimated that 50% of all irrigated systems (totaling approximately 250 million ha) are affected by salinity or shallow water–related problems. Waterlogging and salinization alone represent a significant threat to the world's productivity capacity (Alexandratos 1995).

8.1.4 Need for Site-Specific Irrigation and Salinity Management

Site-specific management (SSM) attempts to manage soil, pests, and crops based upon spatial variation within a field (Larson and Robert 1995). In contrast, conventional farming treats a field uniformly, ignoring the naturally inherent variability of soil and crop conditions between and within fields. There is well-documented evidence that spatial variation within a field is highly significant and amounts to a factor of 3–4 or more for crops (Birrel et al. 1995; Verhagen et al. 1995) and up to an order of magnitude or more for soil variability (Corwin et al. 2003a). SSM is the management of agricultural crops at a spatial scale smaller than the whole field that takes local variation into account to cost-effectively balance crop productivity and quality, detrimental environmental impacts, and the use of resources (e.g., water, fertilizer, pesticides) in an economically optimal way by applying them when, where, and in the amount needed. One of the most promising approaches for attaining sustainability in water-limited agricultural areas is site-specific irrigation management (SSIM). SSIM has the potential to conserve precious freshwater resources by applying irrigation water when, where, and in the amounts needed to optimize yield, which in arid and semiarid climates is often influenced most by salinity and water distributions. To manage within-field variation, georeferenced areas (or units) that are similar with respect to a specified characteristic (e.g., salinity, water content, etc.) must be identified (van Uffelen et al. 1997). Site-specific management units (SSMUs) have been proposed as a means of dealing with the spatial variation of edaphic properties that affect crop productivity (or quality) to achieve the goals of SSM. A SSMU is simply a mapped unit within a field that could be based on soil properties, landscape units, past yield, etc., that is managed to achieve the goals of SSM.

8.1.5 Advanced Information Technologies for Site-Specific Irrigation and Salinity Management

The delineation of site-specific irrigation management units (SSIMUs) is not a trivial task and requires advanced information technologies including proximal sensors, geographic information system (GIS), Global Positioning System (GPS), spatial statistics, and design- or model-based sampling (Corwin and Lesch 2010). Corwin and colleagues (Corwin et al. 2003b; Corwin and Lesch 2010) have developed the methodology for defining SSIMUs based on protocols and guidelines developed by Corwin and Lesch (2003, 2005a) for characterizing the field-scale spatial variability of soil salinity (and other soil properties, including water content and texture) with apparent soil electrical conductivity (EC_a) directed soil sampling.

8.2 FIELD-SCALE SALINITY MEASUREMENT AND MAPPING

The measurement of soil salinity has a long history prior to its assessment with EC_a. Soil salinity refers to the presence of major dissolved inorganic solutes in the soil aqueous phase, which consist of soluble and readily dissolvable salts including charged species (e.g., Na^+, K^+, Mg^{+2}, Ca^{+2}, Cl^-, HCO_3^-, NO_3^-, NO_4^{-2}, and CO_3^{-2}), nonionic solutes, and ions that combine to form ion pairs. The need to measure soil salinity stems from its detrimental impact on plant growth. Effects of soil salinity are manifested in loss of stand, reduced plant growth, reduced yields, and, in severe cases, crop failure. Salinity limits water uptake by plants by reducing the osmotic potential, making it more difficult for the plant to extract water. Salinity may also cause specific ion toxicity or upset the nutritional balance of plants. In addition, the salt composition of the soil water influences the composition of cations on the exchange complex of soil particles, which influences soil permeability and tilth.

8.2.1 Historical Approaches

Historically, six methods have been developed for determining soil salinity at field scales: (1) visual crop observations; (2) electrical conductance of soil solution extracts or extracts at higher than normal water contents; (3) in situ measurement of electrical resistivity (ER); (4) noninvasive measurement of electrical conductance with electromagnetic induction (EMI); (5) in situ measurement of electrical conductance with time domain reflectometry (TDR); and, most recently, (6) multispectral and hyperspectral imagery.

Visual crop observation is the oldest method of determining the presence of soil salinity in a field. It is a quick method, but it has the disadvantage in that salinity development is detected after crop damage has occurred, making it the least desirable method for obtaining soil salinity information. However, remote sensing, including multispectral and hyperspectral imagery, plays an increasing role in agriculture management practices and represents a quantitative approach to visual observation that may offer a potential for early detection of the onset of salinity damage to plants. Even so, multispectral and hyperspectral remote imagery technologies are currently unable to differentiate osmotic from matric or other stresses. This distinction is key to the successful application of remote imagery as a tool to map salinity and/or water content.

The determination of salinity through the measurement of electrical conductance has been well established for decades (U.S. Salinity Laboratory Staff 1954). Electrical conductivity (EC) of water is a function of its chemical composition. McNeal et al. (1970) were among the first to establish the relationship between the EC and molar concentrations of ions in the soil solution. Soil salinity is quantified in terms of the total concentration of the soluble salts, as measured by the EC of the solution in dS m^{-1}. To determine the EC, the soil solution is placed between two electrodes of constant geometry and distance of separation (Bohn et al. 1979). At a constant potential, the current is inversely proportional to the solution's resistance. The measured conductance is a consequence of the solution's salt concentration and the electrode geometry whose effects are embodied in a cell constant. The electrical conductance is a reciprocal of the resistance, as shown in Equation 8.1:

$$EC_T = k/R_T \tag{8.1}$$

where EC_T is the electrical conductivity of the solution in dS m^{-1} at temperature T (°C), k is the cell constant, and R_T is the measured resistance at temperature T. Customarily, EC is expressed at a reference temperature of 25°C for purposes of comparison. The EC measured at a particular temperature T (°C), EC_T, can be adjusted to a reference EC at 25°C, EC_{25}, using the following equations from Handbook 60 (U.S. Salinity Laboratory Staff 1954):

$$EC_{25} = f_T \times EC_T \tag{8.2}$$

where f_T is a temperature conversion factor approximated by Sheets and Hendrickx (1995):

$$f_T = 0.4470 + 1.4034e^{-T/26.815} \tag{8.3}$$

Customarily, soil salinity has been defined in terms of laboratory measurements of the EC of the saturation extract (EC_e), because it is impractical for routine purposes to extract soil water from samples at typical field water contents. Partitioning of solutes over the three soil phases (i.e., gas, liquid, solid) is influenced by the soil-to-water ratio at which the extract is made, so the ratio must be standardized to obtain results that can be applied and interpreted universally. Commonly used extract ratios other than a saturated soil paste are 1:1, 1:2, and 1:5 soil-to-water mixtures.

Developments in the measurement of soil EC to determine soil salinity shifted away from extractions to the measurement of EC_a because the time and cost of obtaining soil solution extracts prohibited their practical use at field scales. Moreover, the high local-scale soil variability rendered salinity sensors and small-volume soil core samples of limited quantitative value. The use of EC_a to measure salinity has the advantage of increased volume of measurement and quickness of measurement but suffers from the complexity of measuring EC for the bulk soil rather than restricted to the solution phase.

8.2.2 Geospatial EC_a Measurements

To measure soil salinity, the electrical conductance of only the soil solution is required; consequently, EC_a measures more than just soil salinity. In fact, EC_a is a measure of anything conductive within the volume of measurement and is influenced, whether directly or indirectly, by any edaphic property that affects bulk soil conductance.

At present, no other measurement provides a greater level of spatial soil information for soil salinity assessment than the geospatial measurements of EC_a. These measurements are particularly useful for directed soil sampling to characterize soil spatial variability of salinity, texture, water content, and other soil properties that are indirectly measured by EC_a (e.g., organic matter [OM], cation exchange capacity [CEC]) (Corwin and Lesch 2005b). The rational for characterizing soil spatial variability with EC_a measurements is based on the hypothesis that this information can be used to develop a directed soil sampling plan that identifies sites that adequately reflect the range and variability of soil salinity and/or other soil properties that are correlated with EC_a. This hypothesis has repeatedly held true for a variety of agricultural applications (Lesch et al. 1992, 2005; Johnson et al. 2001; Corwin and Lesch 2003, 2005b,c; Corwin et al. 2003a,b; Corwin 2005).

The EC_a measurement is particularly well suited for establishing within-field spatial variability of soil properties because it is a quick and dependable measurement that integrates within its measurement the influence of several soil properties that contribute to the electrical conductance of the bulk soil. The EC_a measurement serves as a means of defining spatial patterns that indicate differences in electrical conductance due to the combined conductance influences of soil water content, texture, and bulk density. Therefore, maps of the variability of EC_a provide the spatial information to direct the selection of soil sample sites in order to characterize the spatial variability of those soil properties correlating, either for direct or indirect reasons, to EC_a.

8.2.2.1 Factors Influencing EC_a

The characterization of the spatial variability of various soil properties with EC_a is a consequence of the physicochemical nature of the EC_a measurement. Three parallel pathways of current flow contribute to the EC_a measurement: (1) a liquid phase pathway via salts contained in the soil water occupying the large pores; (2) a solid pathway via soil particles that are in direct and continuous contact with one another; and (3) a solid–liquid pathway primarily via exchangeable cations associated with clay minerals (Rhoades et al. 1989, 1999). Rhoades et al. (1989) formulated an electrical conductance model that describes the three conductance pathways of EC_a:

$$EC_a = \left[\frac{(\theta_{ss} + \theta_{ws})^2 \cdot EC_{ws} \cdot EC_{ss}}{\theta_{ss} \cdot EC_{ws} + \theta_{ws} \cdot EC_s} \right] + (\theta_{sc} \cdot EC_{sc}) + (\theta_{wc} \cdot EC_{wc}) \tag{8.4}$$

where θ_{ws} and θ_{wc} are the volumetric soil water contents in the soil–water pathway ($cm^3\ cm^{-3}$) and in the continuous liquid pathway ($cm^3\ cm^{-3}$), respectively; θ_{ss} and θ_{sc} are the volumetric contents of the surface conductance ($cm^3\ cm^{-3}$) and indurated solid phases of the soil ($cm^3\ cm^{-3}$), respectively; EC_{ws} and EC_{wc} are the specific electrical conductivities of the soil–water pathway ($dS\ m^{-1}$) and continuous liquid pathway ($dS\ m^{-1}$); and EC_{ss} and EC_{sc} are the electrical conductivities of the surface conductance ($dS\ m^{-1}$) and indurated solid phases ($dS\ m^{-1}$), respectively. Equation 8.4 was reformulated by Rhoades et al. (1989) into Equation 8.5:

$$EC_a = \left[\frac{(\theta_{ss} + \theta_{ws})^2 \cdot EC_{ws} \cdot EC_{ss}}{(\theta_{ss} \cdot EC_{ws}) + (\theta_{ws} \cdot EC_s)}\right] + (\theta_w - \theta_{ws}) \cdot EC_{wc} \tag{8.5}$$

where $\theta_w = \theta_{ws} + \theta_{wc}$ = total volumetric water content ($cm^3\ cm^{-3}$), and $\theta_{sc} \cdot EC_{sc}$ was assumed to be negligible. The following simplifying approximations are also known:

$$\theta_w = \frac{(PW \cdot \rho_b)}{100} \tag{8.6}$$

$$\theta_{ws} = 0.639\theta_w + 0.011 \tag{8.7}$$

$$\theta_{ss} = \frac{\rho_b}{2.65} \tag{8.8}$$

$$EC_{ss} = 0.019(SP) - 0.434 \tag{8.9}$$

$$EC_w = \left[\frac{EC_e \cdot \rho_b \cdot SP}{100 \cdot \theta_w}\right] \tag{8.10}$$

where PW is the percent water on a gravimetric basis; ρ_b is the bulk density ($Mg\ m^{-3}$); SP is the saturation percentage; EC_w is the average electrical conductivity of the soil water assuming equilibrium (i.e., $EC_w = EC_{sw} = EC_{wc}$); and EC_e is the electrical conductivity of the saturation extract ($dS\ m^{-1}$).

Because of the pathways of conductance, EC_a is influenced by a complex interaction of edaphic properties including EC_e (soil salinity), texture (quantitatively approximated by SP), θ_w (water content), ρ_b (bulk density), and temperature. The SP and the ρ_b are both directly influenced by clay content and OM. Furthermore, the exchange surfaces on clays and OM provide a solid–liquid phase pathway primarily via exchangeable cations; consequently, clay content and mineralogy, CEC, and OM are recognized as additional factors that influence EC_a measurements. EC_a is a complex property that must be interpreted with these influencing factors in mind.

The interpretation of EC_a measurements is not trivial because of the complexity of current flow in the bulk soil. Numerous EC_a studies have been conducted that have revealed the site specificity and complexity of geospatial EC_a measurements with respect to the particular property or properties that influence the EC_a measurement at the study site. Corwin and Lesch (2005b) provide a compilation of EC_a studies and the associated dominant soil property or properties that are measured by EC_a for that study.

8.2.2.2 Techniques for Measuring EC_a

There are three primary geophysical techniques for measuring EC_a: ER, EMI, and TDR. ER and EMI are easily mobilized and are well suited for field-scale applications because of the ease and low cost of measurement with a volume of measurement that is sufficiently large (>1 m^3) to reduce the influence of local-scale variability. Developments in agricultural applications of ER and EMI have occurred along parallel paths, with each filling a needed niche based upon inherent strengths and limitations. Even though TDR is a useful and well-studied technique for measuring EC_a, it has lagged behind ER and EMI as an "on-the-go" proximal sensor because it does not provide a continuous stream of georeferenced measurements. Rather, TDR requires the user to go from one location to the next, stopping at each location to take discrete measurements; consequently, it is less rapid and less appealing for mapping EC_a at field scales and larger spatial extents. ER and EMI are the current methods of choice for mapping soil salinity and other soil properties that are related to EC_a, so they, and not TDR, will be subsequently discussed. For greater details regarding ER, EMI, and TDR, refer to Hendrickx et al. (2002).

8.2.2.2.1 Electrical Resistivity

ER was originally used by geophysicists to measure the resistivity of the geological subsurface. ER methods involve the measurement of the resistance to current flow across four electrodes inserted in a straight line on the soil surface at a specified distance between the electrodes. The electrodes are connected to a resistance meter that measures the potential gradient between the current and the potential electrodes. These methods were developed in the second decade of the 1900s by Conrad Schlumberger in France and Frank Wenner in the United States for the evaluation of near-surface ER (Rhoades and Halvorson 1977; Burger 1992). Although two electrodes (one current and one potential electrode) could be used, the stability of the reading is greatly improved with the use of four electrodes.

The resistance is converted to EC using Equation 8.1, where the cell constant, k, in the equation is determined by electrode configuration and distance. The depth of penetration of the electrical current and the volume of measurement increase as the interelectrode spacing increases. The four-electrode configuration is referred to as a "Wenner array" when the four electrodes are equidistantly spaced (interelectrode spacing = a). For a homogeneous soil, the depth of measurement of the Wenner array is equal to a, and the soil volume measured is roughly equal to πa^3.

There are additional four-electrode configurations that are frequently used, as discussed by Burger (1992), Telford et al. (1990), and Dobrin (1960). The influence of the interelectrode configuration and distance on EC_a is given by (Equation 8.11):

$$EC_{a,25^\circ C} = \left(\frac{1000}{2\pi R_t}\right)\left\{\frac{f_t}{\dfrac{1}{\dfrac{1}{r_1}-\dfrac{1}{r_2}-\dfrac{1}{R_1}+\dfrac{1}{R_2}}}\right\} \tag{8.11}$$

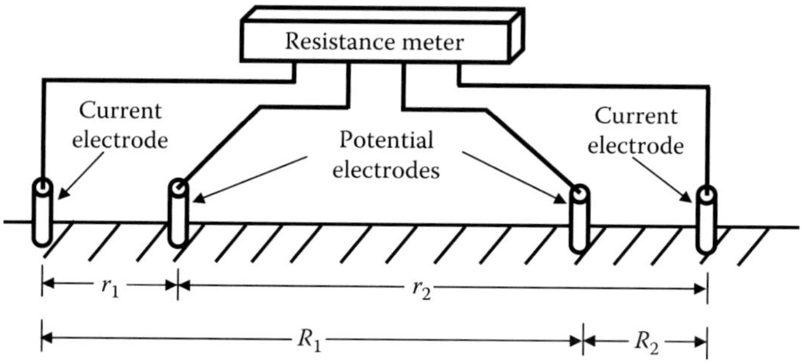

FIGURE 8.1 Schematic of four-electrode probe ER used to measure EC_a. (From Corwin, D.L. and J.M.H. Hendrickx, Solute content and concentration—Indirect measurement of solute concentration: Electrical resistivity—Wenner array. In *Methods of Soil Analysis. Part 4—Physical Methods, Agronomy Monograph No. 9*, eds. J. H. Dane and G.C. Topp, pp. 1282–1287, 2002. Madison, WI: Soil Science Society of America. With permission.)

where $EC_{a,25°C}$ is the EC_a temperature corrected to a reference of 25°C (dS m^{-1}), and r_1, r_2, R_1, and R_2 are the distances in centimeters between the electrodes, as shown in Figure 8.1. For the Wenner array, where $a = r_1 = r_2 = R_1 = R_2$, Equation 8.11 reduces to $EC_a = 159.2f_t/aR_t$ and $159.2/a$ represents the cell constant (k).

ER is an invasive technique that requires good contact between the soil and the four electrodes inserted into the soil; consequently, it produces less reliable measurements in dry or stony soils than a noninvasive measurement such as EMI. Nevertheless, the ER has a flexibility that has proven to be advantageous for field application, i.e., the depth and volume of measurement can be easily changed by altering the spacing between the electrodes. A distinct advantage of the ER approach is that the volume of measurement is determined by the spacing between the electrodes, which makes a large volume of measurement possible. For example, a 1-m interelectrode spacing for a Wenner array results in a volume of measurement of more than 3 m^3. This large volume of measurement integrates the high level of local-scale variability often associated with EC_a measurements.

8.2.2.2.2 Electromagnetic Induction

EC_a can be measured noninvasively with EMI. A transmitter coil located at one end of the EMI instrument induces circular eddy-current loops in the soil, with the magnitude of these loops directly proportional to the EC in the vicinity of that loop. Each current loop generates a secondary electromagnetic field that is proportional to the value of the current flowing within the loop. A fraction of the secondary induced electromagnetic field from each loop is intercepted by the receiver coil of the instrument, and the sum of these signals is amplified and formed into an output voltage, which is related to a depth-weighted EC_a. The amplitude and phase of the secondary field will differ from those of the primary field as a result of soil properties (e.g., clay content, water content, salinity), spacing of the coils and their orientation, frequency, and distance from the soil surface (Hendrickx et al. 2002).

The most commonly used EMI conductivity meters in soil science and in vadose zone hydrology are the Geonics EM-31 and EM-38 (Geonics Limited, Mississauga,

Ontario, Canada) and the DUALEM-2 (DUALEM Inc., Milton, Ontario, Canada). The EM-38 has had considerably greater application for agricultural purposes because the depth of measurement corresponds roughly to the root zone (i.e., generally 1–1.5 m). When the instrument is placed in the vertical coil configuration (EM_v, coils perpendicular to the soil surface), the depth of measurement is approximately 1.5 m, and in the horizontal coil configuration (EM_h, coils parallel to the soil surface), the depth of the measurement is 0.75–1.0 m. The perpendicular planar coils in the DUALEM-2 measure to a depth of 1.0 m and are analogous to the horizontal position in the EM-38, while the horizontal coplanar coils in the DUALEM-2 measure to a depth of 3 m and are analogous to the vertical position in the EM-38. The EM-31 has an intercoil spacing of 3.66 m, which corresponds to measurement depths of 3 and 6 m in the horizontal and vertical dipole orientations, respectively, which extends well beyond the root zone of agricultural crops. However, the EM-38 has one major pitfall, which is the need for calibration. The DUALEM-2 does not require calibration. Further details about and operation of the EM-31 and EM-38 equipment are discussed in Hendrickx et al. (2002). Documents concerning the DUALEM-2 can be found online at http://www.dualem.com/documents.html (accessed January 16, 2014).

EC_a measured by EMI at $EC_a < 1.0$ dS m^{-1} is given by Equation 8.12 from McNeill (1980):

$$EC_a - \frac{4}{2\pi\mu_0 f s^2}\left(\frac{H_s}{H_p}\right) \tag{8.12}$$

where EC_a is measured in S m^{-1}; H_p and H_s are the intensities of the primary and secondary magnetic fields at the receiver coil (A m^{-1}), respectively; f is the frequency of the current (Hz); μ_0 is the magnetic permeability of air ($4\pi 10^{-7}$ H m^{-1}); and s is the intercoil spacing (m).

8.2.2.2.3 *Advantages and Disadvantages of ER and EMI*

Both ER and EMI are rapid and reliable technologies for the measurement of EC_a, each with its advantages and disadvantages. The primary advantage of EMI over ER is that EMI is noninvasive, so it can be used on dry and stony soils that would not be amenable to invasive ER equipment. The disadvantage relates to the response function. Both EMI and ER have nonlinear response functions, but EC_a measured with EMI is a depth-weighted value that is more nonlinear than ER. More specifically, EMI concentrates its measurement of conductance over the depth of measurement at shallow depths, whereas ER is more nearly uniform with depth. Because of the greater linearity of the response function of ER, the EC_a for a discrete depth interval of soil, EC_x, can be determined with the Wenner array by measuring the EC_a of successive layers by increasing the interelectrode spacing from a_{i-1} to a_i and using Equation 8.13 from Barnes (1952) for resistors in parallel:

$$EC_x = EC_{a_i - a_{i-1}} - \frac{(EC_{a_i} \cdot a_i) - (EC_{a_{i-1}} - a_{i-1})}{(a_i - a_{i-1})} \tag{8.13}$$

where a_i is the interelectrode spacing, which equals the depth of sampling, and a_{i-1} is the previous interelectrode spacing, which equals the depth of the previous sampling. Measurements of EC_a by ER and EMI at the same location and over the same volume of measurement are not comparable because of the dissimilarity of their response functions. An advantage of ER over EMI is the ease of instrument calibration. Calibrating the EM-31 and EM-38 is more involved than for ER equipment. However, there is no need to calibrate the DUALEM-2.

8.2.2.3 Field-Scale Mapping of Soil Salinity and EC_a

An understanding and interpretation of geospatial EC_a data can only be obtained from ground-truth measures of soil properties that correlate with EC_a from either a direct influence or indirect association. For this reason, geospatial EC_a measurements are used as a surrogate of soil spatial variability to direct soil sampling when mapping soil salinity at field scales and larger spatial extents. They are not generally used as a direct measure of soil salinity, particularly at EC_a < 1–2 dS m^{-1} where the influence of conductive soil properties other than salinity can have an increased influence on the EC_a reading. At high EC_a values (i.e., EC_a > 1–2 dS m^{-1}), salinity most likely dominates the EC_a reading; consequently, geospatial EC_a measurements most likely map soil salinity.

8.2.2.3.1 Approach and Protocols for EC_a-Directed Soil Sampling

Scientists at the U.S. Salinity Laboratory have developed an integrated system for the measurement of field-scale salinity, which consists of (1) mobile EC_a measurement equipment (Rhoades 1993), (2) protocols for EC_a-directed soil sampling (Corwin and Lesch 2005a), and (3) sample design software (Lesch et al. 2000). The integrated system for mapping soil salinity is schematically illustrated in Figure 8.2.

The protocols of an EC_a survey for measuring soil salinity at field scale include nine basic elements: (1) EC_a survey design, (2) georeferenced EC_a data collection, (3) soil sample design based on georeferenced EC_a data, (4) soil sample collection, (5) physical and chemical analysis of pertinent soil properties, (6) calibration of EC_a to EC_e, (7) statistical analysis to determine dominant soil properties influencing the EC_a measurements at the study site, (8) GIS development, and (9) graphic display of spatial data. The basic steps for each element are provided in Table 8.1. A detailed discussion of the protocols can be found in Corwin and Lesch (2005a), and an update of the protocols specific to mapping soil salinity can be found in Corwin and Lesch (2013). Corwin and Lesch (2005c) provide a case study demonstrating the use of the protocols.

8.2.2.3.2 Factors to Consider during an EC_a-Directed Survey

There are a number of considerations that must be followed when conducting a geospatial EC_a survey to map soil salinity. Each of these considerations can influence the EC_a measurement leading to a potential misinterpretation of the salinity distribution. These considerations account for temporal, moisture, surface roughness, and surface geometry effects.

Temporal comparisons of geospatial EC_a measurements to determine spatiotemporal changes in salinity patterns of distribution can only be made from EC_a survey

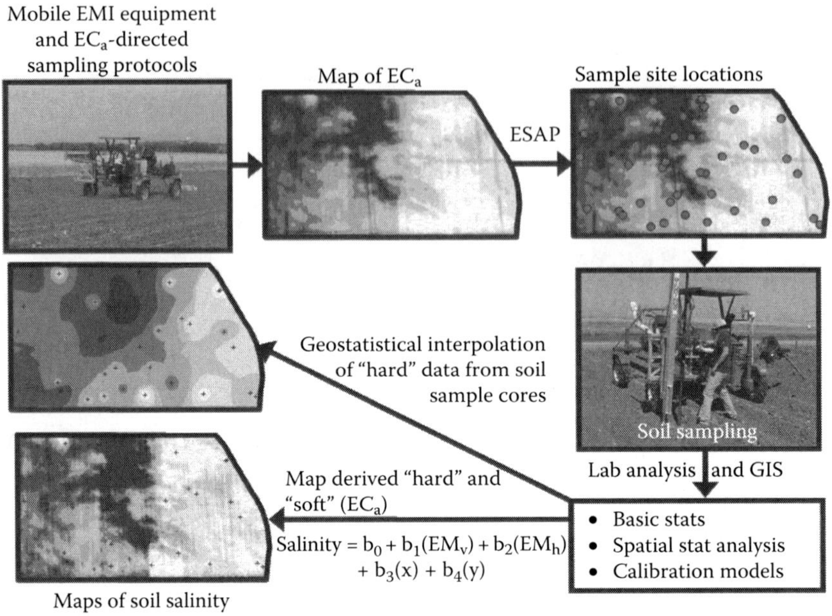

FIGURE 8.2 Schematic of the integrated system for field-scale salinity assessment using EC_a-directed sampling protocols, mobile EMI equipment, ESAP software, and GIS. EM_v refers to EMI measurement in the vertical coil configuration, and EM_h refers to EMI measurement in the horizontal coil configuration. (Modified from Corwin, D.L. et al., Laboratory and field measurements. In *Agricultural Salinity Assessment and Management*, eds. K.K. Tanji and W. Wallender, pp. 295–341, 2012. New York: American Society of Civil Engineers. With permission.)

data that have been obtained under similar water content and temperature conditions. Surveys of EC_a should be conducted when the water content is at or near field capacity and the soil profile temperatures are similar. For irrigated fields, EC_a surveys should be conducted roughly 2–4 days after irrigation or longer if the soil is high in clay content and additional time is needed for the soil to drain to field capacity. For dry-land farming, the survey should occur 2–4 days or longer after a substantial rainfall, depending on soil texture. The effects of temperature can be addressed by taking soil profile temperatures at the time of the EC_a survey and temperature-correcting the EC_a measurements, or by conducting the surveys roughly at the same time during the year so that the temperature profiles are the same for each survey.

The type of irrigation used can influence the within-field spatial distribution of water content and should be kept in mind as a factor that influences EC_a spatial patterns. Sprinkler irrigation has a high level of application uniformity, whereas flood irrigation and drip irrigation are highly nonuniform. In general, flood irrigation results in higher water contents and overleaching at the "head" end of the field, whereas underleaching and lower water contents can occur at the "tail" end of the field. This general across-the-field trend is observed for both flood irrigation with basins and flood irrigation with beds and furrows, but the beds and furrows

TABLE 8.1
Outline of Steps to Conduct an EC_a Field Survey to Map Soil Salinity

1. Site description and EC_a survey design
   a. Record site metadata
   b. Define the project's/survey's objective (e.g., inventorying, spatiotemporal monitoring, site-specific management, etc.)
   c. Establish site boundaries
   d. Select GPS coordinate system
   e. Establish EC_a measurement intensity (i.e., number and location of traverses and space between EC_a measurements with careful consideration of edge effects)
   f. Minimize secondary influences on EC_a (e.g., compaction, surface roughness and geometry, metal)
   g. Special EC_a survey design considerations
      i. Presence of beds and furrows: perform separate surveys for the beds and for the furrows
      ii. Vineyards with metal trellising
         A. Maximize distance from metal for surveys with EMI
         B. Place an insulator between metal posts and trellis wires to break the conductance loop from the soil to the posts along the wires and back into the soil (this applies to both ER and EMI surveys)
      iii. Presence of drip lines: perform separate EC_a surveys over and between drip lines
      iv. Variations in surface geometry or roughness: perform separate surveys with separate sampling designs for each area differing in surface roughness or surface geometry
      v. Temporal studies
         A. Reference all EC_a measurement to 25°C or
         B. Conduct EC_a surveys at the same time of the day and the same day of the year
2. EC_a data collection with mobile GPS-based equipment
   a. Conduct drift runs when using EMI to determine the effect of ambient temperature on EMI instrumentation
   b. Geo-reference site boundaries and significant physical geographic features with GPS
   c. Assure that water content at the study site is at or near field capacity ($\geq$70% field capacity) throughout the field (if water content is <70%, then do not conduct EC_a survey)
   d. Measure geo-referenced EC_a data at the predetermined spatial intensity and record associated metadata
   e. Keep the speed of mobile GPS-based equipment < 10 km h^{-1} to reduce GPS positional errors
3. Soil sample design based on geo-referenced EC_a data
   a. Statistically analyze EC_a data using an appropriate statistical sampling design (i.e., model- or design-based sampling design) to establish the soil sample site locations
   b. Establish site locations, depth of sampling, sample depth increments, and number of cores per site (>100 soil samples are desirable but the total number of samples is largely determined by the resources available to analyze the soil properties of concern)
4. Soil core sampling at specified sites designated by the sample design
   a. Obtain measurements of soil temperature through the profile at selected sites
   b. At randomly selected locations, obtain duplicate soil cores within a 1-m distance of one another to establish local-scale variation of soil salinity (and other soil properties) for 20% or more of the sample locations
   c. Record soil core observations (e.g., temperature, color, $CaCO_3$, gleying, organic matter, mottling, horizonation, textural discontinuities, etc.)

(Continued)

TABLE 8.1 (CONTINUED)
Outline of Steps to Conduct an EC_a Field Survey to Map Soil Salinity

5. Laboratory analysis of soil salinity and other EC_a-correlated soil properties relevant to the project objectives
6. Stochastic and/or deterministic calibration of EC_a to EC_e (and to other soil properties, e.g., water content, SP, etc.)
7. Statistical analysis to determine the soil properties influencing EC_a
   a. Perform a basic statistical analysis of soil salinity (and other relevant soil properties) by depth increment and by composite depth over the depth of measurement of EC_a
   b. Determine the correlation between EC_a and salinity (and between EC_a and other soil properties) by composite depth over the depth of measurement of EC_a
8. GIS database development
9. Graphic display of spatial distribution of soil salinity (and other properties correlated to EC_a) using various interpolation methods (e.g., inverse distance weighting, cubic spline, geostatistics)

Source: Modified from Corwin, D.L. and S.M. Lesch, *Computers and Electronics in Agriculture* 46(1–3): 103–134, 2005, specifically for mapping soil salinity.

introduce an added level of localized complexity resulting from localized variations in water contents. Higher water contents and greater leaching occur under the furrows, whereas beds will typically show lower water contents and accumulations of salinity.

The presence or absence of beds and furrows is a significant factor during a geospatial EC_a survey. Measurements taken in furrows will differ from measurements taken in beds due to water flow and salt accumulation patterns. In addition, the physical presence of the bed influences the conductivity pathways, particularly when using EMI. These surface geometry effects are in addition to the effects of moisture and salinity distribution patterns that are present in beds and furrows. To assess salinity in a bed–furrow irrigated field, it is probably best to take the EC_a measurements in the bed. Above all, the EC_a measurements must be consistent either entirely in the furrow or entirely in the bed.

Surveys of drip-irrigated fields are even more complicated than EC_a surveys of bed–furrow irrigated fields. Drip irrigation produces complex local- and field-scale three-dimensional patterns of water content and salinity that are very difficult to spatially characterize with geospatial EC_a measurements or any salinity measurement technique for that matter. The easiest approach is to run EC_a transects both over and between drip lines to capture the local-scale variation.

The roughness of the soil surface can also influence the spatial EC_a measurements. Geospatial conductance measurements taken on a smooth field surface will be higher than the same field with a rough surface from disking. This is due to the fact that the disturbed, disked soil acts as an insulated layer to the conductance pathways, thereby reducing its conductance. When conducting a geospatial EC_a survey of a field, the entire field must have the same surface roughness.

The above factors, if not taken into account when conducting an EC_a survey, will likely produce a "banding" effect. For example, if an EC_a survey is conducted

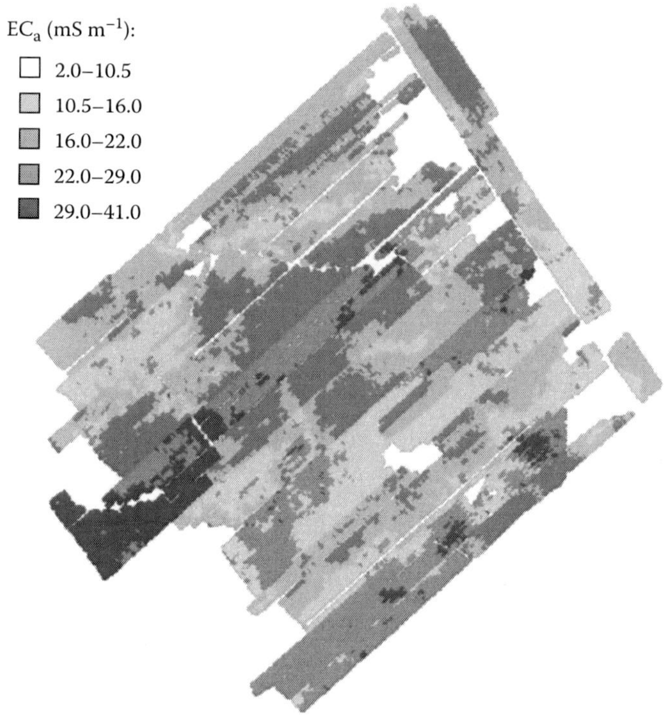

FIGURE 8.3 A poorly designed EC_a survey showing the banding that occurs when surveys are conducted at different times under varying water contents, temperatures, surface roughnesses, and surface geometry conditions. (From Corwin, D.L. and S.M. Lesch, *Journal of Environmental and Engineering Geophysics*, 18:1–25, 2013. With permission.)

on a field that has real differences in water content, soil profile temperature, surface roughness, and surface geometry, then bands of EC_a such as those found in Figure 8.3 will result. These bands reflect the variations in soil moisture, temperature, roughness, and surface geometry, which must be uniform across a field to produce a reliable EC_a survey that can be used to direct soil sampling to spatially characterize the distribution of salinity.

8.2.2.3.3 Model- and Design-Based Sampling

Once a georeferenced EC_a survey is conducted, the data are used to establish the locations of the soil core sample sites for (1) calibration of EC_a to soil sample EC_e and/or (2) delineation of the spatial distribution of soil properties correlated to EC_a within the field surveyed. To establish the locations where soil cores are to be taken, either design-based or prediction-based (i.e., model-based) sampling schemes can be used.

Arguably, the most significant element of the protocols is the EC_a-directed soil sampling design. Design-based sampling schemes have historically been the most commonly used and hence are more familiar to most research scientists. An excellent review of design-based methods can be found in Thompson (1992).

Design-based methods include simple random sampling, stratified random sampling, multistage sampling, cluster sampling, and network sampling schemes. The use of unsupervised classification by Fraisse et al. (2001) and Johnson et al. (2001) is an example of design-based sampling. Prediction-based sampling schemes are less common, although significant statistical research has recently been performed in this area (Valliant et al. 2000). Prediction-based sampling approaches have been applied to the optimal collection of spatial data by Müller (2001), the specification of optimal designs for variogram estimation by Müller and Zimmerman (1999), the estimation of spatially referenced linear regression models by Lesch (2005) and Lesch et al. (1995), and the estimation of geostatistical mixed linear models by Zhu and Stein (2006). Conceptually similar types of nonrandom sampling designs for variogram estimation have been introduced by Bogaert and Russo (1999), Warrick and Myers (1987), and Russo (1984). Both design-based and prediction-based sampling methods can be applied to geospatial EC_a data to direct soil sampling as a means of characterizing soil spatial variability (Corwin and Lesch 2005a).

The prediction-based sampling approach was introduced to EC_a-directed sampling by Lesch et al. (1995). This sampling approach attempts to optimize the estimation of a regression model (i.e., minimize the mean square prediction error produced by the calibration function) while simultaneously ensuring that the independent regression model residual error assumption remains approximately valid. This in turn allows an ordinary regression model to be used to predict soil property levels at all remaining (i.e., nonsampled) conductivity survey sites. The basis for this sampling approach stems directly from traditional response surface sampling methodology (Box and Draper 1987).

There are two main advantages to the response surface approach. First, a substantial reduction in the number of samples required for effectively estimating a calibration function can be achieved in comparison to more traditional design-based sampling schemes. Second, this approach lends itself naturally to the analysis of EC_a data. Indeed, many types of ground-, airborne-, and/or satellite-based remotely sensed data are often collected specifically because one expects these data to correlate strongly with some parameter of interest (e.g., crop stress, soil type, soil salinity, etc.), but the exact parameter estimates (associated with the calibration model) may still need to be determined via some type of site-specific sampling design. The response surface approach explicitly optimizes this site-selection process.

A user-friendly software package (ESAP) developed by Lesch et al. (2000), which uses a response surface sampling design, has proven to be particularly effective in delineating spatial distributions of soil properties from EC_a survey data (Corwin and Lesch 2003, 2005c; Corwin et al. 2003a,b, 2006; Corwin 2005). The ESAP software package, which is available online at http://www.ars.usda.gov/services/software/download.htm?softwareid=94, identifies the optimal locations for soil sample sites from the EC_a survey data. These sites are selected based on spatial statistics to reflect the observed spatial variability in EC_a survey measurements. Generally, 6 to 20 sites are selected depending on the level of variability of the EC_a measurements for a site. The optimal locations of a minimal subset of EC_a survey sites are identified to obtain soil samples.

Once the number and location of the sample sites have been established, the depth of soil core sampling, the sample depth increments, and the number of sites where duplicate or replicate core samples should be taken are established. The depth of sampling should be the same at each sample site and should extend over the depth of measurement by the EC_a measurement equipment used. For instance, the Geonics EM-38 measures to a depth of roughly 0.75–1.0 m in the horizontal coil configuration (EM_h) and 1.2–1.5 m in the vertical coil configuration (EM_v). Composite soil cores to the depth of interest can be taken, but generally, cores are taken at depth increments. Sample depth increments are flexible and depend to a great extent on the study objectives. A depth increment of 0.3 m has been commonly used at the USDA-ARS Salinity Laboratory because it provides sufficient soil profile information over the root zone (i.e., 0–1.2 to 1.5 m) for statistical analysis without an overly burdensome number of samples to conduct physicochemical analyses. Typically, core samples are taken at 0–0.3, 0.3–0.6, 0.6–0.9, 0.9–1.2, and 1.2–1.5 m depth increments. Depth increments should be the same from one sample site to the next. The number of duplicates or replicates taken at each sample site is determined by the desired accuracy for characterizing soil properties and the need for establishing the level of local-scale variability at the site. Duplicates or replicates are not necessarily needed at every sample site to establish local-scale variability.

8.3 SSIMUs: CASE STUDY

Geospatial measurements of EC_a are a powerful tool in SSM when combined with GIS, spatial statistics, and crop-yield monitoring. It is hypothesized that in instances where EC_a correlates with crop yield, spatial EC_a information can be used to direct a soil sampling plan that identifies sites that adequately reflect the range and variability of various soil properties thought to influence crop yield. The objective is to integrate spatial statistics, GIS, EC_a-directed soil sampling, and a crop-yield response model to (1) identify edaphic properties that influence cotton yield and (2) use this spatial information to delineate SSMUs with associated management recommendations for an irrigated crop (i.e., cotton in the subsequent case study) to increase productivity. The following case study summarizes the work conducted and published by Corwin et al. (2003b). For an in-depth discussion of the delineation of SSM units using proximal sensors, such as ER and EMI, refer to Corwin and Lesch (2010).

8.3.1 EC_a-Directed Sampling Methodology

A 32.4-ha field located in the Broadview Water District on the west side of California's San Joaquin Valley was used as the study site. Broadview Water District is located approximately 100 km west of Fresno, California. The soil at the site is slightly alkaline and has good surface and subsurface drainage (Harradine 1950). The subsoil is thick, friable, calcareous, and easily penetrated by roots and water.

Spatial variation of cotton yield was measured at the study site in August 1999 using a four-row cotton picker equipped with a yield sensor and a GPS. A total of 7706 cotton yield readings were collected (Figure 8.4a). Each yield observation represented a total area of approximately 42 m^2. From August 1999 to April 2000, the field was fallow.

On March 2000, an intensive EC_a survey (Figure 8.4b) was collected using mobile fixed-array ER (Figure 8.5) and mobile EMI (Figure 8.6) equipment, developed by

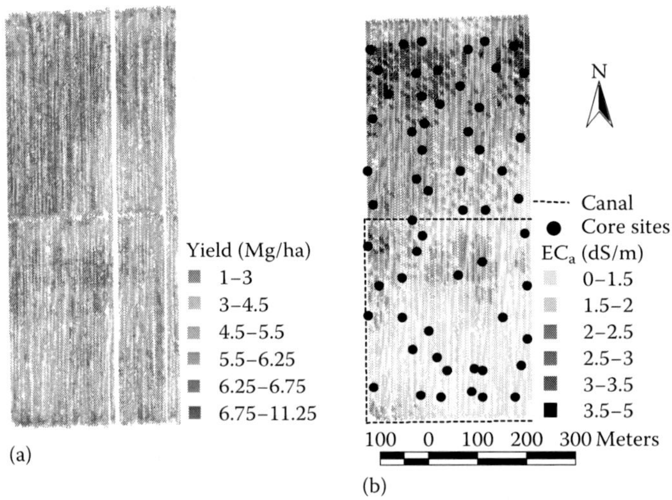

FIGURE 8.4 **(See color insert.)** Maps of (a) cotton yield and (b) EC_a measurements including the locations of the 60 soil core sites. (Modified from Corwin, D.L. et al., *Agronomy Journal* 95(2):352–364, 2003b. With permission.)

FIGURE 8.5 **(See color insert.)** Mobile GPS-based ER equipment showing (a) fixed-array tool bar holding four ER electrodes and (b) a close-up of one of the ER electrodes.

FIGURE 8.6 Mobile GPS-based EMI equipment showing (a) a side view of the entire rig and (b) a close-up of the sled holding the EMI unit.

Rhoades and colleagues at the U.S. Salinity Laboratory (Rhoades 1992a,b; Carter et al. 1993).

The methods and materials used in the EC_a survey were those subsequently published as a set of guidelines and protocols by Corwin and Lesch (2003, 2005a). The fixed-array ER electrodes were spaced to measure EC_a to a depth of 1.5 m. Over 4000 EC_a measurements were collected (Figure 8.4b).

Following the EC_a survey, soil samples were collected at 60 locations. The data from the EC_a survey were used to direct the selection of soil sample sites. The ESAP-95 version 2.01 software package developed by Lesch et al. (1995, 2000) at the U.S. Salinity Laboratory was used to establish the locations where soil cores were taken based on the EC_a survey data. The software used a model-based response surface sampling strategy to locate the 60 sites. These sites reflected the observed spatial variability in EC_a while simultaneously maximizing the spatial uniformity of the sampling design across the study area. Figure 8.4b visually displays the distribution of EC_a survey data in relation to the locations of the 60 core sites. Soil core samples were taken at each site at 0.3-m increments to a depth of 1.8 m: 0–0.3, 0.3–0.6, 0.6–0.9, 0.9–1.2, 1.2–1.5, and 1.5–1.8 m. The soil samples were analyzed for soil properties thought to influence cotton yield, including pH, boron (B), nitrate–nitrogen (NO_3–N), Cl^-, salinity (EC_e), LF, gravimetric water content (θ_g), bulk density, % clay, and saturation percent. All samples were analyzed following the methods outlined in Agronomy Monograph No. 9 Part 1 (Blake and Hartge 1986) and Part 2 (Page et al. 1982).

Statistical analyses were conducted using SAS software (SAS Institute 1999). The statistical analyses consisted of three stages: (1) determination of the correlation between EC_a and cotton yield using data from the 60 sites, (2) exploratory statistical analysis to identify the significant soil properties that influence cotton yield, and (3) development of a crop-yield response model based on ordinary least squares regression adjusted for spatial autocorrelation with restricted maximum likelihood.

Because the location of EC_a and cotton yield measurements did not exactly overlap, ordinary kriging was used to determine the expected cotton yield at the 60 sites. The spatial correlation structure of yield was modeled with an isotropic variogram. The following fitted exponential variogram was used to describe the spatial structure at the study site:

$$v(\delta) = (0.76)^2 + (1.08)^2[1 - \exp(-D/109.3)] \quad (8.14)$$

where D is the lag distance.

All spatial data were compiled, organized, manipulated, and displayed within a GIS. Kriging was selected as the preferred method of interpolation because in all cases, it outperformed inverse distance weighting based on comparisons using jackknifing.

8.3.2 Development of a Crop Yield Response Model Based on Edaphic Properties

8.3.2.1 Correlation between Cotton Yield and EC_a

The fitted variogram model of Equation 8.14 was used in an ordinary kriging approach to estimate cotton yield at the 60 sites. The correlation of EC_a to yield at the 60 sites was 0.51. The moderate correlation between yield and EC_a suggests that some soil properties that influence EC_a measurements also influence cotton yield, making an EC_a-directed soil sampling strategy a viable approach for this site. The similarity of the spatial distributions of EC_a measurements and cotton yield in Figure 8.4 visually confirms the reasonably close relationship of EC_a to the yield.

8.3.2.2 Exploratory Statistical Analysis

Exploratory statistical analyses were conducted to determine the significant soil properties influencing cotton yield and to establish the general form of the cotton yield response model. The exploratory statistical analysis consisted of three stages: (1) a preliminary multiple linear regression (MLR) analysis, (2) a correlation analysis, and (3) scatter plots of yield versus potentially significant soil properties. The preliminary MLR analysis and the correlation analysis were used to establish the significant soil properties that influence cotton yield, while the scatter plots were used to formulate the general form of the cotton yield response model. Both the preliminary MLR analysis and the correlation analysis showed that the 0–1.5 m depth increment resulted in the best correlations and best fit of the data; consequently, the 0–1.5 m depth increment was considered to correspond to the active root zone.

The preliminary MLR analysis indicated that the following soil properties were most significantly related to cotton yield: EC_e, LF, pH, % clay, θ_g, and ρ_b. The correlation between cotton yield and soil properties indicated that the highest correlation occurred with EC_e.

A scatter plot of EC_e and yield indicates a quadratic relationship where the yield increased up to a salinity of 7.17 dS m^{-1} and then decreased (Figure 8.7a). The scatter plot of LF and the yield shows a negative, curvilinear relationship (Figure 8.7b). The yield shows a minimal response to an LF below 0.4 and falls off rapidly for an LF > 0.4. Clay percentage, pH, θ_g, and ρ_b appear to be linearly related to yield to various degrees (Figure 8.7c–f, respectively). Even though there was clearly no correlation between the yield and the pH ($r = -0.01$; see Figure 8.7d), the pH became significant in the presence of the other variables, which became apparent in both the preliminary MLR analysis and in the final yield response model.

Based on the exploratory statistical analysis, it became evident that the general form of the cotton yield response model was

$$Y = \beta_0 + \beta_1(EC_e) + \beta_2(EC_e)^2 + \beta_3(LF)^2 + \beta_4(pH) + \beta_5(\%\ clay) + \beta_6(\theta_g) + \beta_7(\rho_b) + \varepsilon \quad (8.15)$$

where, based on the scatter plots of Figure 8.7, the relationships between cotton yield (Y) and pH, percentage clay, θ_g, and ρ_b are assumed linear; the relationship between yield and EC_e is assumed to be quadratic; the relationship between the yield and the LF is assumed to be curvilinear; $\beta_0, \beta_1, \beta_2, \ldots, \beta_7$ are the regression model parameters; and ε represents the random error component.

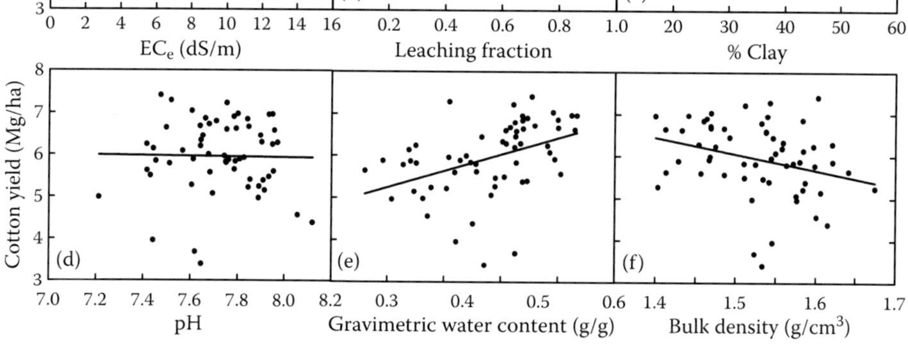

FIGURE 8.7 Scatter plots of soil properties and cotton yield: (a) electrical conductivity of the saturation extract (EC_e, dS m^{-1}), (b) LF, (c) percentage clay, (d) pH, (e) gravimetric water content, and (f) bulk density. (From Corwin, D.L. et al., *Agronomy Journal* 95(2):352–364, 2003b. With permission.)

8.3.2.3 Cotton Yield Response Model

Ordinary least squares regression based on Equation 8.15 resulted in the following response model:

$$Y = 20.90 + 0.38(EC_e) - 0.02(EC_e)^2 - 3.51(LF)^2 - 2.22(pH) + 9.27(\theta_g) + \varepsilon \quad (8.16)$$

where the nonsignificant *t*-test for % clay and ρ_b indicated that these soil properties did not contribute to the yield predictions in a statistically meaningful manner and dropped out of the regression model, while all other parameters were significant near or below the 0.05 level. The R^2 value for Equation 8.16 was 0.61, indicating that 61% of the estimated spatial yield variation was successfully described by Equation 8.16. However, the residual variogram plot indicated that the errors were spatially correlated, which implied that Equation 8.16 must be adjusted for spatial autocorrelation.

Using a restricted maximum likelihood approach to adjust for spatial autocorrelation, the most robust and parsimonious yield response model for cotton was:

$$Y = 19.28 + 0.22(EC_e) - 0.02(EC_e)^2 - 4.42(LF)^2 - 1.99(pH) + 6.93(\theta_g) + \varepsilon \quad (8.17)$$

A comparison of measured and simulated cotton yields at the locations where EC_a-directed soil samples were taken showed close agreement, with a slope of 1.13, a *y*-intercept of −0.70, and an R^2 value of 0.57. A visual comparison of the measured and simulated spatial yield distributions of cotton (Figure 8.8) shows a spatial association between interpolated measured (Figure 8.8b) and predicted (Figure 8.8c) maps.

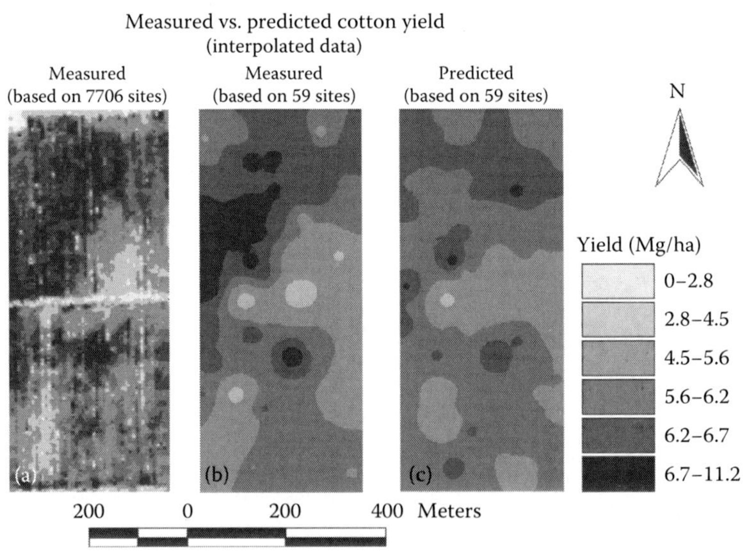

FIGURE 8.8 Comparison of (a) measured cotton yield based on 7706 yield measurements, (b) kriged data at 59 sites for measured cotton yield, and (c) kriged data at 59 sites for predicted cotton yields based on Equation 8.4. (From Corwin, D.L. et al., *Agronomy Journal* 95(2):352–364, 2003b. With permission.)

Sensitivity analysis revealed that LF was the single most significant factor influencing cotton yield with the degree of predicted yield sensitivity to one standard deviation change resulting in a percentage yield reduction for EC_e, LF, pH, and θ_g of 4.6%, 9.6%, 5.8%, and 5.1%, respectively.

8.3.3 Delineating SSIMUs

Based on Equation 8.17, Figure 8.7, and the knowledge of the interaction of the significant factors influencing cotton yield in the Broadview Water District, four recommendations can be made to improve cotton productivity at the study site:

1. Reduce the LF in highly leached areas (i.e., areas where LF > 0.5)
2. Reduce salinity by increased leaching in areas where the average root zone (0–1.5 m) salinity is >7.17 dS m^{-1}
3. Increase the plant-available water (PAW) in coarse-textured areas by more frequent irrigation
4. Reduce the pH where pH > 7.9

Figure 8.9 maps the areas pertaining to the above recommendations. All four recommendations can be accomplished by improving water application scheduling and distribution and by site-specific application of soil amendments. The use of variable-rate irrigation technology at this site would enable the site-specific application of irrigation water at the times and locations needed to optimize yield.

Hypothetically, when crop yield correlates with EC_a, then spatial distributions of EC_a provide a means of determining edaphic properties that influence yield. A yield map could potentially provide the same capability as an EC_a map; but an EC_a map provides information specific to the spatial distribution of edaphic properties, whereas a yield map reflects the influence of numerous additional factors.

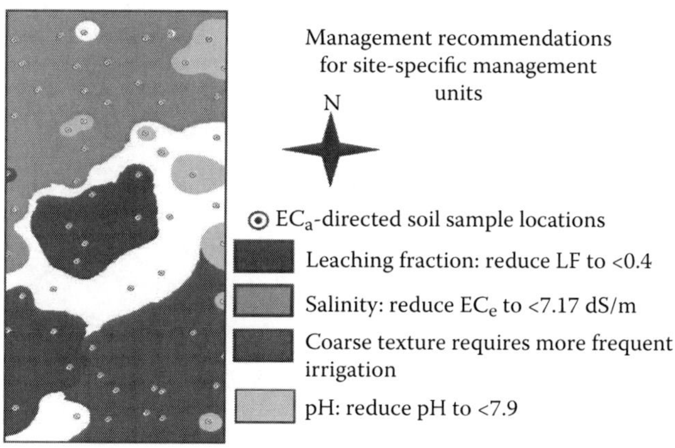

FIGURE 8.9 (See color insert.) Site-specific management units for a 32.4-ha cotton field in the Broadview Water District of central California's San Joaquin Valley. Recommendations are associated with the SSMUs for LF, salinity, texture, and pH. (From Corwin, D.L. and S.M. Lesch, *Computers and Electronics in Agriculture*, 46:11–43, 2005. With permission.)

8.4 OTHER CONSIDERATIONS FOR WATER CONSERVATION IN WATER-SCARCE, SALT-AFFECTED AREAS

Maps of SSIMUs indicate where to apply irrigation water, but knowledge of when to apply and how much to apply is also needed. When to apply is determined primarily by matric and osmotic stress to the plant. Instruments such as a tensiometer to measure the matric potential or a neutron probe or gypsum block to measure water content, which relates to matric potential, are commonly used. Time domain reflectometry and capacitance probes can also be used to measure water content. Irrigation targets are usually set as a percent depletion of the PAW, which is the difference between field capacity (−0.1 bar) and permanent wilting point (−15 bars). The bulk of irrigation research recommends irrigating row crops such as grain or cotton at 50% of the PAW and at 40% for vegetable crops that are more sensitive to water stress. Osmotic stress due to the presence of salts in the soil requires an increase in the frequency of water application because the plant has a combined osmotic and matric stress, which makes it difficult for the plant to imbibe water through the roots. The osmotic potential is generally obtained from a measurement of the dissolved salt concentration in the soil solution, since the osmotic potential in bars is approximately equal to the EC in dS m^{-1} multiplied by a factor of −0.36. The EC of the soil solution is obtained from techniques outlined by Corwin et al. (2012). The positioning of the instrumentation to measure matric and osmotic potentials within a field can be obtained from maps of texture and salinity obtained from EC_a-directed sampling, as explained in Corwin and Lesch (2003, 2005a). From a more practical standpoint, knowing when to irrigate can be determined simply by feel. By squeezing the soil between the thumb and forefinger, or squeezing the soil in the palm of a hand, a fairly accurate estimate of soil moisture can be determined, but this requires considerable experience.

Determining the amount of water to apply is less straightforward than knowing when to apply. To prevent the accumulation of excessive soluble salts in irrigated soils, more water than is required to meet the ET needs of the crops must pass through the root zone to leach soluble salts. This additional irrigation water has typically been expressed as the leaching requirement (LR). LR was originally defined as the fraction of infiltrated water that must pass through the root zone to keep soil salinity from exceeding a level that would significantly reduce crop yield under steady-state conditions, with associated good management and uniformity of leaching (U.S. Salinity Laboratory Staff 1954).

8.4.1 LR: Steady-State vs. Transient Approach

As published in Handbook 60 (U.S. Salinity Laboratory Staff 1954), the original LR model is based on the concept of LF for steady-state conditions with no precipitation or dissolution and good drainage:

$$\mathrm{LF} = \frac{V_{\mathrm{dw}}}{V_{\mathrm{inf}}} = \frac{\mathrm{EC}_{\mathrm{iw}}}{\mathrm{EC}_{\mathrm{dw}}} \tag{8.18}$$

where V_{dw} (mm) and V_{inf} (mm) are the volumes of drainage water and infiltrating irrigation water, respectively, and EC_{iw} (dS m^{-1}) and EC_{dw} (dS m^{-1}) are the electrical conductivities of the irrigation and drainage water, respectively. The LR was originally defined by the U.S. Salinity Laboratory Staff (1954) as the lowest value of LF that could be allowed without EC_{dw} (and thus, inferentially, soil salinity) becoming excessive for optimum plant growth. Thus, the minimum value of LF (i.e., LR) would be given when the maximum permissible salinity level of EC_{dw} (i.e., EC^*_{dw}) was inserted into Equation 8.18 resulting in Equation 8.19, which is considered the original LR model:

$$LR = \frac{EC_{iw}}{EC^*_{dw}} \tag{8.19}$$

The LR is an estimate of what the LF must be to keep soil water salinity within tolerable limits for crop production.

The determination of the LR, as originally formulated in Equation 8.19, required the selection of the appropriate value of EC^*_{dw} for the crop in question. These crop-related values were not known and would be expected to vary with irrigation water salinity and management. However, data obtained from controlled test-plot studies utilizing relatively uniform soil conditions and optimal irrigation and crop management were available (Bernstein 1974; Maas and Hoffman 1977). These controlled studies related the response of many crops to average root zone soil salinity in terms of the EC_e (dS m^{-1}), which is approximately half that of the soil water salinity at field capacity (U.S. Salinity Laboratory Staff 1954). The nearly uniform root zone EC_e values that resulted in 50% yield decreases in forage, field, and vegetable crops, and 10% yield decreases in fruit crops were originally substituted for EC^*_{dw} in Equation 8.19 to estimate the LR. No direct evidence or clear reasoning was given to support the appropriateness of this substitution or the corresponding LR values. Another inherent assumption in the original approach used to determine the LR is that plants respond primarily to average root zone soil salinity. This assumption is not always true. Some evidence for this conclusion is given in Rhoades and Merrill (1976). In addition, the traditional LR model assumes uniform water applications and does not adjust for salt precipitation or dissolution nor does it account for irrigation frequency effects, upward water flow, preferential flow, water chemical composition, and salt removal in surface runoff. Several, but not all, of these inherent weaknesses are accounted for in many of the transient solute transport models that have been developed since 1980, as a consequence of increased computational speeds and memory capabilities of computers.

Work by Letey and Feng (2007) and Corwin et al. (2007) showed that traditional steady-state models calculated higher LRs than more sophisticated and mechanistically rigorous transient models. For instance, Corwin et al. (2007) ran simulations for a typical 6-year crop rotation in California's Imperial Valley and found a reduction in the LR from 0.13 to 0.08. Reducing the estimated LR from 0.13 to 0.8 will reduce irrigation water needs that deplete scarce surface water supplies and will reduce drainage volumes that impact the environment when disposed. To put this into perspective for water conservation, each year, an estimated 2.46 × 109 m^3

(2 million ac-ft) of water infiltrates into the cropped soil of California's Imperial Valley; consequently, reducing the LR from 0.14 to 0.08 would reduce the drainage volume by approximately 1.23 × 108 m^3 (100,000 ac-ft).

Combining site-specific irrigation management with LRs determined from transient solute transport models will unquestionably conserve significant volumes of water and reduce drainage loads that are costly and difficult to dispose without impacting the environment. In response to the research of Letey and Feng (2007) and Corwin et al. (2007), the University of California Center for Water Resources appointed a workgroup to determine whether the current recommended guidelines for the LR based on traditional steady-state analyses need to be revised. The workgroup concluded that the present guidelines overestimate the LR and the negative consequences of irrigating with saline waters (Letey et al. 2011). This error is most significant at low LFs, which is a fortuitous finding because irrigating to achieve low LFs provides a more efficient use of limited water supplies (Letey et al. 2011).

8.5 CAVEAT

Even though EC_a-directed soil sampling provides a viable means of identifying some soil properties that influence within-field variation of yield and of delineating SSIMUs, it is only one piece of a complicated puzzle of interacting factors that result in observed within-field crop variation. Water conservation is just one aspect of sustainable agriculture. Crop yield is influenced by complex interactions of meteorological (e.g., temperature, humidity, wind, etc.), biological (e.g., pests, earthworms, etc.), anthropogenic (management related), and edaphic (e.g., salinity, soil pH, water content, etc.) factors. Furthermore, sustainability requires more than just a myopic look at crop productivity. Sustainability must balance profitability, crop productivity, optimization of resource inputs (e.g., water, fertilizers, pesticides), and minimization of environmental impacts.

ACKNOWLEDGMENTS

The author acknowledges the past collaboration of numerous colleagues on field research that resulted in the techniques discussed in this chapter regarding EC_a-directed soil sampling. In particular, collaboration with Scott Lesch and Jim Rhoades has been invaluable. The technical support of Kevin Yemoto, Clay Wilkinson, Harry Forster, Nahid Vishteh, and Jack Jobes is deeply appreciated. Kevin Yemoto's significant contributions to the problem set and answers are especially appreciated. Product identification is provided solely for the benefit of the reader and does not imply the endorsement of the USDA.

REFERENCES

Alexandratos, N. (ed.). 1995. *World Agriculture: Towards 2010*. Chichester: John Wiley & Sons and Rome: FAO.

Barnes, H.E. 1952. Soil investigation employing a new method of layer-value determination for earth resistivity interpretation. *Highway Research Board Bulletin* 65:26–36.

Bernstein, L. 1974. Crop growth and salinity. In *Drainage for Agriculture, Agronomy Monograph No. 17*, ed. J. van Schilfgaarde, 39–54. Madison, WI: Soil Science Society of America.

Birrel, S.J., S.C. Borgelt, and K.A. Sudduth. 1995. Crop yield mapping: Comparison of yield monitors and mapping techniques. In *Proceedings of 2nd International Conference on Site-Specific Management for Agricultural Systems*, eds. P.C. Robert, R.H. Rust and WE. Larson, 15–32. Madison WI: ASA-CSSA-SSSA.

Blake, G.R., and K.H. Hartge. 1986. Bulk density. In *Methods of Soil Analysis, Part 1, Physical and Mineralogical Methods, 2nd Edition, Agronomy Monograph No. 9A*, ed. A. Klute, 363–375. Madison, WI: ASA-CSSA-SSSA.

Bogaert, P., and D. Russo. 1999. Optimal spatial sampling design for the estimation of the variogram based on a least squares approach. *Water Resources Research* 35:1275–1289.

Bohn, H.L., B.L. McNeal, and G.A. O'Connor. 1979. *Soil Chemistry*. New York: Wiley.

Box, G.E.P., and N.R. Draper. 1987. *Empirical Model-Building and Response Surfaces*. New York: John Wiley & Sons.

Burger, H.R. 1992. *Exploration Geophysics of the Shallow Subsurface*. Upper Saddle River, NJ: Prentice-Hall.

Carter, L.M., J.D. Rhoades, and J.H. Chesson. 1993. Mechanization of soil salinity assessment for mapping. Paper No. 931557. *Proc. 1993 ASAE Winter Meetings*. St. Joseph, MI: ASAE.

Commission on Sustainable Development (CSD). 1997. *Comprehensive Assessment of the Freshwater Resources of the World. Report of the Secretary-General*. United Nations Economic and Social Council (UNESCO). Available at http://www.un.org/esa/documents/ecosoc/cn17/1997/ecn171997-9.htm (accessed January 12, 2014).

Corwin, D.L. 2005. Geospatial measurements of apparent soil electrical conductivity for characterizing soil spatial variability. In *Soil–Water–Solute Characterization: An Integrated Approach*, eds. J. Álvarez-Benedí and R. Muñoz-Carpena, 640–672. Boca Raton, FL: CRC Press.

Corwin, D.L., and J.M.H. Hendrickx. 2002. Solute content and concentration—Indirect measurement of solute concentration: Electrical resistivity—Wenner array. In *Methods of Soil Analysis. Part 4—Physical Methods, Agronomy Monograph No. 9*, eds. J.H. Dane and G.C. Topp, 1282–1287. Madison, WI: Soil Science Society of America.

Corwin, D.L., and S.M. Lesch. 2003. Application of soil electrical conductivity to precision agriculture: Theory, principles and guidelines. *Agronomy Journal* 95(3):455–471.

Corwin, D.L., and S.M. Lesch. 2005a. Characterizing soil spatial variability with apparent soil electrical conductivity: I. Survey protocols. *Computers and Electronics in Agriculture* 46(1–3):103–134.

Corwin, D.L., and S.M. Lesch. 2005b. Apparent soil electrical conductivity measurements in agriculture. *Computers and Electronics in Agriculture* 46:11–43.

Corwin, D.L., and S.M. Lesch. 2005c. Characterizing soil spatial variability with apparent soil electrical conductivity: II. Case study. *Computers and Electronics in Agriculture* 46(1–3):135–152.

Corwin, D.L., and S.M. Lesch. 2010. Delineating site-specific management units with proximal sensors. In *Geostatistical Applications in Precision Agriculture*, ed. M. Oliver, 139–166. New York: Springer.

Corwin, D.L., and S.M. Lesch. 2013. Protocols and guidelines for field-scale measurement of soil salinity distribution with EC_a-directed soil sampling. *Journal of Environmental and Engineering Geophysics* 18:1–25.

Corwin, D.L., S.R. Kaffka, J.W. Hopmans, Y. Mori, S.M. Lesch, and J.D. Oster. 2003a. Assessment and field-scale mapping of soil quality properties of a saline–sodic soil. *Geoderma* 114:231–259.

Corwin, D.L., S.M. Lesch, P.J. Shouse, R. Soppe, and J.E. Ayars. 2003b. Identifying soil properties that influence cotton yield using soil sampling directed by apparent soil electrical conductivity. *Agronomy Journal* 95(2):352–364.

Corwin, D.L., S.M. Lesch, J.D. Oster, and S.R. Kaffka. 2006. Monitoring management-induced spatio-temporal changes in soil quality through soil sampling directed by apparent electrical conductivity. *Geoderma* 131:369–387.

Corwin, D.L., J.D. Rhoades, and J. Šimůnek. 2007. Leaching requirement for soil salinity control: Steady-state versus transient models. *Agricultural Water Management* 90:165–180.

Corwin, D.L., S.M. Lesch, and D.B. Lobell. 2012. Laboratory and field measurements. In *Agricultural Salinity Assessment and Management*, eds. K.K. Tanji and W. Wallender, 295–341. New York: American Society of Civil Engineers.

Dobrin, M.B. 1960. *Introduction to Geophysical Prospecting*. New York: McGraw-Hill.

Fraisse, C.W., K.A. Sudduth, and N.R. Kitchen. 2001. Delineation of site-specific management zones by unsupervised classification of topographic attributes and soil electrical conductivity. *Transactions of the ASAE* 44(1):155–166.

Harradine, F.F. 1950. *Soils of Western Fresno County California*. Berkeley, CA: University of California.

Hendrickx, J.M.H., J.M. Wraith, D.L. Corwin, and R.G. Kachanoski. 2002. Solute content and concentration. In *Methods of Soil Analysis: Part 4—Physical Methods, SSSA Book Series No. 5*, eds. J.H. Dane and G.C. Topp, 1253–1321. Madison, WI: Soil Science Society of America.

Intergovernmental Panel on Climate Change (IPCC). 2007. *Climate Change 2007: The Physical Science Basis*. Contribution of Working Group I to the Fourth Assessment Report of the Intergovernmental Panel on Climate Change, eds. S. Solomon, D. Qin, M. Manning, M. Marquis, K. Averyt, M.M.B. Tignor, H.L. Miller, and Z. Chen. Cambridge: Cambridge University Press.

Johnson, C.K., J.W. Doran, H.R. Duke, B.J. Weinhold, K.M. Eskridge, and J.F. Shanahan. 2001. Field-scale electrical conductivity mapping for delineating soil condition. *Soil Science Society of America Journal* 65:1829–1837.

Jury, W.A., and H.J. Vaux, Jr. 2007. The emerging global water crisis: Managing scarcity and conflict between water users. *Advances in Agronomy* 95:1–76.

Larson, W.E., and P.C. Robert. 1995. Farming by soil. In *Soil Management for Sustainability*, eds. R. Lal and F.J. Pierce, 103–112. Ankeny, IA: Soil and Water Conservation Society.

Lesch, S.M. 2005. Sensor-directed response surface sampling designs for characterizing spatial variation in soil properties. *Computers and Electronics in Agriculture* 46(1–3):153–179.

Lesch, S.M., J.D. Rhoades, L.J. Lund, and D.L. Corwin. 1992. Mapping soil salinity using calibrated electromagnetic measurements. *Soil Science Society of America Journal* 56:540–548.

Lesch, S.M., D.J. Strauss, and J.D. Rhoades. 1995. Spatial prediction of soil salinity using electromagnetic induction techniques: 2. An efficient spatial sampling algorithm suitable for multiple linear regression model identification and estimation. *Water Resources Research* 31:387–398.

Lesch, S.M., J.D. Rhoades, and D.L. Corwin. 2000. *ESAP-95 Version 2.10R: User and Tutorial Guide. Research Rpt. 146*. Riverside, CA: USDA-ARS, U.S. Salinity Laboratory.

Lesch, S.M., D.L. Corwin, and D.A. Robinson. 2005. Apparent soil electrical conductivity mapping as an agricultural management tool in arid zone soils. *Computers and Electronics in Agriculture* 46(1–3):351–378.

Letey, J., and G.L. Feng. 2007. Dynamic versus steady-state approaches to evaluate irrigation management of saline waters. *Agricultural Water Management* 91:1–10.

Letey, J., G.J. Hoffman, J.W. Hopmans, S.R. Grattan, D. Suarez, D.L. Corwin, J.D. Oster, L. Wu, and C. Amrhein. 2011. Evaluation of soil salinity leaching requirement guidelines. *Agricultural Water Management* 98:502–506.

Maas, E.V., and G.J. Hoffman. 1977. Crop salt tolerance—Current assessment. *Journal of Irrigation and Drainage Engineering* 103(IR2):115–134.

McNeal, B.L., J.D. Oster, and J.T. Hatcher. 1970. Calculation of electrical conductivity from solution composition data as an aid to in-situ estimation of soil salinity. *Soil Science* 110:405–414.

McNeill, J.D. 1980. *Electromagnetic Terrain Conductivity Measurement at Low Induction Numbers. Technical Note TN-6*. Mississauga, Ontario: Geonics Limited.

Müller, W.G. 2001. *Collecting Spatial Data: Optimum Design of Experiments for Random Fields*, 2nd Edition. Heidelberg: Physica-Verlag.

Müller, W.G., and D.L. Zimmerman. 1999. Optimal designs for variogram estimation. *Environmetrics* 10:23–37.

Page, A.L., R.H. Miller, and D.R. Kenney (eds.). 1982. *Methods of Soil Analysis, Part 2—Chemical and Microbiological Properties, 2nd Edition, Agronomy Monograph No. 9*. Madison, WI: ASA-CSSA-SSSA.

Rhoades, J.D. 1992a. Instrumental field methods of salinity appraisal. In *Advances in Measurement of Soil Physical Properties: Bring Theory into Practice, SSSA Special Publ. No. 30*, eds. G.C. Topp, W.D. Reynolds and R.E. Green, 231–248. Madison, WI: Soil Science Society of America.

Rhoades, J.D. 1992b. Recent advances in the methodology for measuring and mapping soil salinity. *Proc. Int'l. Symposium on Strategies for Utilizing Salt-Affected Lands*, ISSS Meeting, 17–25 February, Bangkok, Thailand.

Rhoades, J.D. 1993. Electrical conductivity methods for measuring and mapping soil salinity. In *Advances in Agronomy, vol. 49*, ed. D.L. Sparks, 201–251. San Diego, CA: Academic Press.

Rhoades, J.D., and A.D. Halvorson. 1977. *Electrical Conductivity Methods for Detecting and Delineating Saline Seeps and Measuring Salinity in Northern Great Plains Soils, ARS W-42*. Berkeley, CA: USDA-ARS Western Region.

Rhoades, J.D., and J. Loveday. 1990. Salinity in irrigated agriculture. In *Irrigation of Agricultural Crops, Agronomy Monograph 30*, eds. B.A. Stewart and D.R. Nielson, 1089–1142. Madison, WI: ASA-CSSA-SSSA.

Rhoades, J.D., and S.D. Merrill. 1976. Assessing the suitability of water for irrigation: Theoretical and empirical approaches. In *Prognosis of Salinity and Alkalinity, Soils Bulletin 31*, 69–109. Rome: FAO.

Rhoades, J.D., N.A. Manteghi, P.J. Shouse, and W.J. Alves. 1989. Soil electrical conductivity and soil salinity: New formulations and calibrations. *Soil Science Society of America Journal* 53:433–439.

Rhoades, J.D., F. Chanduvi, and S. Lesch. 1999. *Soil Salinity Assessment: Methods and Interpretation of Electrical Conductivity Measurements. FAO Irrigation and Drainage Paper #57*. Rome: Food and Agriculture Organization of the United Nations.

Russo, D. 1984. Design of an optimal sampling network for estimating the variogram. *Soil Science Society of America Journal* 48:708–716.

SAS Institute. 1999. *SAS Software, Version 8.2*. Cary, NC: SAS Institute.

Sheets, K.R., and J.M.H. Hendrickx. 1995. Non-invasive soil water content measurement using electromagnetic induction. *Water Resources Research* 31:2401–2409.

Shiklomanov, I.A., and J.C. Roda (eds.). 2003. *World Water Resources at the Beginning of the 21st Century*. Cambridge: Cambridge University Press.

Telford, W.M., L.P. Gledart, and R.E. Sheriff. 1990. *Applied Geophysics*, 2nd Edition. Cambridge: Cambridge University Press.

Thompson, S.K. 1992. *Sampling*. New York: Wiley.

United Nations Educational, Scientific and Cultural Organization (UNESCO). 2009. *Water in a Changing World. The United Nations World Water Development Report #3*. Paris: UNESCO.

United Nations FAO. 2001. *AQUASTAT—FAO's Information System on Water and Agriculture*. Available at http://www.fao.org/nr/water/aquastat/main/index.stm (accessed April 28, 2014).

United Nations FAO, Natural Resources and Environment Department. 2011. *Hot Issues: Water Scarcity*. FAO Water. Available at http://www.fao.org/nr/water/issues/scarcity.html (accessed January 12, 2014).

U.S. Salinity Laboratory Staff. 1954. Diagnosis and improvement of saline and alkali soils. *U.S. Department of Agriculture Handbook. No. 60.* Washington, DC: U.S. Printing Office.

Valliant, R., A.H. Dorfman, and R.M. Royall. 2000. *Finite Population Sampling and Inference: A Prediction Approach*. New York: John Wiley.

van Uffelen, C.G.R., J. Verhagen, and J. Bouma. 1997. Comparison of simulated crop yield patterns for site-specific management. *Agricultural Systems* 54:207–222.

Verhagen, A., H.W.G. Booltink, and J. Bouma. 1995. Site-specific management: Balancing production and environmental requirements at farm level. *Agricultural Systems* 49:369–384.

Warrick, A.W., and D.E. Myers. 1987. Optimization of sampling locations for variogram calculations. *Water Resources Research* 23:496–500.

World Resources Institute. 1998. *1998–99 World Resources—A Guide to the Global Environment*. New York: Oxford University Press.

World Water Council. 2000. *World Water Vision Commission Report: A Water-Secure World*. Vision for Water, Life and the Environment. Water World Council. Available at http://www.worldwaterforum5.org/fileadmin/wwc/Library/Publications_and_reports/Visions/CommissionReport.pdf (accessed January 12, 2014).

Zhu, Z., and M.L. Stein. 2006. Spatial sampling design for prediction with estimated parameters. *Journal of Agricultural, Biological, and Environmental Statistics* 11:24–44.

9 Resource Management in Rangeland

D. Terrance Booth and Samuel E. Cox

CONTENTS

EXECUTIVE SUMMARY

Adequate assessments of vast expanses of rangeland—a primary prerequisite to effective conservation planning—require landscape-scale evaluations that accurately represent the resources (e.g., soil, vegetation, wildlife, water), the structure and function of the resource-providing systems, and the natural range of variation in measured resource condition indicators. We discuss why conventional rangeland survey methods are inadequate for this task and how geographic information systems, ground survey data, and high-altitude and/or satellite imagery can be used with 1- to 50-mm-resolution digital imagery and new software programs to make rangeland surveys more objective, repeatable, and cost effective. We then review example applications of rangeland surveys that we have conducted to answer specific landscape-scale management questions. The online exercise adds to the illustrated utility of these kinds of aerial surveys with hands-on examples of survey planning, database queries, and how to use commonly available software in data evaluation and analyses.

KEYWORDS

Aerial photography, image analysis, landscape resource monitoring, sampling, scientific rigor

9.1 INTRODUCTION

Rangeland resource conservation requires accurate information on the ecology of the systems of interest and the benefits (goods and services) that we value, followed by action to sustainably use and enhance those benefits. Action may be taken to optimize some benefits over others, but usually the challenge is to balance resource use in ways that sustain all benefits. Inventory and monitoring (i.e., surveys) are the basis for assessing ecological status; however, deciding how, what, when, and where to measure the patterns and processes of rangeland systems has been one of the foremost challenges to developing optimal rangeland conservation plans (Friedel et al. 2000).

9.1.1 Conventional Rangeland Surveys

The earliest rangeland surveys were "ocular" estimates (West 2003a). Such estimates are subjective. They are opinions, judgments, or guesses that are based on experience and feelings—the "art" of rangeland management. Nowadays, the emphasis is on measureable indicators with repeatable measurements (NRC 1994). The goal is "consistent, uniform, and standard vegetation attribute sampling that is economical, repeatable, statistically reliable, and technically adequate" (ITT 1996). We believe that meeting this goal has been obstructed by

1. Variation in data collected among observers and through time by single observers.
2. Unverifiable data—conventional rangeland vegetation surveys capture information at points in space and time under ever-changing conditions; exactly relocating and duplicating point-sampling and other nonimage survey observations (see Section 9.1.2) is not economically or practically feasible, and thus repeatability/verifiability of these types of vegetation survey data does not exist.
3. High costs of conventional survey methods resulting in
   a. Limited sampling, thus a high risk of false-negative results (finding no change when change has occurred).

   b. Subjective selection of relatively small "representative" sample areas with a focus on specific resources rather than ecological systems (ITT 1996).
4. Lack of application of the scientific method including
   a. Lack of a comprehensive, thorough, consistent, and complete analysis of key ecological and management questions.
   b. Lack of presurvey statistical design.
   c. Lack of rigor in survey efforts.
5. Time constraints.

9.1.2 Data Variation in Rangeland Inventory and Monitoring

"Human stress is a significant factor in most botanical analysis techniques and may easily invalidate the results obtained" (Walker 1970, p. 415). This 1970 conclusion is among the first of many reports that deal with the causes and consequences of variation in conventional rangeland survey data. Stress, age, and aging effects on eyes (Booth et al. 2005; Cagney et al. 2011), and likely a number of unstudied factors like experience, changing light conditions and related effects on color perception and object identification, and varying interpretations of definitions and protocol rules, are all noteworthy because rangeland measurements are not simple readings of rulers, gauges, or dials, where simplicity would overcome interpersonal and situational differences. Moreover, across-year comparisons needed to test for ecologically important change (see Section 9.1.4) must account for differences in plant–biology–weather relationships (phenology). However, the number of sample sites that can be surveyed using conventional ground-based methods during phenologically comparable times of a given year is limited. In numerous ways, conventional ground-based point and plot-sampling methods—alone—are inadequate for a landscape-scale approach to the management of a vast and variable resource (West 1999).

9.1.3 Need for Landscape-Scale Management

Historically, rangeland management focused on livestock grazing (USDA 1940). That focus changed with increases in societal demands. Multiple-use management—the attempt to appropriately balance and sustain major resources and their uses—became the standard land management concept in much of the world; failure to achieve sustainable balances brought the realization that ecological systems, and not just single resources like forages, must be sustained for multiple-use management to succeed (Sidle and Sharpley 1991). As W. Muiruri of the University of Nairobi, Kenya, states, "The vital conceptual factor in seeking to develop...natural resources is to understand the structure and function of the ecosystem... The renewable natural resources...soils, water, forest and wildlife... are so intertwined that the effect of man's use on any of them immediately sets in motion a chain reaction and has repercussions on all the others" (Muiruri 1978, p. 328). Muiruri (1978, p. 328) advocated regional monitoring and concluded that an effective land management program must have a "framework with which the… society ecosystem relationship may be successfully regulated." These concepts have been reiterated by many authors and have been accepted in principle by most land management agencies.

A modern understanding of ecosystem management goes beyond the "intertwined" effects. For example, we now recognize the need to capture the natural range of variation with respect to (1) variation in measured indicators due to location, i.e., spatial heterogeneity; (2) variation in species pools and identities at landscape versus local scales; and (3) trait-influenced indirect effects that are evident only at landscape scales. This includes behaviors or other ecological characteristics that change in response to some environmental influence—perhaps another organism—that can in turn change the spatial distribution of organisms in the area of interaction (Zirlewagen and von Wilpert 2004, 2010; Underwood et al. 2005). Consequently, Zirlewagen and von Wilpert (2004, 2010) maintained that the management should be judged at the landscape level. They supported their assertion by showing that regional land management models had a lower risk of misinterpreting relationships than did small-area models. It seems probable that regional models were more successful because they included more of the natural range of variation than the small-area models. Thus, the importance of assessing resource management at the landscape scale is established by the evidence of interrelated ecological services; by the consequential need to understand the structure, function, and interactions within whole systems (Muiruri 1978; Sidle and Sharpley 1991); and by the importance of large-area monitoring to capture the range of natural variation in measured ecological indicators (Zirlewagen and von Wilpert 2004, 2010; Underwood et al. 2005).

9.1.4 Cost and Scientific Rigor

Rangeland managers test for change in plant communities to assess the effects of weather and management on resource conditions and trends. The implied null hypothesis is no ecologically important change over time. Effective monitoring, i.e., monitoring that has the potential to detect change in selected indicators with acceptable statistical error, must be stable, powerful, robust, and cost-effective (Brady et al. 1995). Stable and powerful refer to the risk of error. A statistical type I error (also called a false positive or false-change rate) is controlled by using the appropriate significance level (usually $\alpha = 0.05$). Type II error is a false negative, e.g., finding no change when an ecologically important change has occurred. Type II error is controlled by the number and spatial distribution of samples; that is, sample size and distribution determine if a survey has the statistical power to detect change. If sampling inadequately represents the resource and its natural variation, then the risk of type II error increases. Most rangeland surveys continue to be at high risk of false-negative results because of the cost of conventional methods. Fortunately, new tools now greatly extend sampling capacity over that of conventional methods, reduce subjectivity in data collection, and thus provide the means for more reliable hypothesis testing of landscape-scale management questions.

9.1.5 Objective—What Can This Chapter Do for You?

We have briefly highlighted the major resource-survey challenges that have confronted rangeland professionals for the past 100 years. Erroneous, incomplete, or undone surveys result in land management decisions based on incorrect perceptions of resource conditions. The consequences in the worst case scenarios are inadequate forage for planned grazing, uncontrolled invasive weed populations, gullied meadows and wetlands, watersheds trending toward higher peak stream flows and lower late-season flows, silted reservoirs, and reduced wildlife populations. Expert judgments of rangeland professionals have often preserved ecological function. However, climate change; increasing human populations and their multiple-use demands; and desk-bound managers harried by financial, administrative, and litigative obligations are aspects of the present that make accurate, objective, and inexpensive rangeland surveys more important than anytime during the previous 100 years. We now describe fundamental survey concepts, methods, and tools that address many limitations in rangeland resource surveys that hindered past assessments. We illustrate these concepts, tools, and techniques with examples from our cooperative work with land management agencies and other entities in which we have obtained high-density, objective, verifiable, rangeland survey data at landscape scales.

9.2 DIGITAL IMAGERY, GIS, AND IMAGE ANALYSIS SOFTWARE

Remote sensing is usually the most logical approach to acquiring information over large rangeland areas in short time periods and on random sites removed from easy ground access. Satellite and high-altitude sensors can provide extensive, small-scale views that aid landscape-level inventories of plant communities, thus facilitating plant–community mapping showing distributions and—over time—providing data on large patch dynamics. Because of the high temporal resolution of many satellites (Landsat 8 covers the globe every 16 days), these tools capture the seasonal and phenological changes of dominant plant communities. The National Agricultural Imagery Program (NAIP) acquires 1-m pixel aerial photography of the United States on a 3-year cycle. Both NAIP and Landsat data sets provide important historical information at no cost to the end user and should be utilized whenever relevant. Often, however, the small scale of these image products is insufficient to gather data, such as ground cover, necessary to address rangeland management questions. In such cases, higher-resolution imagery (1–50 mm ground sample distance [GSD]) provides detailed views, though typically over smaller areas and at reduced temporal resolution.

9.2.1 Image-Based Ecological Assessments

Ecological investigators have long sought to use image-based data collection (Cooper 1924). With the advent of digital cameras, the use of fine-scale information from images began replacing conventional point, plot, transect, and ocular estimates for the collection of ecological data. Because an image is a permanent record of conditions at a specific place in time, images provide a critical advantage for measurement verification and change detection. A number of tools exist for obtaining images of different resolutions for ecological measurements. Ground-based nadir (vertical) digital images can be acquired using staffs, stands, booms, and gantries, including mounts for all terrain vehicles and automobiles (Booth et al. 2004; Booth and Cox 2011), or by using only one's hands (Cagney et al. 2011). For areas greater than approximately 200 ha, aerial image acquisition is the most efficient means for acquiring images as samples (Booth and Cox 2011). But even where sampling will be done primarily from aircraft, we recommend acquisition of at least some ground images as an important part of resource surveys. Ground acquisition provides the opportunity to capture sharper nadir images than can be obtained from the air and to make notes that will help with species identification and the interpretation of other fine details during image analysis. The use of vehicle platforms—aerial or wheeled—may be regarded as an economical means of supplementing, but not replacing, stationary ground sampling and observation.

Aerial sampling with image resolutions ranging from 1- to 50-mm GSD (a measure of image resolution) can be done with piloted conventional and light sport airplanes (LSAs), manned and unmanned helicopters, and unmanned fixed-wing aircraft (Booth and Cox 2011). LSAs and helicopters are the only manned aircraft that can fly slowly enough to obtain low-blur, 1-mm images (i.e., capable of resolving a blade of grass).* Helicopters (manned and unmanned) can fly in weather and mountainous conditions where LSAs cannot safely operate, but contract costs for manned helicopters are more expensive than for LSAs. Manned helicopters and LSAs operate over longer distances than unmanned aerial vehicles and carry larger payloads, allowing for multiple cameras producing nested, multiresolution image sets showing wide fields of view and high resolution. The acquisition of images in the range of 1- to 50-mm GSD multispatial imagery has been useful for a number of rangeland survey applications (Booth and Cox 2009; Booth et al. 2010). Aerial surveys acquiring thousands of samples overcome the need for subjective selection of study areas and cost less than extensive ground sampling for similar-size areas.

Image resolution is a function of sensor resolution, lens focal length, and aircraft altitude; these influence the choice of camera and lens. We found that, generally, if an LSA was at 100-m altitude above ground level (AGL), a camera with at least 11 megapixels and an 840-mm lens was required to achieve 1-mm GSD in our aerial images. Although an 11-megapixel camera will suffice at 100-m AGL, we recommend using cameras that have at least 16 megapixels (Booth et al. 2006a). Once the images are acquired, they can be examined and interpreted. The tools of photo-interpretation have rapidly evolved since the advent of the computer, and today, most image analysis is conducted with the aid of computer software.

---

* What level of resolution is necessary to allow rangeland plants to be identified at the species level? It is not possible to make an overall guiding recommendation on image resolution for plant identification. Obviously, larger plants such as juniper, sagebrush and other large shrubs, and robust forbs at maximum growth will not require the same resolution as low-growing grasses, sedges, and forbs. Factors besides resolution, including motion blur, light conditions, plant phenology, visibility of identifying characteristics, and user field experience, have major influences on the potential for photo interpretation to the species level. Where identification to life form is inadequate for user purposes and species identification is critical, we recommend preliminary work with all must-identify species to determine the image-acquisition conditions needed to accomplish the species-identification need.

9.2.2 Image-Based Measurements

To establish efficient, repeatable methods for image analysis, we developed SamplePoint (Booth et al. 2006b) and ImageMeasurement (Booth et al. 2006a), software applications that facilitate measurements of key ecological indicators from digital imagery. These programs are effective and accurate but only when used with image resolutions that are suited to the tasks and by people with adequate experience on the ground in the areas of interest (AOIs). For measuring ground cover, image resolutions should be approximately 1-mm GSD or greater (Booth et al. 2006b). Attempts have been made to obtain accurate ground-cover measurements from 10-, 20-, and 50-mm images. However, image pixels at these resolutions capture correspondingly greater ground areas than are captured by 1-mm pixels (Cook and Stubbendieck 1986; Weber et al. 2013). This violates point sampling theory and does not give accurate measurements (Booth and Cox 2009).

Image analysts must also consider the acquisition season and the related environmental conditions. For example, basal cover is a less variable and more robust measurement than canopy cover, and basal cover can be measured if images are acquired after green-up and before canopy development.

9.2.2.1 SamplePoint Software

SamplePoint is a tool that facilitates manual point sampling of digital images when measuring ground cover or other percent-of-image indicators (Booth et al. 2006b; Booth and Cox 2011). The software uses a crosshair to allow the reader to locate the sample point. That point is always a *single pixel* of the image and is the center pixel of a 9-pixel array at the center of the crosshair; only the center pixel is read, but the 9-pixel array helps users to clearly separate the center pixel from the crosshair. The software automatically moves the crosshair from point to point and will find the same pixels each time the image database is opened. By returning to the same pixel, the software facilitates comparisons of SamplePoint readings (i.e., "hits" on ground-cover characteristics) for any given image, thus allowing data verification among users. Compared with conventional on-the-ground point sampling, image capture and SamplePoint analysis reduces survey time and cost and increases user comfort and data verifiability (Cagney et al. 2011).

9.2.2.2 ImageMeasurement Software

ImageMeasurement can be used to manually measure the distance (length) or area of any feature in digital images (Booth et al. 2006a,c) and is unique in utilizing the resolution information for every image in a multiresolution image data set. The need for image-specific information results from topographic variability in the AOI and from altitude variation by the light airplanes used to achieve the slow flight speeds that avoid excessive motion blur in low-altitude aerial images. Having image-specific information is particularly important when linear and area measurements are being made, as previously discussed (Booth et al. 2006a). ImageMeasurement facilitates up to 50 measurements per image that are automatically saved to a spreadsheet. Some common feature measurements we have used are stream width, tree and shrub canopy diameters and area, and bare-ground patch length along a linear transect.

9.2.3 Scientific Method and Use of GIS in Planning Aerial and Ground Surveys

Images are samples. They mean little without a carefully designed plan that incorporates critical thinking and experimental design appropriate to testing one or more hypotheses. A hypothesis is a precise, testable statement about something. "Testable" denotes that there is a means by which probable truth can be established by evidence. Rangeland science deals with the study of natural phenomena. It is often impractical to manipulate ecosystems, or natural phenomena within ecosystems; and in such cases, hypothesis testing depends on careful, systematic methods of observation (measurements) that can be used as evidence for supporting or rejecting a hypothesis. We agree with West's (2003b, p. 335) emphasis on the importance to rangeland management of "science-based tests…used to fine tune the management and put it on a more objective basis." He stressed that the "crucial" first step in planning is writing down the questions to be tested. We stress that image-based rangeland surveys bring the power of true analytical research to landscape-scale questions and that plans should not only be written down but also be reviewed by a consulting statistician to protect the investment in, and maximize the information return from, landscape-scale rangeland surveys.

Steps in developing a survey plan for an AOI are given as follows:

1. State the question(s) as completely, clearly, and concisely as possible—carefully defining key words to avoid ambiguity.
2. State what is known and what is unknown about the question and then use a critical and rigorous logic sequence to progress to a clearly testable statement (hypothesis). "If/then" statements are often useful in formulating a hypothesis. For example, a manager might write that if grazing in a damaged AOI is now at sustainable levels, then future surveys will show no new stream channel headcuts or any migration of existing headcuts. (A headcut is a form of channel erosion resulting in an abrupt step in a stream channel profile.)
3. Use a geographic information system (GIS) to draft sampling plans for aerial and ground image acquisition. This allows for consideration of the components of a desired sampling design: number, spatial distribution and resolution of samples with the realities of budget, access for ground-image acquisition, and logistical complications of aerial acquisition (e.g., mountains, prevailing winds, operational and emergency landing sites, aerial platform capabilities/limitations and crew safety). It facilitates planning for data storage, considerations of costs for alternative platforms and plans, and the trade-offs of what management questions can be legitimately addressed with the available budget. Given all this, a GIS is essential in crafting robust sampling plans and in using resulting samples to examine landscape-scale questions. In the remaining portion of this chapter, we describe studies used to address common land-management questions.

9.3 CASE STUDIES

9.3.1 Has Stocking Rate Affected Ground Cover?

Our experimental question was whether two image sets, ground and aerial, could be used to detect differences among stocking rates on a shortgrass prairie (Booth and Cox 2008). We used land that had been grazed from May to October for 14 years by steers at 10 to 25 animal days per hectare prior to a 3-year experiment in which stocking rates were adjusted. The land was divided into three pastures: a 63-ha pasture grazed with 10 steers to obtain 35%–45% utilization (moderate), a 34-ha pasture grazed with 10 steers to obtain 70%–80% utilization (heavy), and a 33-ha pasture left ungrazed. Grazing began in late May and continued for 3 to 5 months. During the fourth year of the stocking rate treatments, we used ArcGIS 9.0 (ESRI, Redlands, California) to plan 200 systematically located sample sites across all of the pastures. Each sample was a color image (red, green, blue) with a resolution of 1-mm GSD. We acquired aerial images using a Canon 1Ds 11.1-megapixel camera with 840-mm-focal-length lens arrangement. The camera was mounted in a light sport airplane (FAA 2004) equipped with a system for navigation and programmed camera triggering, plus a laser range finder for measuring altitude AGL (Booth and Cox 2006). Target flight altitude was 100-m AGL, and the camera was triggered at 80-m intervals. The images had a 3 × 4 m field of view (FOV).

Ground images were acquired 2 days after the aerial survey by a single person who located the aerial photo center points on the ground using a Garmin eTrex Venture WAAS-enabled GPS (Olathe, Kansas) and used an Olympus E20, 5-megapixel camera mounted on an aluminum camera frame that positioned the camera for nadir (vertical perspective) image collection at 2-m AGL (Booth et al. 2004). Ground-cover measurements were made from TIF ground and aerial images using SamplePoint software (Booth et al. 2006b). We used seven ground-cover classes: brown grass, green grass, cactus, shrub, litter, cow manure, and bare ground. We used PROC MIXED in SAS V 9.1 (Nashville, Tennessee) to test means (without fitting a mixed model to the data), a method that accommodated the unequal sample sizes among different-sized pastures. We then tested mean differences using F-protected ($P < 0.05$) t-tests.

The cover of brown grass, green grass, bare ground, litter, and cow manure differed among pastures ($P < 0.05$) when measured using either ground or aerial images. Shrub cover was similar among pastures when measured from ground images (1-m^2 FOV) but differed between ungrazed and the two grazed pastures when measured from aerial (12-m^2 FOV) images ($P < 0.05$). We conclude that because shrubs are larger in size and fewer in number than grasses and forbs, the larger FOV would have been the more appropriate spatial scale for testing grazing effects on shrubs. Measurements from both aerial and ground images showed the same grazing-related trend for five of the six cover classes (Figure 9.1). In both image sets, the heavily grazed pasture showed more bare ground than the ungrazed and moderately grazed pastures (Figure 9.1). We conclude that both the ground and aerial imagery facilitate detection of differences among grazing treatments on shortgrass prairie and are thus effective tools for monitoring rangeland condition as influenced by stocking rate

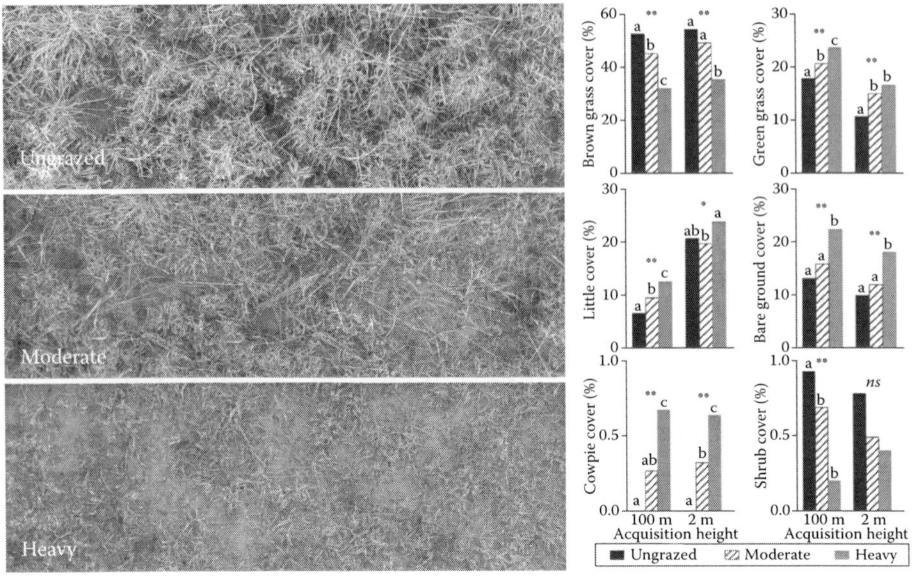

FIGURE 9.1 Bare ground differences by stocking rate (ungrazed, moderate, heavy) are illustrated in the above 100 × 33 cm (ground distance) panels taken from images with bare ground equal to the mean value for the pasture represented (10% ungrazed, 12% moderate, and 18% heavy). Images were acquired from 2-m AGL and have a ground sample distance of 1 mm. Note that dead organic matter is not in contact with the soil surface in the ungrazed pasture. This material, because it was not in contact with the soil, was not defined as litter. This resulted in the heavily grazed pasture having greater litter cover because the reduced grass canopy allowed dead organic matter to reach the soil surface. The right panel shows cover (%) means for every cover type showing significant differences ($P < 0.05$) among pastures as the result of a one-way ANOVA with F-protected t-test separation of data collected using SamplePoint from images captured from both 100- and 2-m AGL. $^*P < 0.05$; $^{**}P < 0.0001$. (From Booth, D.T., and S.E. Cox, *Frontiers in Ecology and the Environment* 6:185–190, 2008.)

and other management practices. This conclusion is further supported by data from a subsequent larger study in the same area (Augustine et al. 2012). The choice of whether to collect aerial or ground images is then a matter of which data set best fits with the study design and budget.

9.3.2 Does Revegetation Meet the Standards of Performance?

To determine if revegetation of two pipeline rights-of-way (ROWs) on public land met permitting requirements, we flew aerial surveys of a 32-km ROW segment of the Lost Creek pipeline south of Jeffrey City, Wyoming (42°14′N, 107°54′W), and a 35-km ROW segment of the Howell pipeline north of Casper, Wyoming (43°23′N, 106°, 26′W) (Booth and Cox 2009). Both segments are in Wyoming big sagebrush (*Artemisia tridentata* Nutt. ssp. *wyomingensis* Beetle & Young) vegetation types, but the Lost Creek segment is in a 250–300 mm precipitation zone compared to

300–360 mm for the segment of the Howell pipeline. We used manual camera triggering for both pipelines. The equipment was similar to that described in Section 9.3.1 except that the airplane was a Moyes–Bailey Dragonfly and was equipped with two SLR cameras (1Ds 11.1 megapixels with a 100-mm, f/2.8 lens; 1DsMarkII 16.7 megapixels with an 840-mm, image-stabilized, f/5.6 lens; Canon USA, Lake Success, New York; shutter speed = 1/4000 s).

The Lost Creek survey was flown in June 2006. We acquired 258 pairs of images from an average AGL of 150 m with each image pair consisting of one 13-mm GSD image (36 × 54 m FOV) and a nested 1.3-mm GSD image (4 × 6 m FOV). Distances between aerial sample locations along the Lost Creek Pipeline ranged between 54 and 917 m with an average separation of 140 m (±76 SD). The Howell survey was flown in June 2007 and acquired 152 pairs of images from an average AGL of 240 m. The greater altitude was used to increase the FOV of the images so that more lower-resolution images would span the full width of the disturbed area. Each image pair consisted of one 21-mm GSD image (57 × 86 m FOV) and one 2.1-mm GSD image (7 × 10 m FOV). Sample separation distance ranged from 90 to 1500 m and averaged 293 m (±175 SD).

Reclamation success for both pipelines was defined by seven indicators: (1) vegetative cover in the ROW at least 75% of the vegetative cover on adjacent land; (2) species composition is composed of predominately seeded species; (3) seeded species are those adapted to established grazing; (4) reproduction of seeded species; (5) evidence that seeded or planted woody species had >50% survival; (6) noxious weeds are controlled; and (7) vegetative cover mitigates visual impacts. We assessed reclamation success in the ROW by measuring vegetation cover by life form (grass, forb, or shrub) in and adjacent to the ROW (within 300 m) using SamplePoint with 100 systematically arranged sample points classified as vegetation by life form, soil, litter, rock, and unknown (Booth et al. 2006b). Images taken over the ROW that included adjacent area were not utilized in calculations of revegetation adequacy and vice versa.

Standard *t*-tests were used for the Howell pipeline data, and Welch's *t*-tests were used for the Lost Creek pipeline data, to compare plant cover within the ROW versus adjacent to the ROW. Where means differed, we tallied and mapped those locations within the ROW that failed to meet reclamation standards. Comparing the vegetative cover inside and adjacent to the ROW was a measure of reclamation success defined by indicator 1. Measurements of vegetation cover also addressed indicators 3–7 since adequate cover is an indication of revegetation survival—and over the long term, of reproduction in the face of grazing and other environmental stressors. Each image was examined for weedy species (indicator 6). The ease with which the ROW was distinguished from adjacent land addressed visual effects (indicator 7).

Only 20% of the Lost Creek images, and 40% of the Howell images, had total vegetation cover ≥ 75% of the vegetation cover of the adjacent land. Similarly, 14% and 28%, respectively, of the images in the Lost Creek and Howell pipeline ROWs had bare ground values equal to, or less than, the values for the adjacent land, and mean bare ground at both ROWs was higher than adjacent land ($P < 0.005$; see Figure 9.2). Neither of these pipelines met revegetation standards, and while noxious weeds were not detected at either site, the high percentage of bare ground implies that the ROWs are at risk for weed invasion.

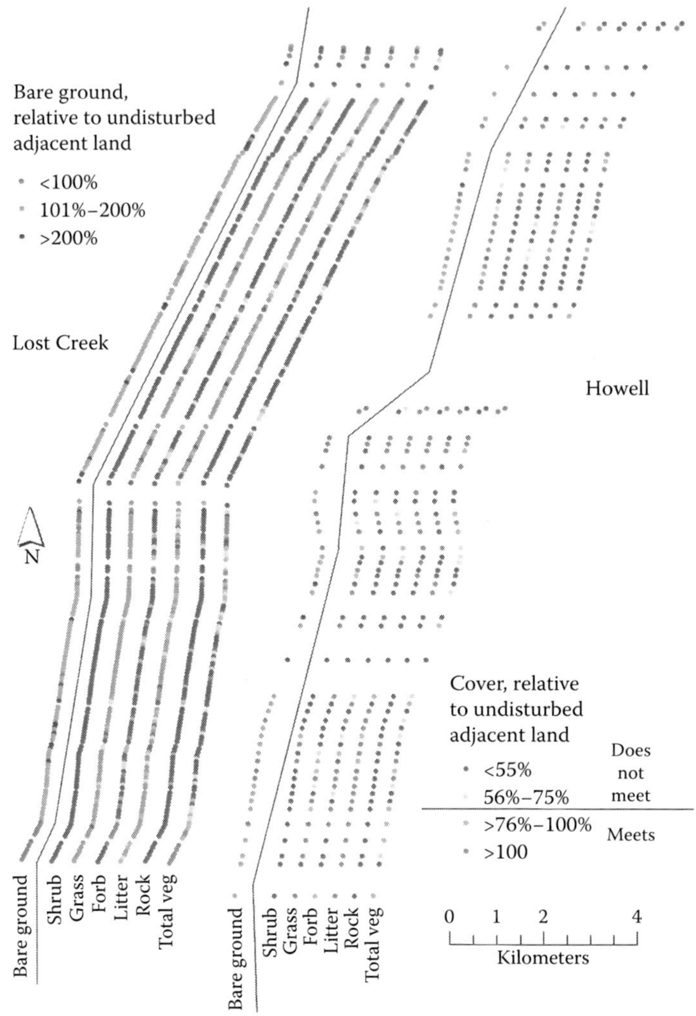

FIGURE 9.2 (See color insert.) Cover values for bare ground, shrub, grass, forb, litter, rock, and total vegetation (shrub, forb, and grass) as a percentage of adjacent, undisturbed land cover values. Values were measured from ~5 × 8 m FOV aerial images and plotted to scale, providing a quick and comprehensive view of reclamation progress along two ~35-km stretches of pipeline ROWs in central Wyoming. Cover breaks indicate areas that meet or do not meet the vegetative cover requirements established by the Plans of Development. Skips in the Howell pipeline sequence are the result of technical problems encountered during that survey. (From Booth, D.T., and S.E. Cox, *Environmental Monitoring and Assessment* 158:23–33, 2009.)

We found that the wider FOV of the lower-resolution imagery acquired for the Howell Pipeline (2 vs. 1 and 20 vs. 13 mm GSD) is less useful than anticipated; thus, we recommend acquisition of high-resolution images without concern for obtaining a wider FOV. We also recommend that a survey flight line be directly over the ROW with a second flight line targeted ~30 m adjacent to the ROW to provide control imagery.

Energy extraction has put unprecedented pressure on western public land managers, but if inadequacies in remediation of disturbances are not identified, they will not be rectified. Aerial surveys and image analyses can increase agency land management efficiency and effectiveness by providing representative high-density, affordable sampling along the length of the ROW disturbances. We recommend these surveys for obtaining the large sample numbers and accurate ground-cover measurements needed to accomplish actionable environmental monitoring. The images provide spatial evidence identifying unmitigated environmental problems for each pipeline segment, allowing fact-based decisions by responsible companies and agencies when allocating funds for protecting or rehabilitating disturbed surface resources.

9.3.3 What Is the Status of Leafy Spurge?

Unmanaged invasive weed populations can be expected to increase exponentially (Maxwell et al. 2009), making weed invasion in difficult-to-access wildlands an especially difficult land management problem. We tested the utility of 1- and 10-mm GSD images for detecting small, dispersed populations of leafy spurge (*Euphorbia esula* L.) in a rough country by conducting a mid-July 2006 aerial survey of land known to have leafy spurge infestations (Booth et al. 2010). We used the aerial survey methods described in Section 9.3.2, except that the navigation system was programmed to automatically trigger two cameras at 804-m intervals along 25 flight lines, resulting in image acquisitions for 1383 locations from 121-m AGL. Our survey area was centered at 44°20′52″N, 112°33′45″W, northwest of Dubois, Idaho, and consisted of over 22,130 ha of public land extending north from the Upper Snake River Plain to the Continental Divide on the Montana–Idaho border. The land ranges in elevation from 1710 to 2650 m and has a generally south-facing slope of 10.9% ± 8.1%. Average annual precipitation ranges from 305 to 508 mm; nine perennial streams run through the survey area.

Key historical events in the study area include the following: (1) the establishment of leafy spurge on the study site; (2) the subsequent release of approximately 297,000 spurge-eating beneficial insects at 32 locations within the study area prior to 2003; (3) a 2003 wildfire that burned 15,378 ha of mountain big sagebrush/Idaho fescue (*Artemisia tridentata* Nutt. ssp. *vaseyana* (Rydb.)/*Festuca idahoensis* Elmer) along with various dispersed infestations of leafy spurge; (4) the postburn release of another 93,000 insects among 29 locations; and (5) chemical spurge control was also used in various postburn applications. This history, and the diverse topography, contributed to considerable variation in leafy spurge populations on the study site.

We examined both 1- and 10-mm GSD images for leafy spurge and counted the number of positive images. Images were classified as riparian leafy spurge if a stream channel was in the image FOV. Leafy spurge cover was determined by measuring ground cover, including a canopy cover of leafy spurge and other identifiable species, from 1-mm GSD images using SamplePoint with a 100-point grid (Booth et al. 2006b).

Leafy spurge was detected in 139 upland and 18 riparian, 10-mm GSD images (10%) and 105 of the 1-mm GSD images (8%). During the 8-year span from 1998 to 2005, agency ground crews detected 214 separate infestations. Although leafy spurge plants showed up more clearly in the 1-mm GSD imagery (Figure 9.3), detection capability was lower at that resolution, owing to its smaller FOV ($P = 0.03$, $n = 1383$). In images where leafy spurge was detected, canopy cover averaged > 16%. Cover categories of native vegetation (brown grass, brown forb, sagebrush [*Artemisia* spp.], bitterbrush [*Purshia tridentate* (Pursh) DC.], and green forb) had significantly greater canopy cover when leafy spurge was absent. Leafy spurge presence did not affect cover of the native snakeweed (*Gutierrezia* spp.) or rabbitbrush (*Chrysothamnus* spp.; Table 9.1), but because of the similarity of color of these species with leafy spurge, the finding should be confirmed on the ground.

Systematic sampling of wildland vegetation is recognized as usually superior to random sampling (Maxwell et al. 2009), but it may require greater sampling intensity for equal accuracy (Wei and Chen 2004). A comparison of leafy spurge detection

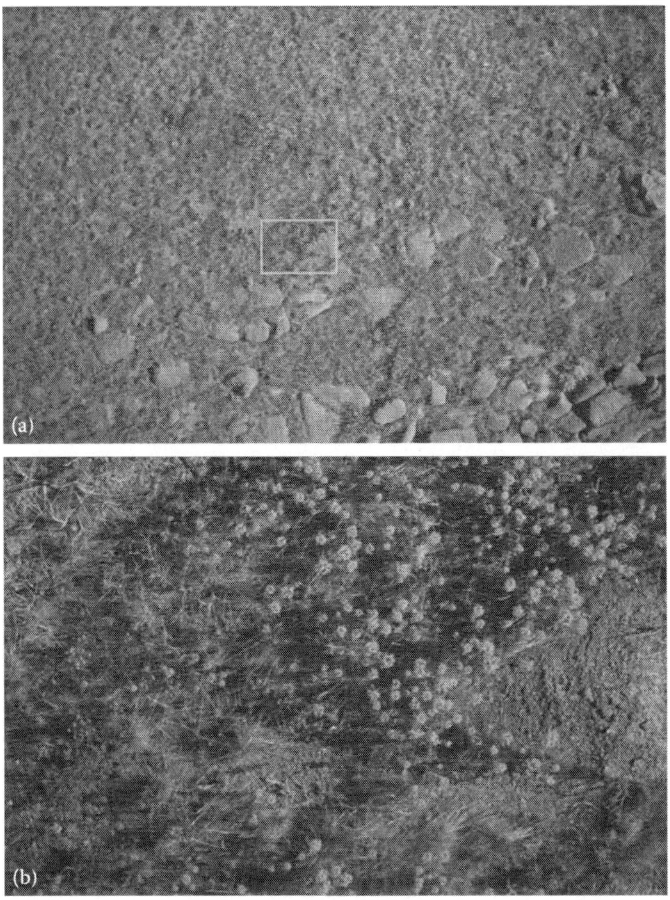

FIGURE 9.3 **(See color insert.)** Leafy spurge captured in 10-mm GSD (a) and 1-mm GSD (b) aerial images. The small rectangle shows coverage of the nested 1-mm GSD image. (From Booth, D.T. et al., *Native Plant Journal* 11:327–339, 2010.)

TABLE 9.1
Cover Means (Percentage) and Standard Deviations for Select Categories Measured from Upland Aerial-Survey Images with Leafy Spurge (LS+) and without Leafy Spurge (LS−)

Cover Category	LS+	LS−	*P*
Bare ground	5.4 ± 6.6	6.6 ± 8.3	0.228
Litter	6.8 ± 7.0	2.4 ± 2.6	<0.001
Brown grass	7.0 ± 8.0	11.4 ± 9.4	<0.001
Brown shrub	1.5 ± 3.4	1.9 ± 2.9	0.402
Brown forb	1.0 ± 5.4	3.2 ± 3.7	<0.001
Green grass	22.8 ± 15.4	23.2 ± 12.4	0.806
Gravel	3.4 ± 5.2	2.4 ± 3.6	0.137
Rock	10.1 ± 11.0	3.1 ± 6.0	<0.001
Snakeweed	0.3 ± 1.0	0.2 ± 0.6	0.108
Sagebrush	5.5 ± 8.3	9.7 ± 12.6	<0.001
Rabbitbrush	0.5 ± 2.6	0.2 ± 1.1	0.275
Bitterbrush	1.2 ± 3.1	0.1 ± 0.3	<0.0001
Green forb	12.9 ± 14.3	24.3 ± 14.0	<0.001
Leafy spurge	16.6 ± 24.6	0 ± 0	n.a.

Source: Booth, D.T. et al., *Native Plant Journal* 11:327–339, 2010.
Note: *P* values from Welch's *t*-test are shown. LS + *n* = 99; LS − *n* = 122. Ground cover was obtained by measuring the selected cover categories from 1-mm GSD samples using the SamplePoint software.

points from 8 years of agency ground monitoring with our one-time aerial survey illustrates both the advantage of a systematic aerial survey (>30 new populations identified) and a need for greater aerial-sampling intensity. This single aerial survey detected 73% of ground-detected populations found during an 8-year period of ground monitoring. Although an unknown number of ground-detected populations may have been eliminated before the aerial survey, we doubt that the number of eliminated spurge populations is great enough to significantly influence our aerial detection rate. Blumenthal et al. (2007) recommended capturing images of 0.5% of the area of interest as appropriate to detect toadflax (*Linaria dalmatica* [L.] P. Mill.) in a mixed-grass prairie. This survey, conducted a year before the Blumenthal recommendation, captured only 0.05% of the area of interest in 10-mm GSD images. We conclude that our sampling density was too low, and we recommend that future aerial surveys intended to determine invasive-species status in scattered populations capture at least 0.5% of a survey area in ~10-mm GSD images that include a nested 1-mm image to aid in plant identification.

9.3.4 Is Your Conservation Grazing Plan Producing a Benefit?

In 2003, the Squaw Valley Ranch (SVR) of Elko County, Nevada, began a conservation grazing management program to improve riparian conditions of streams affected by ranch operations. The goal was to regain lost habitat for the Lahonton cutthroat trout (*Oncorhynchus clarki henshawi* [Richardson]). We assessed the effectiveness of SVR's reduced hot-season riparian grazing by using willow cover as an indicator of stream recovery after conducting aerial surveys in 2003, 2004, and 2006 over the 330,000-ha Rock Creek watershed (41°17′N, 116°23′W) in the area's Tuscarora Mountains (Booth et al. 2012). Watershed elevation is 1500 to 2400 m with precipitation ranging from 250 to 300 mm (PRISM 2014). The riparian zones contain coyote and yellow willows (*Salix exigua* [Nutt.] and *S. lutea* [Nutt.], syn. *rigida*). From 2003 to 2006, there was no intentional hot-season livestock grazing on riparian areas, although there was some nonpermitted use in 2003 and 2004. Three major wildfires in 2005 and 2006 burned upland areas surrounding 15 of the 27 stream reaches surveyed, inflicting damage to some riparian vegetation.

The aerial surveys used methods previously described. Although images were acquired at two resolutions, only data obtained from images of ~20-mm GSD were used to measure willow cover. Segments of 11 streams (27 reaches), totaling 170 km, were surveyed using manual camera triggering, and each stream was sampled at approximately 100-m intervals. The camera equipment was the same as in the case study discussed in Section 9.3.2, but target flight altitude AGL was 200 m in 2003 and 2004 and 250 m in 2006—a change made to increase image FOV. We used SamplePoint (Booth et al. 2006b) with images of nine streams that represent the watershed's geomorphological variation and used 100 points per image to measure cover for (1) nonriparian, (2) water, (3) willow, (4) riparian vegetation, and (5) others. Dark green vegetation defined the riparian area. Points outside the riparian area were classified as nonriparian. Only points falling inside the riparian area were classified into the other four categories. This method required subjective delineation of riparian boundaries, as do all methods that measure riparian indicators. Cover percentages were converted into actual area (m^2) and were then normalized for interyear comparison by dividing the actual area by the stream length within the image. Thus, cover is reported in $m^2 \cdot m^{-1}$ stream. Stream length was used for normalization because it shows higher annual consistency than the riparian width or area. It was measured on a line with a minimum of 20 segments placed down the center of the bank full channel using ImageMeasurement (Booth et al. 2006a). Additionally, the distance from each sample to the upstream end of the stream reach containing the sample was measured from topographical maps (1:100,000 scale) in ArcMap 9.0.

Willow-cover data were square root-transformed to satisfy the equal variance assumption. The 2005 fire burned two reaches of Upper Rock Creek, reducing willow cover for both and resulting in loss of statistical independence between them. These data were averaged together and treated as one reach. Other reaches burned, but we saw no evidence that they sustained significant damage. Because most data were spatially autocorrelated, we condensed the data into means for stream reaches. Annual change was then tested using *t*-tests paired by years for individual streams.

Willow cover ($m^2 \cdot m^{-1}$ stream) increased 3.1% between 2003 and 2006 (P = 0.02, n = 12 reaches) and 2.0% between 2004 and 2006 (P = 0.004, n = 20 reaches), but there was not a significant increase between 2003 and 2004 (P = 0.18, n = 12 reaches; see Table 9.2). The 3.1% increase in actual cover means that between 2003 and 2006, the willow cover nearly tripled on one stream, more than doubled on three

TABLE 9.2
Willow Cover for Nine Streams by Year

Creek	2003 Willow Cover	*n*	2004 Willow Cover	*n*	2006 Willow Cover	*n*	Relative Increase
Frazer[a]	2.9 ± 3.1	42	3.9 ± 4.5	30	7.3 ± 6.0	36	2.5×
Lewis	9.5 ± 4.8	30	9.9 ± 6.3	21	11.0 ± 6.8	46	1.2×
Middle Rock[a]	0.9 ± 1.1	28	0.2 ± 0.5	36	1.0 ± 2.0	51	1.1×
Nelson	7.2 ± 4.3	21	8.4 ± 6.7	12	16.8 ± 8.8	36	2.3×
Upper Rock[a]	18.2 ± 17.0	28	10.2 ± 10.1	25	10.2 ± 8.0	87	−0.6×
Soldier[a]	—	—	1.2 ± 1.7	21	1.7 ± 2.1	26	1.4×
Toejam	3.8 ± 4.0	71	6.4 ± 6.4	54	8.7 ± 8.8	81	2.3×
Trout	—	—	4.1 ± 3.9	40	4.2 ± 6.2	56	1.1×
Willow[a]	1.8 ± 3.2	64	3.6 ± 5.8	40	5.2 ± 6.9	146	2.9×
All	5.2 ± 7.9	284	4.9 ± 6.3	279	7.1 ± 7.9	565	1.4×

Source: Booth, D.T. et al., *Ecological Indicators* 18:512–519, 2012.
Note: Cover data are $m^2 \cdot m^{-1}$ of stream in image ± S.D. and the relative increase factor over 2 or 3 years is in multiples of 2003 or 2004 cover.
[a] Portions of this stream were within a 2005 burn perimeter.

others, and increased on all but one fire-affected stream. Thus, the aerial surveys recorded major increases in willow cover on most streams between 2003 and 2006. This finding was strongly supported by historical evidence from both ground-based and Landsat images (Booth et al. 2012), illustrating the benefit of using data from different remote sensing platforms. Therefore, we conclude that the SVR conservation grazing plan is producing a biological benefit in terms of improved stream conditions and recovery of lost trout habitat.

9.4 CONCLUSIONS

Large-area natural resource surveys have been hampered by high cost, data variation, and lack of statistical design, rigor, and power. These factors have reduced the information return per dollar of survey cost, often nullifying any value of the survey effort. The use of GIS, 1-mm and other high-detail digital imagery, and image analysis software make valid tests of landscape-scale management questions possible and bring an unprecedented capability to answer ecological questions that are key to optimum land management and sustainable resource use.

9.5 HANDS-ON EXAMPLES

The online exercise contains both an in-depth generalized approach to high-resolution aerial image survey planning that utilizes multiple geographic data sources, and step-by-step instructions on how to plan a specific aerial project. The latter utilizes geographic data available from the CRC website: http://http://www.crcpress.com/product/isbn/9781439867228. The examples show how aerial data sets can be queried and how software tools can be used to assess data adequacy or to analyze selected data that address specific questions.

ACKNOWLEDGMENTS

This chapter is based on work funded as part of the research program at the USDA-ARS, High Plains Grassland Research Station, Cheyenne, Wyoming. Grants from USDI-BLM, Wyoming State Office, and Casper and Lander, Wyoming Field Offices, to D.T. Booth were the primary supplementary funding. Programming for SamplePoint and ImageMeasurement was done by Robert D. Berryman. Aerial images were acquired by CloudStreet Flying Service, Fort Collins, Colorado. Mention of trade names is for information only and does not imply an endorsement of products or services to the exclusion of similar products or services of equal quality. S.E. Cox had the primary responsibility for the online exercise. We especially thank Dr. Laurie Abbott (New Mexico State University), Dr. Lee Vierling (University of Idaho), Dr. Gretchen Sassenrath (editor), and Dr. Tom Mueller (editor) for reviewing this material and for their many excellent suggestions.

REFERENCES

Augustine, D.J., D.T. Booth, S.E. Cox, and J.D. Derner. 2012. Grazing intensity and spatial heterogeneity in bare soil in a grazing-resistant grassland. *Rangeland Ecology and Management* 65:39–46.

Blumenthal, D., D.T. Booth, S.E. Cox, and C. Ferrier. 2007. Large-scale aerial images capture details of plant populations. *Rangeland Ecology and Management* 60:523–528.

Booth, D.T., and S.E. Cox. 2006. Very large scale aerial photography for rangeland monitoring. *Geocarto* 21:27–34.

Booth, D.T., and S.E. Cox. 2008. Image-based monitoring to measure ecological change. *Frontiers in Ecology and the Environment* 6:185–190.

Booth, D.T., and S.E. Cox. 2009. Dual-camera, high-resolution aerial assessment of pipeline revegetation. *Environmental Monitoring and Assessment* 158:23–33.

Booth, D.T., and S.E. Cox. 2011. Art to science: Tools for greater objectivity in resource monitoring. *Rangelands* 33:27–34.

Booth, D.T., S.E. Cox, and R.D. Berryman. 2006a. Precision measurements from very large scale aerial digital imagery. *Environmental Monitoring and Assessment* 112:293–307.

Booth, D.T., S.E. Cox, and R.D. Berryman. 2006b. Point sampling digital imagery with 'SamplePoint'. *Environmental Monitoring and Assessment* 123:97–108.

Booth, D.T., S.E. Cox, and D.E. Johnson. 2005. Detection-threshold calibration and other factors influencing digital measurements of bare ground. *Rangeland Ecology and Management* 58:598–604.

Booth, D.T., S.E. Cox, and G.E. Simonds. 2006c. Riparian monitoring using 2-cm GSD aerial photography. *Journal of Ecological Indicators* 7:636–648.

Booth, D.T., S.E. Cox, and D. Teel. 2010. Aerial assessment of leafy spurge (*Euphorbia esula* L.) on Idaho's deep fire burn. *Native Plant Journal* 11:327–339.

Booth, D.T., S.E. Cox, M. Louhaichi, and D.E. Johnson. 2004. Technical Note: Lightweight camera stand for close-to-earth remote sensing. *Journal of Range Management* 57:675–678.

Booth, D.T., S.E. Cox, G. Simonds, and E. Sant. 2012. Willow cover as a stream-recovery indicator under a conservation grazing plan. *Ecological Indicators* 18:512–519.

Brady, W.W., J.E. Mitchell, C.D. Bonham, and J.W. Cook. 1995. Assessing the power of the point–line transect to monitor changes in plant basal cover. *Journal of Range Management* 48:187–190.

Cagney, J., S.E. Cox, and D.T. Booth. 2011. Comparison of point intercept and image analysis for monitoring rangeland transects. *Rangeland Ecology and Management* 64: 309–315.

Cook, C.W., and J. Stubbendieck. 1986. *Range Research: Basic Problems and Techniques*. Society for Range Management, Denver, CO.

Cooper, W.S. 1924. An apparatus for photographic recording of quadrats. *Journal of Ecology* 12:317–321.

Federal Aviation Administration. 2004. Airworthiness certification of aircraft and related products. Order 8130.2F.

Friedel, M.H., W.A. Laycock, and G.N. Bastin. 2000. Assessing rangeland condition and trend. In *Field and Laboratory Methods for Grassland and Animal Production Research*, eds. L. 't Mannetje, and R.M. Jones, 227–262. CAB International, New York.

Interagency Technical Team. 1996. *Sampling Vegetation Attributes*. Interagency Technical Reference, Report No. BLM/RS/ST-96/002+1730. U.S. Department of the Interior, Bureau of Land Management—National Applied Resources Science Center, Denver, CO. Available at http://www.blm.gov/nstc/library/pdf/samplveg.pdf (accessed January 7, 2014).

Maxwell, B.D., E. Lehnhoff, and L.J. Rew. 2009. The rationale for monitoring invasive plant populations as a crucial step for management. *Invasive Plant Science and Management* 2:1–9.

Muiruri, W. 1978. Bio-economic conflicts in resource use and management: A Kenyan case study. *GeoJournal* 2(4):321–330.

National Research Council. 1994. *Rangeland Health*. National Academy Press, Washington, DC.

PRISM (PRISM Climate Group). 2014. 30-yr normal precipitation: Annual. Available at http://www.prism.oregonstate.edu/normals/ (accessed January 8, 2014.)

Sidle, R.C., and A.N. Sharpley. 1991. Cumulative effects of land management on soil and water resources: An overview. *Journal of Environmental Quality* 20:1–3.

Underwood, N., P. Hamback, and B.D. Inouye. 2005. Large-scale questions and small-scale data: Empirical and theoretical methods for scaling up in ecology. *Oecologia* 145:176–177.

U.S. Dept. Agriculture, Agricultural Adjustment Administration. 1940. *Instructions to Field Range Examiners for Making Range Surveys*. USDA-AAA Western Division Report WD-25, Washington, DC.

Walker, B.H. 1970. An evaluation of eight methods of botanical analysis on grasslands in Rhodesia. *Journal of Applied Ecology* 7:403–416.

Weber, K.T., F. Chen, D.T. Booth, M. Raza, K. Serr, and B. Gokhale. 2013. Comparing two ground-cover measurement methodologies for semiarid rangelands. *Rangeland Ecology and Management* 66:82–87.

Wei, H., and D. Chen. 2004. The effect of spatial autocorrelation on the sampling design in accuracy assessment: A case study with simulated data. *Environmental Informatics Archives* 2:910–919.

West, N.E. 1999. Accounting for rangeland resources over entire landscapes. In *Proceedings VI, People and Rangelands: Building the Future*, eds. D. Eldridge, and D. Freudenberger, 726–736. VI International Rangeland Congress, Inc., Aitkenvale, Queensland, Australia.

West, N.E. 2003a. History of rangeland monitoring in the U.S.A. *Arid Land Research and Management* 17:495–545.

West, N.E. 2003b. Theoretical underpinnings of rangeland monitoring. *Arid Land Research and Management* 17:333–346.

Zirlewagen, D., and K. von Wilpert. 2004. Using model scenarios to predict and evaluate forest-management impacts on soil base saturation at landscape level. *European Journal of Forest Research* 123:269–282.

Zirlewagen, D., and K. von Wilpert. 2010. Upscaling of environmental information: Support of land-use management decision by spatio-temporal regionalization approaches. *Environmental Management* 46:878–893.

10 Landscape Tension Index

Brian D. Lee, James Adams, and Stephen D. Austin

CONTENTS

EXECUTIVE SUMMARY

As human population increases, land resources will become more vital to human survival. Society will expand the emphasis of strategically viewing the landscape for a variety of often conflicting uses. Identifying land resources based on the provisioning of important ecological services provides one landscape perspective. Mapping the factors that contribute to the conversion of forests and agricultural lands to urbanized uses provides another way to understand the landscape. Combining these two landscape views multiplies the understanding of the location and level of landscape tension. This chapter describes a modified land evaluation and site assessment (LESA) approach to identify land that has environmental value while simultaneously having attributes that make it attractive for urbanization. Using expert knowledge, spatial data layer factors and weighting values were developed for a 15-county study area in the Bluegrass Region of central Kentucky, USA. This work provides an approach that builds upon existing uses of LESA described in the literature. The value of this method is that it provides for stakeholder input from multiple perspectives in a landscape utilization process. In addition, this method creates a combined geospatial index of landscape tension utilizing subanalyses of landscape preservation and landscape change through urbanization pressure in the landscape. The combined geospatial landscape tension index utilizes subanalyses that can be valued and informed through stakeholder input using an iterative Delphi process with submodels.

KEYWORDS

Conservation planning, decision support, farmland preservation, GIS, land evaluation and site assessment (LESA), urban growth management

10.1 INTRODUCTION

The goal of this chapter is to provide a method for resolving potential incompatibilities between uses that provide societal ecosystem services and those that offer a physical base for urbanization. This chapter extends the Land Evaluation and Site Assessment (LESA) approach to resolve these conflicts in a manner that is similar to its use to ascertain agricultural land suitability (Wright et al. 1983; Pease and Coughlin 1996). The Landscape Tension Index (LTI) approach helps stakeholders visually distinguish areas for landscape preservation from places that are more suitable for urbanization. By then combining the two analyses, LTI can be created for use in land-planning discussion and decision making.

LESA was developed to provide a flexible tool for integrating many agricultural use factors for land assessment (Soil and Water Conservation Society 2003). The approach has traditionally been used on a parcel basis for rating and ranking properties for agricultural and land conservation programs (e.g., purchase of development rights [PDR] or conservation easements). Over 200 local and state programs are using LESA for land conservation (Pease et al. 1994). In June 2003, LESA expert participants met in Nebraska City, Nebraska, for a series of focused sessions to identify opportunities and constraints relative to enhancing LESA. Among several interest areas, the participants identified incorporating geographic information systems (GISs) into expanded LESA applications that identify lands of critical concern (Soil and Water Conservation Society 2003). In addition, LESA has a long history of support from the Soil and Water Conservation Society as well as through its publications and activities. The extent of implementation of this method development, a case study approach to address a wide range of issues, is documented in more than 15 journal publications over the last several decades (e.g., Dunford et al. 1983; Wright et al. 1983; Little 1984; Richards 1984; Nellis and Nicholson 1985; Reganold 1986; Luckey 1989; Van Horn et al. 1989; Daniels 1990; Ferguson et al. 1991; Freedgood 1991; Coughlin et al. 1994; Hoobler et al. 2003; Dung and Sugumaran 2005; Machado et al. 2006).

The LTI is described here as a case study using a modified LESA approach for identifying landscape preservation value (Figure 10.1) and landscape urbanization change value (Figure 10.2), and then combining the two sub-analyses for a product map of an LTI (Figure 10.3). The factors that comprise the landscape preservation and the landscape change maps can utilize values based on stakeholder input about issues that are specific to the study area as well as author experience. The resulting LTI identified parts of the landscape that could be used by designers, planners, policy makers, and stakeholders to inform land use decisions by identifying the ecologically and culturally important areas within the landscape that may be more susceptible to urbanization pressure.

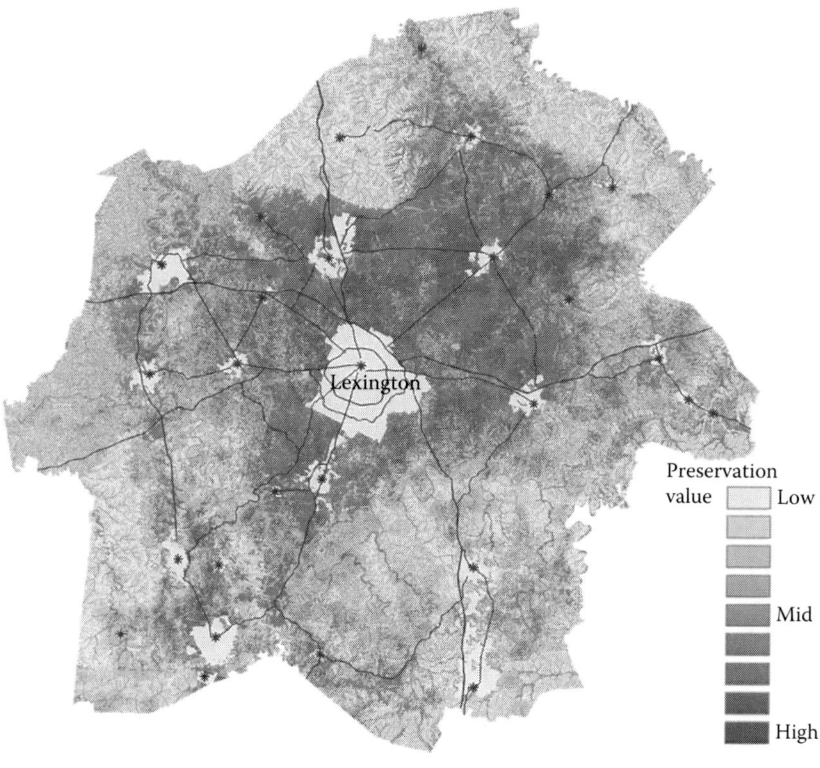

FIGURE 10.1 The 15-county study area with the incorporated cities and Lexington-Fayette County Urban Service Area in white. The darker areas indicate parts of the landscape that are more valuable to ecological and cultural systems.

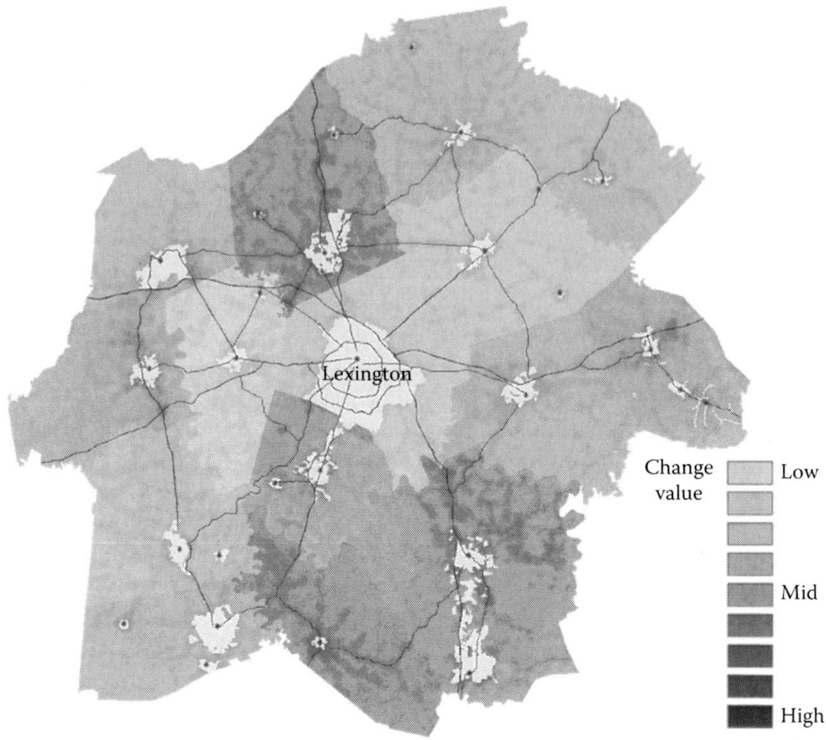

FIGURE 10.2 The 15-county study area indicating relatively higher change (urbanization) in darker color and lower change in lighter color. The blocky pattern is apparent primarily because of the predicted population growth and a review of the land use regulations/restrictions relative to each county.

10.1.1 Kentucky Bluegrass Region Context

The study region, located in the Kentucky Bluegrass, is faced with the daunting challenge of preserving its cultural landscape amidst urbanization change potential. The region is within a day's drive to approximately two-thirds of the US population. Routes connect a north–south and east–west interstate highway system in the region. Lexington-Fayette County is the regional population center with 295,803 people according to the 2010 population census. This is a 13.5% population increase from 2000. According to the 2007 Lexington-Fayette Urban County Government (LFUCG) Comprehensive Plan (LFUCG 2007), Lexington-Fayette County serves as the center of economic, educational, health, and cultural activities for the surrounding area. Lexington has been inhabited continuously since 1775 and has grown from a rural community to the primary urban center of Central Kentucky. The area is noted for its thoroughbred and saddlebred horse farms, bourbon distilleries, and rolling hills with wooden four-plank and rock fences. In 2006, the Bluegrass Cultural Landscape of Kentucky was placed on the watch site list of the World Monuments Fund because of inadequately coordinated regionwide planning that failed to protect this cultural landscape (World Monuments Fund 2006). There are several nearby communities in

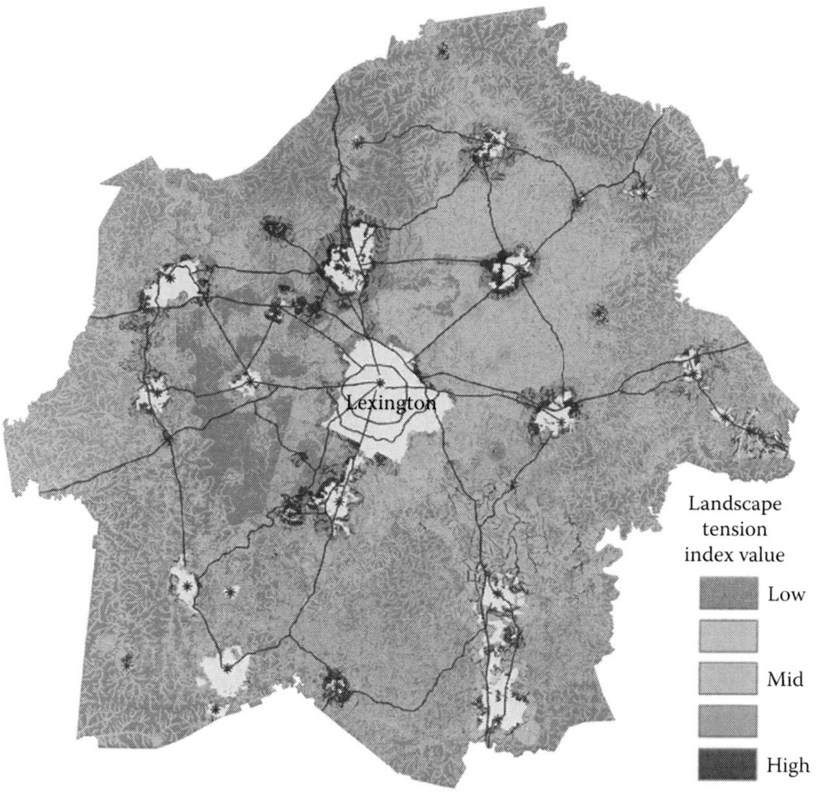

FIGURE 10.3 The 15-county study area indicating parts of the landscape with relative amounts of tension between change and preservation. Darker shades indicate places that have a pressure to be urbanized while at the same time have a relatively high value to be preserved for ecological and/or cultural reasons. The figure is made by combining (multiplying) the two previous analyses in a GIS. In addition, the color scale in the legend has been reclassified down to five categories in order to emphasize places of landscape tension.

six adjacent counties, which are all within commuting distance/time of Lexington. These counties are generally increasing in population with Scott County particularly notable for both its historical and projected growth. In 2000, Scott County had a human population of 33,061; in 2010, that had risen to 47,173, with projected growth for 2020 as 63,984 and 2050 as 117,889 (Kentucky State Data Center 2011).

Kentucky does not require local growth management policies in all communities. The Kentucky Legislature in the 2000–2001 session passed House Bill 524 that would have required urban growth boundaries, but the governor did not sign it into law (Howland and Sohn 2007). Kentucky does require communities to have a comprehensive plan to "serve as a guide for the physical development and economic and social well-being of the planning unit" under Kentucky Revised Statutes (KRS) 100.183 (Kentucky Legislature n.d.a). Therefore, planning activities frequently differ across local jurisdictions. Planning jurisdictions surrounding Lexington vary in their growth management strategies, but all have implemented a number of zoning ordinances. Lexington's urban service area determines where city services, such as city wastewater lines, are available; land outside this boundary is labeled as "rural" with a 40-ac minimum lot size enforced through a traditional zoning approach. In its more than 50 years of existence, this boundary has changed seven times, including one reduction in 1967 (LFUCG 2007, p. 287). The presence of highly desirable soils and many cultural points of interest, coupled with projected population growth and regionally fragmented land use policies, has the potential to create landscape tension. The region is also known worldwide in the thoroughbred horse industry for breeding, training, and related equestrian activities. For example, in 2010, the Alltech FEI World Equestrian Games took place in the study area. This is the first time this event has been held in the United States and outside of Europe.

Within the study region, policies have been enacted in several counties to help preserve the existing open space and rural character. Rural character has been defined as representing "a quality of life based upon traditional rural landscapes, activities, lifestyles, and aesthetic values" (Municipal Research and Service Center of Washington 1999). Rural development and farmland fragmentation are threatening the rural character of the region. Fragmentation is described as the division of a habitat or land types into smaller units (Forman 1995). Farm fragmentation near urban areas decreases the traditional local farming economic base and changes the social and esthetic attributes of rural communities (Heimlich 1989). In addition, human-induced fragmentation is a detriment to "biodiversity with losses to numbers of species, their abundance, and genetic diversity" (Lord and Norton 1990). Fragmentation within the study region may be attributed to differing and conflicting policies among jurisdictions. One means of rural character protection in the region is through a PDR program, as instituted by the LFUCG (2000), which seeks to preserve valuable farmland (Daniels 1991). In general, these rights are typically held in perpetuity regardless of the property owner, which ensures that it will not be developed in the future. PDR programs typically provide more permanent farmland protection than either zoning or property tax advantages (Daniels 1991). Many Bluegrass residents cherish their surrounding farmlands and natural areas not only from an esthetic standpoint but also from an economic standpoint. The agricultural industry plays a vital role in the local economy and contributes to the multimillion dollar annual tourism industry.

In 2000, the PDR program and the Rural Land Management Board, which oversees the PDR program, were adopted by the City Council for only the Lexington-Fayette County through an ordinance (LFUCG 2000). Landowners with potential PDR-eligible properties submit an application to a Rural Land Management Board, which analyzes each property individually based on a point value system. Essentially, the point system is an adaptation of a LESA framework (Pease and Coughlin 1996; LFUCG 2000). This point system rates property parcel factors such as parcel size, road frontage, soil quality, and proximity to federal highways and interstate interchanges, among other things. Higher-scoring properties gain purchase priority of the development rights. As of October 20, 2014, 244 farms totaling 28,169 acres are now permanently protected by conservation easements (LFUCG 2014). PDR is now at 56.34% of the 50,000 acre goal defined in the Ordinance. The PDR program is not antidevelopment; rather, its purpose is to encourage denser development within the existing urban area while maintaining the integrity of the rural character outside of the urban service area's limits.

The PDR program is predated by the adoption of a 40-ac minimum lot size in 1999 from 10 ac as well as the 1958 adoption of the Urban Service Boundary. Lexington-Fayette County is not the only county within the study region that has taken steps to affect its rural heritage by way of PDR, cluster zoning, and/or large lot zoning (Georgetown News-Graphic 2008). These programs are some ways for preserving the rural character of the landscape. Farmland fragmentation and loss of rural character appear to be issues within the study region. Although some preservation policies have been implemented within the study region, it is beneficial for stakeholders to be informed of other options that are available. Identifying parts of the landscape that have specific environmental value and are apt to be urbanized is potentially important to crafting land use policy recommendations.

10.2 MATERIALS AND METHODS

In order to perform an LTI study, it can be conceptually developed in five distinct phases: (1) geospatial and nongeospatial data preparation, (2) stakeholder input and refinement, (3) landscape change preservation sub-analyses, (4) landscape change sub-analyses, and (5) LTI calculation. Of course, phases 2–5 can and likely should be iteratively performed. In the exercise provided on the CRC Press website at http://www.crcpress.com/prod/isbn/9781439867228, the data and basic valuing and weighting approach will be described, allowing the user to perform the essential components of an analysis described here.

10.2.1 Preparation of Spatial Data

Initial descriptive mapping was developed using basic inventory data such as soils, infrastructure, land cover, and demography, as described in the LESA Guidebook (Pease and Coughlin 1996) for process organization and model development. These data from maps and spreadsheets as well as conversations with the stakeholders are essential to complete a full study for use in decision making.

Using Environmental Systems Research Institute's ArcGIS software, and in particular ModelBuilder (ESRI, Redlands, California), the raster and vector data layers were assembled following conventions of map algebra (Tomlin 1990, 1994; DeMers 2002). The GIS model process used ideas from Williams (1985) and DeMers (1994, 2002). This approach was used to create two separate geospatial sub-analyses for visualizing landscape preservation areas and landscape urbanization change areas across the 15-county study region. Subsequently, these two sub-analyses were combined to create the LTI for the nonincorporated areas and outside the Urban Services Area of Lexington-Fayette County, Kentucky. The data used were all publically available from government sources and are described in detail later in Section 10.2.3.

10.2.2 Stakeholders' Input Meeting

Typically, a meeting with stakeholders that functions as described in the LESA Committee (Pease and Coughlin 1996) denotes each project phase. The stakeholders are asked questions relative to geospatial and political factors that they believe drive landscape change (urbanization) in the study area as well as places that are important to maintain for environmental or cultural reasons. The factors that comprise the landscape preservation and the landscape change maps are given values based on stakeholder opinions and issues specific to the study area using meeting survey forms. These expert opinions are used to determine a weight of percent influence for each contributing factor to each sub-analysis. The analyses identify areas, particularly in transportation corridors, that require greater attention from designers, planners, policy makers, and stakeholders. These survey forms explain each contributing factor, their value breakdown, and the overall percent influence. This process follows the process that is described in Pease and Coughlin (1996) using expert feedback and can be combined with a digital audience response system.

The inventory data are valued on a scale of 1 (least) to 10 (most) according to each factor's level of susceptibility to change and/or landscape preservation. For example, prime agricultural soil is given a value of 10 because stakeholders would likely consider this resource as a valuable asset for landscape preservation. Likewise, counties with a high percentage of projected population growth were also given a high value because of susceptibility to landscape change. In general, areas that were already developed received a value of 1 because initial urbanization conversion had already occurred. Each factor is weighted based on interaction with stakeholders during expert panel sessions and public meetings, and/or authors' expert knowledge. It is also possible to start with all factors weighted equally if and when there is no sound reasoning to do otherwise. Refinements can be made to weights as well as factors as a result of stakeholder interactions during the iterative process.

10.2.3 Landscape Preservation Sub-Analysis Description

The landscape preservation analysis uses several factors representing primarily ecological but also cultural locations of importance in the region. Selecting factors for the sub-analysis will depend on a number of issues relative to the environment in which the study is operating. Landscape preservation can be framed from an ecological point of view whereby ecosystem function and services are used for selection or cultural factors such as viewable areas around historic structures or pastoral landscapes. In the work described here, "Conservation Thresholds for Land Use Planners" by Kennedy et al. (2003) was primarily used for giving overall direction on factor selection and valuation. What will follow is a description of the specific factors used along with the factor weighting in this sub-analysis. Each sub-analysis is based on 100% to complete the weighting.

10.2.3.1 Soil Analysis (Weight = 40%)

Soil types were assigned specific values for different soil properties, reflective of their limitations and advantages. For the purposes of this study, soil types were valued in terms of quality for farm production. The values are as follows: Not prime farmland = 1; Limited farmland = 6; Prime farmland and farmland of statewide importance = 10. The soil data were downloaded from the US Department of Agriculture, Natural Resources Conservation Service's—Soil Data Mart at the time. The Soil

Data Mart is no longer in service, but the functionality has been incorporated into the Web Soil Survey at http://websoilsurvey.nrcs.usda.gov/.

10.2.3.2 Contiguous Pasture, Hay, Grassland, and Cultivated Crops (Weight = 15%)

The 2001 National Land Cover Data GRID for Kentucky was produced through a cooperative project conducted by the Multi-Resolution Land Characteristics Consortium that involved a local representative in the process. The consortium successfully completed mapping the United States into the National Land Cover Data thematic detail (Homer et al. 2004). A variety of spatial analysis and metrics can be applied to these data to interpret and understand the landscape in different ways. For this study, the land cover classes of pasture/hay, grassland/herbaceous, and cultivated crops were extracted and valued in terms of contiguous area. The larger contiguous areas were valued highly in this study. The areas were broken into 10 divisions using natural data breaks based on acres. The values are as follows: 0–268 ac = 1; 269–654 = 2; 655–1069 = 3; 1070–1485 = 4; 1486–1961 = 5; 1962–2584 = 6; 2585–3505 = 7; 3506–4901 = 8; 4902–7604 = 9; >7604 a = 10.

10.2.3.3 Points of Interest (Weight = 15%)

Interest points were a layer within the study region from national register properties, historic landmarks, horse farms, and other Bluegrass tourism sites (Kentucky Heritage Council 2008). For the purposes of this study, a proximity value was given to the areas surrounding these points of interest. The rationale was that these locations represent many of the historic and cultural amenities of the landscape experience. The values are as follows: 1+ mi from point = 1; 0.5–1 mi = 6; 0–0.5 mi = 10.

10.2.3.4 Forest Fragmentation (Weight = 15%)

The 2001 National Land Cover Data GRID for Kentucky was also used to define the forest fragmentation types. Riitters et al. (2000) identified categories of fragmentation from the amount of forest and its occurrence as adjacent forest pixels. Knowledge of fragmentation type permits general inferences about probable impacts, even without detailed knowledge of the flora, fauna, and ecological processes of the area (Riitters et al. 2000). In this study, the land cover classes of deciduous forest, evergreen forest, mixed forest, and woody wetland were reclassified and valued in terms of the relationship to their surrounding 9 × 9 cell area or 81-cell neighborhood using the focal sum of Spatial Analyst. This study uses four forest fragmentation types (Riitters et al. 2000): Patch, Edge, Transitional, and Interior forest. The values are as follows: Not forest = 1; 0%–40% of the 81-pixel neighborhood (Patch) = 4; 41%–80% (Edge) = 6; 81%–99% (Transitional) = 8; 100% (Interior) = 10.

10.2.3.5 Stewardship Lands (Weight = 5%)

The Kentucky Fish and Wildlife Information System provided the Kentucky GAP Analysis Program land stewardship layer as part of a national project to assess biodiversity at state and national levels (Kentucky GAP 2001). This stewardship layer provides information about the location, ownership, management, and conservation status of public and semipublic lands in Kentucky (Kentucky GAP 2001). The basic idea is that land close to these areas would be more valuable for ecological land preservation. The stewardship layer also represents a possible landscape amenity and will potentially attract development. For the purpose of this study, the stewardship properties and the surrounding areas were given the following values: 1+ mi from area = 1; 0.5–1 mi = 6; 0–0.5 mi = 8; within property line = 10.

10.2.3.6 Aquatic Systems (Weight = 10%)

This factor layer used streams, springs, and wetlands. The National Hydrography Dataset reports on surface water features (NHD 2003). The Kentucky Groundwater Data Repository compiles the springs in the Kentucky layer that provides information on springs (Kentucky Geological Survey 2003). The Kentucky Environmental and Public Protection Cabinet (2002) compiled the National Wetlands Inventory (NWI) digital layer, which shows the extent of wetlands, as defined by the US Fish and Wildlife Service, and is mapped as part of the NWI program. In general, the aquatic systems are valued by buffer sizes, as described in *Conservation Thresholds for Land Use Planners* by Kennedy et al. (2003). The specific buffer sizes of 50 and 300 ft. are recommended generalizations. The values are as follows: >300 ft. from a stream, spring, and wetland = 1; 50–300 ft. from two of the three = 6; 50–300 ft. from all three = 7; 0–50 ft. from one of the three = 8; 0–50 ft. from two of the three = 9; 0–50 ft. from all three = 10.

10.2.4 Landscape Change Sub-Analysis Description

This sub-analysis sums to 100% in weight just like the landscape preservation sub-analysis and was intended to identify general urban development landscape suitability. The authors sought to identify places in the landscape that could potentially be urbanized more easily given the land and its context. This analysis is similar to the site assessment component of LESA. The essential point is that areas where it would be easier to develop are given relatively high values (for example, low land slope), while the harder areas are given low values (steep land slope). In this example, urbanization is a broad term encompassing housing and commercial land uses generally.

10.2.4.1 Population Growth (Weight = 30%)

The Urban Institute at the University of Louisville compiles the projected population growth data for the state of Kentucky, based on the US Census Bureau data from 2000 (Kentucky State Data Center 2008). These data are projected for each county every 5 years through 2030. In this study, the population growth data are valued in terms of percent growth from 2000 to 2030. The values are as follows: 0%–19% = 1; 20%–31% = 3; 32%–43% = 6; 44%–54% = 8; 55%+ = 10.

10.2.4.2 Land Use Policy (Weight = 28%)

In the Commonwealth of Kentucky, the county or the city government has the ability to determine land use policy (Kentucky Legislature n.d.b). There is a range of regulations that may differ greatly between counties and cities. Counties that have less strict land use policies, such as smaller lot minimums, were considered more susceptible to urbanization change. In this study, land use policy is generalized for each county and based on the minimum lot sizes and the ability to create new roads in relationship to the new development. The values are as follows: 40-ac minimum = 1; 5-ac minimum with 30-ac minimum zone = 2; 5-ac minimum, no new roads = 5; 5-ac minimum = 7; 1-ac minimum with 2.5-ac zone = 8; 1-ac minimum = 10.

10.2.4.3 Median Household Value (Weight = 5%)

The US Census Bureau (2014) gathers demographic information from residents of the United States for publication every 10 years in the decennial census. Median household value is gathered to reflect the respondent's estimate of the current dollar value of the household rather than construction cost or purchase price (US Census Bureau 2008). For the purposes of this study, median household value in 2000 is assessed by census tract in terms of higher property values being more susceptible to change based on work by Alig et al. (2004). The values are as follows: \$0–\$54,420 = 1; \$54,421–\$108,840 = 3; \$108,841–\$163,260 = 5; \$163,261–\$217,680 = 8; >\$217,681 = 10.

10.2.4.4 Municipal Water Infrastructure (Weight = 5%)

The Water Resources Information System (n.d.a) compiles data on waterlines in the Commonwealth of Kentucky, as surveyed by Kentucky's Area Development Districts and as provided by the Division of Water, to illustrate the location of waterlines and reference basic attribute data concerning those features. Land closer to existing or proposed infrastructure was considered more susceptible to change. Specifically, existing and proposed waterline infrastructure that is 6 in. or larger was valued in terms of the distance to the infrastructure. The values are as follows: >0.5 mi away, outside urban service area = 1; 0.25–0.5 mi = 6; 0–0.25 mi = 10.

10.2.4.5 Sanitary Sewer Infrastructure (Weight = 10%)

The Water Resources Information System (n.d.b) also compiles data on wastewater collection lines. Land closer to existing or proposed infrastructure is more susceptible to change. For the purposes of this study, existing and proposed sewer infrastructure were valued in terms of the distance to the infrastructure. The values are as follows: >0.5 mi away, outside urban service area = 1; 0.25–0.5 mi = 6; 0–0.25 mi = 10.

10.2.4.6 Major Road Intersections (Weight = 8%)

The KyVector transportation layers represent all vector datasets published to KyGeonet under the transportation category by the Kentucky Transportation Cabinet, Kentucky GIS Community, and the Kentucky Geography Network (Kentucky Transportation Cabinet 2007). The major road layer is restricted to the larger and more heavily traveled roads. The major roads were valued in terms of distance to their intersections. The values were as follows: >1 mi from point = 1; 0.5–1 mi = 6; 0–0.5 mi = 10.

10.2.4.7 Minor Roads (Weight = 4%)

The local road layer of the KyVector Transportation Layers includes Kentucky's smaller local and county roads. The minor roads were valued in terms of distance to their intersections. The values are as follows: >0.5 mi from point = 1; 0.25–0.5 mi = 6; 0–0.25 mi = 10.

10.2.4.8 Land Slope (Weight = 10%)

The Shuttle Radar Topography Mission of 2000 collected topographic measurements of the earth's surface to provide data to enhance the activities of scientists, military, commercial, and civilian users (National Aeronautics and Space Administration 2002). Using ArcGIS with the Spatial Analyst Extension, these data were utilized to create a percent slope map for the study region. For the purposes of this study, the percent slope map was valued in terms of lesser slopes being more susceptible to landscape change. The values are as follows: >20% = 1; 12–20 = 3; 6–12 = 5; 3–6 = 8; 0–3 = 10.

10.3 RESULTS AND DISCUSSION

In map algebra (Tomlin 1990, 1994) terms, a local operation was used to multiply the landscape preservation and landscape change sub-analyses to produce the LTI. The index can potentially range from 1 (low tension) to 100 (high tension). The resulting LTI (Figure 10.3) indicated areas with the highest (shown in darker shade of gray) and lowest (shown in lighter shade of gray) potential for tension; they have places that are characterized with a high susceptibility to urbanization and a high value for preservation. Areas shown in the lightest shade of gray have the lowest potential for tension, places characterized by a low susceptibility to urbanization, and a low value for preservation. This map may be interpreted and used by stakeholders in many ways.

A particular way that these types of results could be used is in connection with a regionally coordinated PDR program. A PDR program, although successful in the Inner Bluegrass Region, is not the only means by which to preserve rural areas. Other means of preservation include transfer of development rights (TDR) and the implementation of cluster developments. TDR programs allow landowners to transfer the rights to develop one parcel of land to another (Brabec and Smith 2002). While PDR programs are completely voluntary on the landowners' part, TDR programs designate TDR areas. TDR programs often require landowners in these areas to participate in the program. Since development rights are transferred rather than purchased, the program is relatively inexpensive when compared to PDR (Brabec and Smith 2002). Cluster development (also called planned residential development) differs from TDR and PDR programs because it typically deals with development on a site-specific basis. Cluster development has been described as a design approach that focuses buildings and other man-made infrastructure on a small portion of the site while reserving the land in common functional open space to be used for agriculture, recreation, or preservation (Gallatin County 2003). Cluster development usually preserves farmland by reducing minimum lot sizes and requiring open space. A study conducted by Brabec and Smith (2002) analyzes the effectiveness of each of these methods of land preservation. Contiguity, which is the grouping of spatially adjacent farmland, was highest in areas that utilized more than one preservation tool. Therefore, just because one protection strategy may be in place, there can still be opportunities to further protect rural land. Furthermore, it was determined that TDRs have more potential to protect large acreages of farmland as well as a greater amount of land than the other two protection strategies. This leads to a higher rate of contiguity and less fragmentation. PDRs were more successful in maintaining agricultural uses on conserved land. Cluster developments, while maintaining a certain degree of greenspace, were not effective in reducing fragmentation as they worked on a site-specific rather than a regional basis (Brabec and Smith 2002). It should be restated that studies have found that a combination of multiple preservation programs is most effective and valuable for farmland protection (Brabec and Smith 2002). This implies that the study area, as a region, should consider adopting multiple programs of preservation in order to ensure that the region's unique rural character is not diminished and that smarter growth within the region is encouraged.

10.3.1 Limitations

The study as it stands has some limitations due to access to updated material beyond the data and resources compiled by KYGeonet. Maintaining up-to-date, quality information on a public network such as KYGeonet is critical to future landscape studies. For example, one limitation of the LTI was the varying dates of publication of input data. Land cover data from 2001 were used with transportation data from 2007. Consistently updating input data and determining the effect of varying data are issues that need further attention and are beyond the scope of this study. The LTI could benefit from validation to adjust weighting and other assumptions. At the time of this study, no data existed to use for validation. Additional limitations were identified in the land use policy layer of the study. Accurate zoning and land use information was difficult to compile, particularly for the entire 15-county region because of a variety of systems to maintain specific area designations. In addition, variance requests are regularly being heard and granted as well as denied in the region.

10.3.2 Opportunities

Despite these limitations, one of the distinct advantages to the LTI model is its dynamic impact potential and flexibility to accommodate a variety of situations. By using a GIS application to build the model, data can be altered and added, as well as weighting differences. Stakeholder and expert feedback can be efficiently applied to produce a map that quantifies their values of the landscape. Such models could be extremely effective for a planning commission or other organization to better understand their landscape from a regional perspective.

10.4 SUMMARY AND CONCLUSIONS

This chapter has set out to build on a framework originally developed primarily for agricultural land conservation programs by extending the ideas to identify potential areas that are suitable for urbanization. By combining the two suitability analyses in a raster-based geospatial model, a tension index in map form is created in order to visualize and subsequently identify areas in the landscape that are simultaneously suitable for both land use activities. This approach is adaptable to local and regional conditions given the current computing technology and generally available data.

The LTI is limited to areas that are located outside the corporate boundaries of the study region, and further attention should be given to the relationships between inside and outside the boundaries. Understanding the urban–rural relationship is critical in land use planning, and a further adaptation of this study could be used to better understand these issues. It is believed that a great deal of rural landscape tension could be avoided if urban infill development becomes more attractive to developers and other stakeholders.

ACKNOWLEDGMENT

The authors thank Karen Goodlet for her invaluable assistance early in the editing process. The material presented in this article has not been used for actual decision making in the study region.

REFERENCES

Alig, R.J., J.D. Kline, and M. Lichtenstein. 2004. Urbanization on the US landscape: Looking ahead in the 21st century. *Landscape and Urban Planning* 69(2):219–234.

Brabec, E., and C. Smith. 2002. Agricultural land fragmentation: The spatial effects of three land protection strategies in the eastern United States. *Landscape and Urban Planning* 58(2):255–268.

Coughlin, R.E., J.R. Pease, F. Steiner, L. Papazian, J.A. Pressley, A. Sussman, and J.C. Leach. 1994. The status of state and local LESA programs. *Journal of Soil and Water Conservation* 49(1):6–13.

Daniels, T. 1990. Using LESA in a purchase of development rights program. *Journal of Soil and Water Conservation* 45(6):617–621.

Daniels, T. 1991. The purchase of development rights: Preserving agricultural land and open space. *Journal of the American Planning Association* 57(4):421–431.

DeMers, M.N. 1994. Requirements analysis for GIS LESA modeling. In F.R. Steiner, J.R. Pease, R.E. Coughlin, eds., *A Decade with LESA: The Evolution of Land Evaluation and Site Assessment*. Soil and Water Conservation Society, Ankeny, IA.

DeMers, M.N. 2002. *GIS Modeling in Raster.* John Wiley & Sons, New York.

Dunford, R.W., R.D. Roe, F.R. Steiner, W.R. Wagner, and L.E. Wright. 1983. Implementing LESA in Whitman County, Washington. *Journal of Soil and Water Conservation* 38(2):87–89.

Dung, E.J., and R. Sugumaran. 2005. Development of an agricultural land evaluation and site assessment (LESA) decision support tool using remote sensing and geographic information system. *Journal of Soil and Water Conservation* 60(5):228–235.

Ferguson, C.A., R.L. Bowen, and M. Akram Kahn. 1991. A statewide LESA system for Hawaii. *Journal of Soil and Water Conservation* 46(4):263–267.

Forman, R. 1995. *Land Mosaics: The Ecology of Landscapes and Regions.* Cambridge University Press, New York.

Freedgood, J. 1991. PDR programs take root in the Northeast. *Journal of Soil and Water Conservation* 46(5):329–331.

Gallatin County. 2003. *Gallatin County Growth Policy*. Gallatin County, Bozeman, MT. Adopted April 15, 2003. Available at http://www.gallatin.mt.gov/public_documents/gallatincomt_plandept/Plans&Policies/GrowthPolicyComplete05.pdf (accessed February 2, 2014).

Georgetown News-Graphic. 2008. Hearing settles Royal Spring property, PDR. Georgetown News-Graphic. Available at http://www.news-graphic.com/.

Heimlich, R. 1989. Metropolitan agriculture: Farming in the city's shadow. *Journal of the American Planning Association* 55(4):457–466.

Homer, C., C. Huang, L. Yang, B.K. Wylie, and M. Coan. 2004. Development of a 2001 national land-cover database for the United States. *Photogrammetric Engineering & Remote Sensing* 70(7):829–840.

Hoobler, B.M., G.F. Vance, J.D. Hamerlinck, L.C. Munn, and J.A. Hayward. 2003. Applications of land evaluation and site assessment (LESA) and a geographic information system (GIS) in East Park County, Wyoming. *Journal of Soil and Water Conservation* 58(2):105–112.

Howland, M., and J. Sohn. 2007. Has Maryland's priority funding areas initiative constrained the expansion of water and sewer investments? *Land Use Policy* 24(1):175–186.

Kennedy, C., J. Wilkinson, and J. Balch. 2003. *Conservation Thresholds for Land Use Planners*. Environmental Law Institute, Washington, DC. Available at http://www.eli.org/sites/default/files/eli-pubs/d13-04.pdf (accessed February 4, 2014).

Kentucky Environmental and Public Protection Cabinet. 2002. National Wetlands Inventory—Polygons. Available at ftp://data.gis.eppc.ky.gov/shapefiles/nwi_polygons.zip (accessed November 9, 2013).

Kentucky GAP. 2001. Kentucky GAP land stewardship. Kentucky Fish & Wildlife Information System. The exact data used are no longer available. There are newer data available that have replaced the specific data originally used and cited. The newer data is US Geological Survey, Gap Analysis Program (GAP). November 2012. Protected Areas Database of the United States (PADUS), version 1.3 Combined Feature Class. Available at http://gapanalysis.usgs.gov/padus/data/download/ (accessed November 15, 2014).

Kentucky Geological Survey. 2003. Springs in Kentucky. Available at http://kgs.uky.edu/kgsweb/download/rivers/springs.zip (accessed November 9, 2013).

Kentucky Heritage Council. 2008. *National Register/Survey/Landmarks—Overview*. Frankfort, KY. Available at http://heritage.ky.gov/natreg/ (accessed February 4, 2014).

Kentucky Legislature. n.d.a. Kentucky Revised Statutes. 100.183 Comprehensive plan required. Available at http://www.lrc.ky.gov/statutes/statute.aspx?id=26710 (accessed February 4, 2014).

Kentucky Legislature. n.d.b. Kentucky Revised Statutes. Title IX—Counties, cities, and other local units, Chapter 100—Planning and zoning. Available at http://www.lrc.ky.gov/statutes/index.aspx (accessed February 4, 2014).

Kentucky State Data Center. 2008. Population projections. Available at http://ksdc.louisville.edu/index.php/kentucky-demographic-data/projections (accessed November 9, 2014).

Kentucky State Data Center. 2011. Population projections—Total population. Available at http://ksdc.louisville.edu/images/DemographicData/Projections/hmk2011_total%20pop.xls (accessed February 4, 2014).

Kentucky Transportation Cabinet. 2007. KYTC local roads. KYTC Division of Planning, Commonwealth of Kentucky. Available at ftp://ftp.kymartian.ky.gov/trans/ (accessed November 9, 2014).

Lexington-Fayette Urban County Government. 2000. *Purchase of Development Rights Ordinance No. 4-2000*. Clerk of Urban County Council, Lexington, KY. Available at ftp://ftp.lfucg.com/AdminSvcs/PDR/PDROrdinance.pdf (accessed February 4, 2014).

Lexington-Fayette Urban County Government. 2007. The 2007 Comprehensive Plan for Lexington-Fayette County, Kentucky—Appendix. Division of Planning, Lexington-Fayette Urban County Government. Available at http://www.lexingtonky.gov/modules/ShowDocument.aspx?documentid=1795 (accessed January 22, 2008 and February 4, 2014).

Lexington-Fayette Urban County Government (LFUCG). 2014. Division of Purchase of Development Rights. Available at http://www.lexingtonky.gov/index.aspx?page=497 (accessed October 20, 2014).

Little, C.E. 1984. Commentary: Farmland preservation: Playing political hardball. *Journal of Soil and Water Conservation* 39(4):248–250.

Lord, J.N., and D.A. Norton. 1990. Scale and the spatial concept of fragmentation. *Conservation Biology* 4(2):197–202.

Luckey, D. 1989. The impact of rural water districts on the conversion of agricultural land in Douglas County, Kansas. *Journal of Soil and Water Conservation* 44(3):251–255.

Machado, E.A., D.M. Stoms, F.W. Davis, and J. Kreitler. 2006. Prioritizing farmland preservation cost-effectively for multiple objectives. *Journal of Soil and Water Conservation* 61(5):250–258.

Municipal Research and Service Center of Washington. 1999. *Keeping the Rural Vision: Protecting Rural Character & Planning for Rural Development.* Municipal Research and Services Center of Washington. Olympia, Washington. Available at http://www.commerce.wa.gov/Documents/GMS-Keeping-the-Rural-Vision-3-1999.pdf (accessed November 9, 2014).

National Aeronautics and Space Administration. 2002. The Commonwealth from Space Collection—Kentucky Shuttle Radar Topography Mission (SRTM) Elevation Data Set—February 2000—Kentucky Single Zone. Available at ftp://ftp.kymartian.ky.gov/kls/KYSRTM2000/KYSRTM_2000 (accessed January 27, 2008).

Nellis, L., and J.K. Nicholson. 1985. Utah's "learning" approach to farmland protection. *Journal of Soil and Water Conservation* 40(3):271–273.

NHD. 2003. 24k streams of Kentucky. United States Geologic Survey and United States Environmental Protection Agency. Kentucky Geography Network. Available at http://www.uky.edu/KGS/gis/NHD24DOWN.html (accessed January 25, 2008).

Pease, J., and R. Coughlin. 1996. *Land Evaluation and Site Assessment: A Guidebook for Rating Agricultural Lands*, 2nd edition. U.S. Department of Agriculture's Natural Resources Conservation Service and Soil and Water Conservation Society, Ankeny, IA. Available at http://www.nrcs.usda.gov/Internet/FSE_DOCUMENTS/stelprdb1047455.pdf (accessed February 4, 2014).

Pease, J.R., R.E. Coughlin, F.R. Steiner, A.P. Sussman, L. Papazian, J.A. Pressley, and J.C. Leach. 1994. State and local LESA systems: Status and evaluation. In F.R. Steiner, J.R. Pease, and R.E. Coughlin, eds., *A Decade with LESA: The Evolution of Land Evaluation and Site Assessment*. Soil and Water Conservation Society, Ankeny, IA.

Reganold, J.P. 1986. Prime agricultural land protection: Washington State's experience. *Journal of Soil and Water Conservation* 41(2):89–92.

Richards, W.Q. 1984. Viewpoint: Reconstructing the national conservation program. *Journal of Soil and Water Conservation* 39(3):156–157.

Riitters, K., J. Wickham, R. O'Neill, B. Jones, and E. Smith. 2000. Global-scale patterns of forest fragmentation. *Conservation Ecology* 4(2):3. Available at http://www.consecol.org/vol4/iss2/art3/ (accessed November 9, 2014).

Soil and Water Conservation Society. 2003. *Enhancing LESA: Ideas for Improving the Use and Capabilities of the Land Evaluation and Site Assessment System.* Report from workshop held June 3–4, Nebraska City, NE. Soil and Water Conservation Society, Ankeny, IA. Available at http://www.farmlandinfo.org/sites/default/files/Enhancing_LESA_Report_1.pdf (accessed November 9, 2014).

Tomlin, C.D. 1990. *Geographic Information Systems and Cartographic Modeling*. Prentice-Hall, Englewood Cliffs, NJ.

Tomlin, C.D. 1994. Map algebra: One perspective. *Landscape and Urban Planning* 30(1):3–12.

United States Census Bureau. 2008. Home page. Available at http://www.census.gov/ (accessed January 19, 2008).

United States Census Bureau. 2014. Decennial census. Available at http://www.census.gov/history/www/programs/demographic/decennial_census.html (accessed February 4, 2014).

Van Horn, T.G., G.C. Steinhardt, and J.E. Yahner. 1989. Evaluating the consistency of results for the agricultural land evaluation and site assessment (LESA) system. *Journal of Soil and Water Conservation* 44(6):615–620.

Water Resources Information System. n.d.a. Water lines. Available at http://wris.ky.gov/kygeonet/watlin.zip (accessed January 28, 2008).

Water Resources Information System. n.d.b. Sewer lines. Available at http://wris.ky.gov/kygeonet/sewlin.zip (accessed January 28, 2008).

Williams, T.H.L. 1985. Implementing LESA on a geographic information system: A case study. *Photogrammetric Engineering and Remote Sensing* 51(12):1923–1932.

World Monuments Fund. 2006. World monuments watch list of 100 most endangered sites. Available at http://www.wmf.org/project/bluegrass-cultural-landscape-kentucky (accessed February 4, 2014).

Wright, L.E., W. Zitzmann, K. Young, and R. Googins. 1983. LESA—Agricultural land evaluation and site assessment. *Journal of Soil and Water Conservation* 38(2):82–86.

11 Prioritizing Land with Geographic Information System for Environmental Quality Incentives Program Funding

Derya Özgöç Çağlar and Richard L. Farnsworth

CONTENTS

EXECUTIVE SUMMARY

The U.S. Department of Agriculture offers technical, financial, and educational support to farm and ranch operators through a diverse set of conservation programs designed to lessen environmental and natural resource problems caused by agricultural activities. Numerous internal and external program assessments point to inconsistencies in the application selection processes of these programs, the end result of which is the inefficient distribution of several billion dollars of scarce federal funds and missed environmental and natural resource program goals. Spatial Multiple Criteria Decision Analysis (SMCDA) is particularly suitable for federal conservation programs where location matters and recommendations call for a more disciplined and transparent process to reduce questions about each program's implementation and effectiveness. ArcGIS ModelBuilder is an extremely useful application that program managers can use to make their SMCDA models operational, understand likely impacts of their models prior to implementation, manage programs, and modify their models as programs change. The goals of our research were to

(1) develop a formal SMCDA model that represents Indiana's 2005 Environmental Quality Incentives Program (EQIP) application selection process; (2) create a working version of the Indiana SMCDA model in ModelBuilder; (3) populate it with sample data from 2005, analyze the results, draw conclusions, and make recommendations regarding future research; and (4) provide a hands-on case study with a step-by-step guide. Results show that the pairing of SMCDA and ModelBuilder provides consistency, centralizing control and statewide analysis in application selection for enrollment in federal conservation programs.

KEYWORDS

ArcGIS ModelBuilder, Environmental Quality Incentives Program (EQIP), federal conservation programs, geographic information system (GIS), Spatial Multiple Criteria Decision Analysis (SMCDA)

11.1 INTRODUCTION

Agricultural activities contribute to numerous environmental and natural resource problems, including water and air pollution, soil erosion, and loss of habitat for native plants and animals. To improve the environment and the resources, the US Department of Agriculture (USDA) offers technical, financial, and educational support to farm and ranch operators through a diverse set of conservation programs. Participation is voluntary and usually competitive.

The Conservation Reserve Program (CRP), the Environmental Quality Incentives Program (EQIP), and other smaller federal conservation programs follow a similar approach for accepting applications. Consistent with the legislation, agency personnel identify resource concerns; propose suitable conservation practices or systems for mitigating the concerns; link resource concerns and practices using measurable criteria; and adopt a decision rule for scoring, ranking, and selecting applications for funding through these programs.

Numerous internal and external program assessments point to inconsistencies in the application selection process, the end result of which is the inefficient distribution of several billion dollars of scarce federal funds and missed environmental and natural resource program goals. A report prepared by the Government Accountability Office (Bertoni 2006), for example, critiqued and then recommended that EQIP managers document their rationale for the selection of resource concerns and the weights assigned to them. Other authors (Babcock et al. 1997; Hajkowicz et al. 2007; Soil and Water Conservation Society and Environmental Defense 2007) discuss how decision rules directly affect a program's benefits and costs and ultimately the overall program performance.

A key recommendation found in most of the conservation program assessments is for federal agencies to adopt a more coherent enrollment process that maximizes net benefits given each program's funding level (Reichelderfer and Boggess 1988; Ribaudo et al. 2001; Higgins et al. 2008). A second recommendation encourages agencies to incorporate geographic information system (GIS) into selection processes because environmental and resource problems and solutions are location-specific. The field of SMCDA is useful for developing and diagramming the much-needed formal selection process, and ArcGIS ModelBuilder is a useful tool to make the formal model operational.

Multiple Criteria Decision Analysis (MCDA) is the general field of study that includes decision making in the presence of two or more conflicting objectives and/or decision analysis processes involving two or more attributes (Tecle and Duckstein 1994). SMCDA combines or transforms geographical data (the input) into a decision (the output) (Thinh et al. 2004). An SMCDA process aggregates multidimensional geographical data and information into one-dimensional values for the alternatives (Sharifi and Retsios 2004). MCDA and SMCDA are transparent and formal approaches that are capable of increasing objectivity and consistency, thus generating repeatable, reviewable, revisable, and easy-to-understand results (Smith and Theberge 1987; Guikema and Milke 1999; Lahdelma et al. 2000; Butler et al. 2001; Janssen 2001; Linkov et al. 2004; Machado et al. 2006). SMCDA is particularly suitable for federal conservation programs where location matters and recommendations call for a more disciplined and transparent process to reduce questions about the validity of the analysis (Baker et al. 2001).

ArcGIS ModelBuilder (Esri, Redlands, California) is an extremely useful application that program managers can use to make their SMCDA models operational, understand the likely impacts of their models prior to implementation, manage programs, and modify their models as programs change. ModelBuilder employs a graphical interface that allows users to make their models operational, add data, run them, and summarize results in tables and maps. The integration of SMCDA and ModelBuilder addresses some of the critiques and affords the opportunity to more fully investigate decision rules.

The goals of our research were to develop a formal SMCDA model that represents Indiana's 2005 EQIP selection process, create a working version of the Indiana spatial MCDA model in ModelBuilder, populate it with sample data from 2005, analyze the results, draw conclusions, and make recommendations regarding future research. Because most of the federal conservation programs follow a similar approach, once the first SMCDA model has been developed and then constructed within ModelBuilder, it can be easily modified to fit other similar conservation programs.

11.2 METHODS

Congress enacts conservation legislation to mitigate environmental and natural resource problems. Using this guiding legislation, federal agencies create programs and develop application selection processes for enrolling applicants and distributing federal funds. Though these programs target different resource and environmental problems, they share a common approach. Program managers identify important environmental and resource concerns (an objective); select criteria that are capable of detecting and measuring changes in concerns; assign weights to the concerns to capture each one's relative importance; identify conservation practices that mitigate the concerns; assess the effectiveness of the practices in mitigating the resource concerns; quantify the mitigating impacts of the conservation practices across the selected criteria; and apply a decision rule that allows them to score, rank, and select a subset of submitted applications for funding based on benefits, costs, acres treated, practices applied, or some combination of benefit and cost factors.

SMCDA provides a formal, structured process for solving this decision problem. Broadly defined, SMCDA is concerned with making decisions over the available alternatives that are explicitly described in terms of multiple, usually conflicting, criteria (Yoon and Hwang 1995; Triantaphyllou and Baig 2005).

In 2005, the Indiana Natural Resources Conservation Service (NRCS) created an application selection process that resulted in dividing EQIP applications into two categories: enroll and do not enroll. The selection process started with the identification of resource concerns and ended with a decision rule for ranking submitted applications.

Using the SMCDA process, we framed the Indiana EQIP application selection process and structured the components into a decision hierarchy (Figure 11.1). First-level objectives—water quality, air quality, soil erosion, and species at risk—represent 2005 national priorities that were required to be addressed in every state's EQIP in a cost-effective manner (USDA, NRCS 2002a). Each of the four first-level national priorities was developed into second-level program objectives, resulting in 12 resource concerns that are specific for Indiana (Figure 11.1). A point-allocation method was then used to assess the relative importance of the national priorities and the 12 second-tier objectives for the 2005 Indiana NRCS EQIP. The water quality priority was divided among surface water (10 points), ground water (10 points), and lakes (10 points) concerns. Air quality was assigned 15 points. The national soil erosion priority was addressed by sheet and rill erosion (10 points), ephemeral classic gully (10 points), and a combined soil condition and wind erosion concern (15 points). The species at risk national priority was divided equally among critical aquatic habitat, critical woodland habitat, critical grassland habitat, and critical wetland habitat concerns (6.25 points each). As shown in Figure 11.1, 100 points were distributed among the four national priorities.

Resource concerns vary across the landscape. To improve efficiency and to avoid paying for conservation practices that did not mitigate Indiana's 12 primary resource concerns, Indiana NRCS identified its resource concerns across the landscape and created individual resource concern maps that are displayed in Figure 11.2. Each resource concern map represents the geographical locations of environmentally sensitive areas. These maps correspond to constraint maps since points for applying conservation practices are only awarded if practices are applied in sensitive areas. Maps of three resource concerns—surface water, sheet and rill erosion, and ephemeral and gully erosion—are not shown because they apply to the entire state.

By definition, conservation practices mitigate one or more of the resource concerns. Indiana NRCS turned to its Conservation Practice Physical Effects (CPPE) matrix (USDA, NRCS 2006a) to provide some indication of increased benefits associated with applying one or more conservation practices. The CPPE matrix summarizes each conservation practice's impact on environmental and natural resource problems according to the qualitative scale that ranges from "significant increase in the problem" to "significant decrease in the problem" (Lawrence et al. 1997; USDA, NRCS 2002b, 2006b). As a last step, this categorical scale was converted

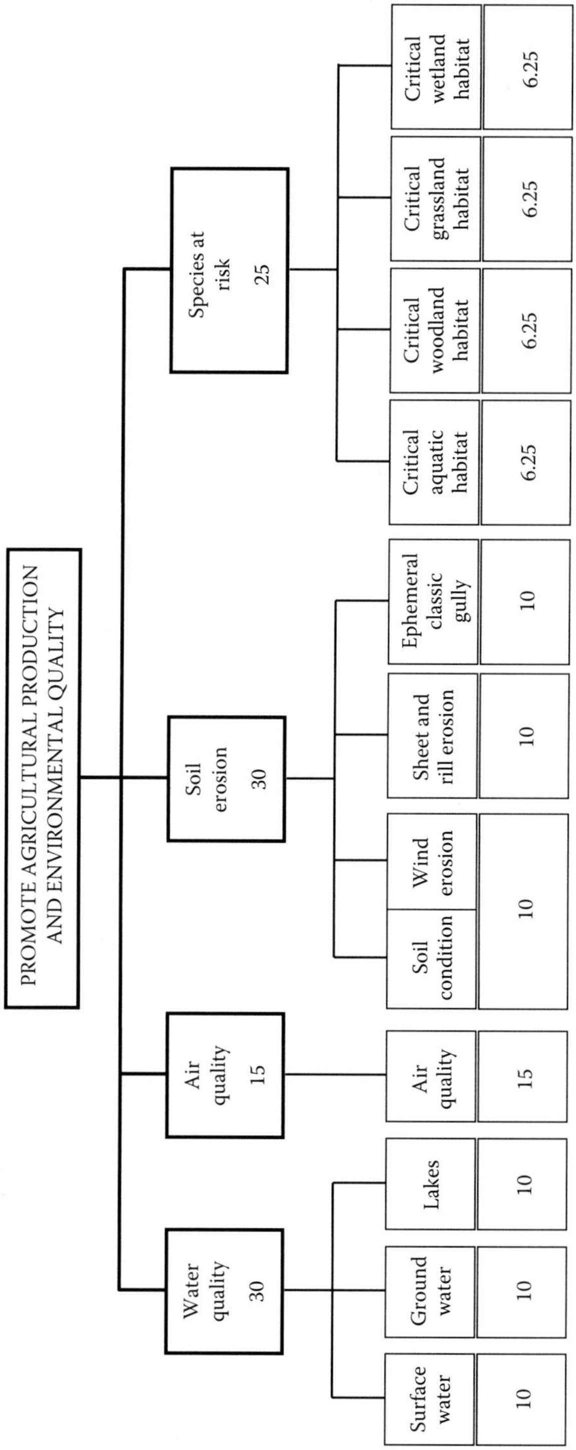

FIGURE 11.1 2005 Indiana EQIP resource concerns hierarchy and weights.

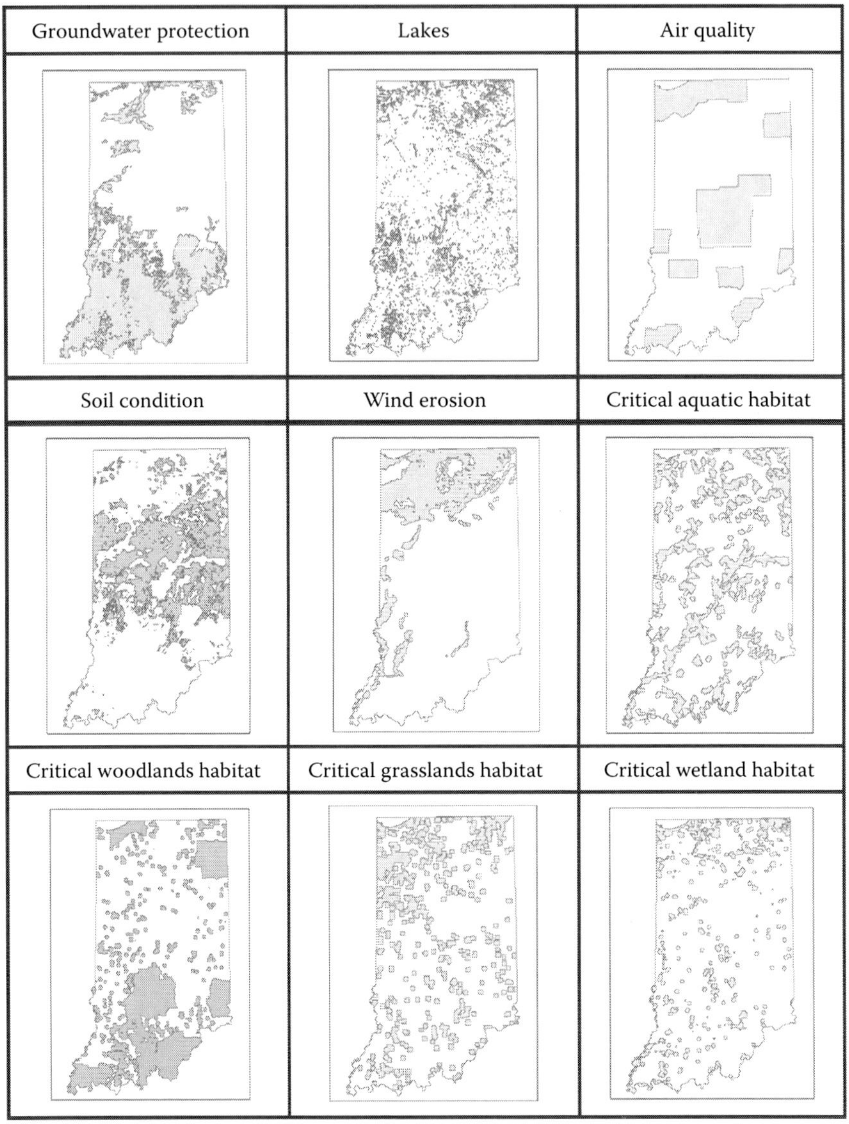

FIGURE 11.2 2005 Indiana EQIP resource concern maps.

to a numerical scale that was judged to be a suitable estimate of expected benefits. Landowners submitting applications that mitigated as many of these resource concerns as possible by applying conservation practices would increase their chances of being accepted for enrollment and funding.

An application's benefit score is determined by summing all weighted outcomes. The weighted outcomes are calculated by multiplying the CPPE-derived benefit scores weighted by each resource concern's relative importance:

$$\mathrm{BS}_i = \left(\left(\sum_{j=1}^{n} w_j s_{ij} y_j \right) \times \mathrm{PL} \right) + \mathrm{LS}, \quad i = 1, \ldots, 800 \tag{11.1}$$

where BS_i = benefit score of application i; j = the resource concerns, j = 1, …, 12; w_j = the weight of the resources concern j; s_{ij} = outcome score of application i against resource concern j; PL = practice lifespan; LS = local score; and y_j = a binary variable (0, 1) denoting the jth resource concern.

Because the benefits increase the longer a practice remains, Indiana NRCS added the multiplicative term PL to boost benefits. If the practice lifespan is 10 years or greater for 50% or more of practices proposed in an application, estimated benefits are multiplied by 1.1. The term LS denotes the points earned from each county's separate benefit scoring rule. The purpose of LS is to recognize that resource concerns and priorities vary from one county to the next and to account for these differences.

Estimation of practice costs is relatively straightforward. The costs of all practices proposed in an application are summed to obtain an estimate of the total costs of implementing an EQIP contract.

$$\mathrm{TC}_i = \sum_{k=1}^{m} c_k, \quad k = 1, \ldots, 86 \tag{11.2}$$

where TC_i = total cost of application i; k = the practices proposed by the application, k = 1, …, 86; and c_k = cost of practice k.

The last step in the completion of the SMCDA model is the decision rule for selecting and funding applications. Benefit ranking, cost ranking, and benefit-to-cost ratio ranking are the three decision rules commonly used. Indiana NRCS adopted a decision rule that is a weighted combination of benefit maximization and benefit–cost maximization. For our purposes, we will use each of these three standard decision rules; these could also be adopted for federal conservation programs.

To illustrate the usefulness and flexibility of SMCDA and ModelBuilder in designing and implementing EQIP and other federal conservation programs, we conducted three analyses using 25 hypothetical EQIP applications. In 2005, Indiana NRCS received 800 applications for the EQIP. To illustrate the model, we decreased this number to 25 and relocated and reshaped the parcels to keep the applications anonymous (Figure 11.3). These hypothetical EQIP applications are from various locations throughout the state and propose a unique set of conservation practices for each location. The first weighted analysis entailed changing the initial weights assigned to the 12 resource concerns, making them equal across all resource concerns, and running the model again. In the second weighted analysis, we increased the weight of the water quality objective to 60% and then equally distributed the remaining 40% among the rest of the objectives. Finally, in the third decision rule analysis, the 25 representative applications were entered, and results for each of the three decision rules (benefit ranking, cost ranking, and benefit-to-cost ratio ranking) embedded in the existing ModelBuilder application selection tool were generated.

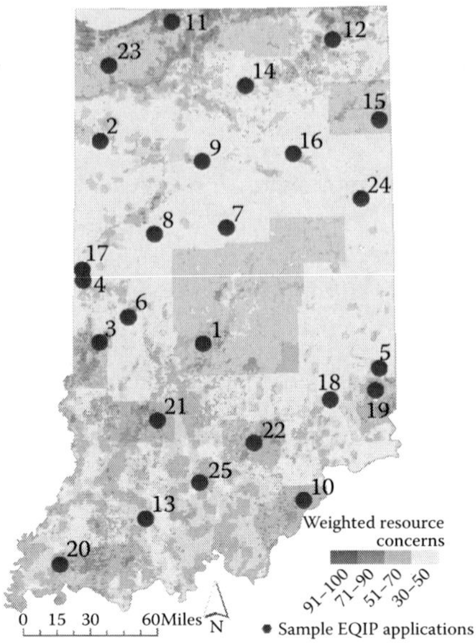

FIGURE 11.3 2005 Indiana EQIP resource concern concentration map using the initial 2005 selection process weights. Areas of highest priority are valued between 91 and 100. Locations of 25 hypothetical applications used in this study are shown as large dots.

11.3 RESULTS AND DISCUSSION

11.3.1 Weighting Analyses

Figure 11.3 illustrates resource concern area concentration with the initial 2005 selection process weightings. In Figure 11.4, the resource concern concentration areas are compared for equal weighting of all objectives and for heavily weighting the water quality objective (surface water, ground water, and lakes). Higher values indicate that the resource concern concentrations are higher in those areas, and hence, they hold higher priority. Comparison of the three maps shows how resource concern concentrations change when the relative importance of those concerns is changed. Air quality, for example, tends to make certain counties more prominent because its resource concern map is based on county boundaries in the majority of cases. The impact of air quality becomes less noticeable and disappears as its relative importance decreases steadily from equal weighting of resource concerns (Figure 11.4a) to heavily weighting water quality (Figure 11.4b). Subtle changes in weights assigned to resource concerns can significantly change the scoring and ranking of applications.

The use of ModelBuilder allows users to change the weights and quickly view the results. This could be very useful when a committee of decision makers works together to value resource concerns. Table 11.1 indicates how the rank of the 25 hypothetical applications changed when the resource concern weights changed. For example, Application 17 was ranked fifth under the 2005 application selection weighting scheme, fourth with equal weighting, and first when the water quality objective

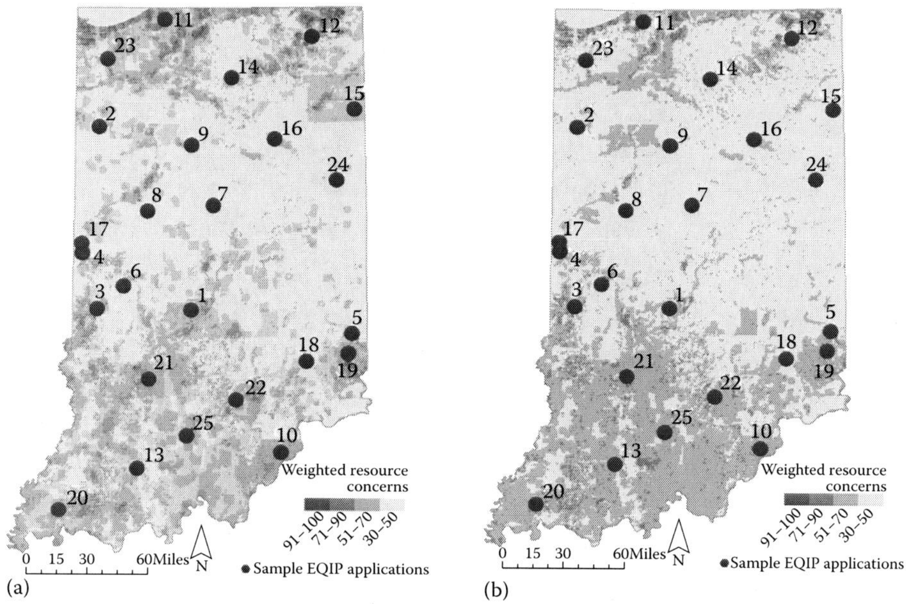

FIGURE 11.4 2005 Indiana EQIP resource concern concentration maps with (a) equal weighting and (b) water quality priority (surface water, ground water, and lakes) heavily weighted. Areas of highest priority are valued between 91 and 100. Locations of 25 hypothetical applications used in this study are shown as large dots.

was weighted heavily. Similarly, application 10 is ranked thirteenth under the 2005 application selection and equal weighting scheme, whereas its rank decreased to twentieth when the water quality objective weight is increased.

The above comparisons of weighting results served two purposes. First, the weights assigned to resource concerns can significantly influence the selection of applications for enrollment. Second, ModelBuilder is extremely flexible, making it easy to incorporate and quickly assess the implications of a different weighting scheme, more or fewer resource concerns, different decision rules, and other tweaks to the SMCDA model. The weights of resource concerns affect the conceptualized environmental problem concentration areas and indicate where limited resources need to be directed.

11.3.2 Decision Rules Analysis

This illustrative analysis includes each of the three commonly used decision rules: benefit ranking, cost ranking, and benefit-to-cost ratio ranking. This provides us the opportunity to test how ranking changes with various decision rules and to analyze how well the enrolled applications target the intense resource concern areas.

The selection of a particular decision rule defines how conservation programs integrate benefits and costs into the selection process. In theory, applications should be ranked based on the total benefits provided, the total project costs, or the benefits

TABLE 11.1
Application Ranks under the Benefit Ranking Decision Rule for Three Different Weighting Schemes: Initial 2005 Selection Process Weighting Scheme, Equal Weighting Scheme, and Heavily Weighted Water Objective Scheme

	2005 Selection Process Weighting	Equal Weighting	Heavy Water Quality Objective Weighting
Application	Rank Order	Rank Order	Rank Order
1	25	25	25
2	24	24	23
3	21	21	19
4	15	14	14
5	19	20	22
6	23	23	24
7	18	17	16
8	22	22	17
9	20	18	21
10	13	13	20
11	16	16	18
12	17	19	15
13	11	11	10
14	14	15	13
15	12	12	12
16	10	10	8
17	5	4	1
18	8	8	6
19	7	7	7
20	2	2	3
21	6	6	9
22	3	3	4
23	9	9	11
24	1	1	2
25	4	5	5

per dollar expended. Rankings of applications by the decision rule and the 2005 selection process guidelines and weighting scheme are shown in Table 11.2.

The top-ranked applications for each decision rule and weighting scheme combination are illustrated in Table 11.3. This table summarizes the comparisons between three decision rules and three weighting schemes. Different applications are ranked higher for different combinations. When the weight of the water quality objective is increased heavily, the ranking of the applications has changed, and application 17 replaces application 24 in the benefit ranking by attaining a higher benefit score. Similarly, application 2 replaces application 18 as the highest benefit-to-cost ratio ranking when the water quality objective is weighted more heavily. When the

TABLE 11.2
Application Ranks under Initial 2005 Selection Process Weighting Scheme and with Three Decision Rules: Benefit Ranking, Cost Ranking, and Benefit-to-Cost Ratio Ranking

	Benefit Ranking	Cost Ranking	Benefit-to-Cost Ratio Ranking
Application	Rank Order	Rank Order	Rank Order
1	25	1	2
2	24	2	3
3	21	5	6
4	15	15	20
5	19	18	23
6	23	4	7
7	18	13	19
8	22	19	25
9	20	6	5
10	13	7	10
11	16	8	11
12	17	14	22
13	11	25	24
14	14	11	12
15	12	16	18
16	10	24	21
17	5	20	14
18	8	3	1
19	7	17	13
20	2	23	16
21	6	22	17
22	3	9	4
23	9	10	8
24	1	12	9
25	4	21	15

TABLE 11.3
Top-Ranked Applications under Each Weighting Scheme and Decision Rules

	Decision Rules		
	Benefit Ranking (Table 11.1)	Cost Ranking	Benefit-to-Cost Ratio Ranking
Weighting			
2005 selection process weighting (Table 11.2)	Application 24	Application 1	Application 18
Equal weighting	Application 24	Application 1	Application 18
Heavy water quality objective weighting	Application 17	Application 1	Application 2

decision rule has changed from benefit maximization to benefit-to-cost ratio ranking, application 24 is replaced by application 18. The rank of applications under cost ranking has not changed since the total costs of applications remained the same. These results indicate the benefit of combining SMCDA and ModelBuilder.

11.4 GIS APPLICATION

The objective of this modeling effort is the development of a ModelBuilder application selection tool that incorporates the critical components found in the application selection processes of federal agricultural conservation programs. With minor modifications, this ModelBuilder tool may be used in multiple federal conservation programs and at a variety of scales to select applications that provide the greatest environmental benefits. To illustrate the tool's general applicability and robustness, Indiana's 2005 EQIP application selection process is modeled. The tool and the input data are available on the CRC Press website at http://www.crcpress.com/prod/isbn/9781439867228 to try the model.

The existing Indiana ModelBuilder application-selection tool has considerable flexibility that can be used to conduct sensitivity analyses or to modify the existing EQIP.

- *Sensitivity analysis—Modification of the weights of resource concerns*: From the dialog box, the weight of each resource concern could be changed to determine if certain resource concerns significantly change the ranking and selection of applications for funding in Indiana's EQIP.
- *Sensitivity analysis—Replacement of existing resource concern maps with more detailed maps*: Are binary resource concern maps sufficient? Will more detailed resource concern maps significantly change the ranking and selection of applications for funding? These questions can be answered by replacing the existing maps with more detailed maps and running ModelBuilder again for the Indiana EQIP.
- *Sensitivity analysis—Replacement of qualitative practice scores with quantitative scores*: A major criticism of EQIP has been the use of expert opinion and a limited number of categories (0 to 4). More categories could be added, and the application selection tool could be rerun to determine if the ranking and selection of applications changed significantly for the Indiana 2005 EQIP. If quantitative impacts such as tons of soil saved are available for conservation practices, the qualitative rankings could be replaced. Again, a sensitivity analysis could be conducted to determine if the quantitative variables improved program performance. For both of these analyses, you would need to update the tabular data of EQIP applications.
- *Update modifications*: If the program does not change from year to year, but the basic data do, updating is relatively straightforward. New applications could be added, or some applications could be deleted from the EQIP application file. The local scores, costs, and practice scores in the table of the EQIP application data layer could be replaced or updated; for example, costs can be updated by replacing the existing "TotalCost" field with the new data.

The ModelBuilder application selection tool is extremely flexible and can be modified to fit EQIPs in other states or new conservation programs. The key concept here is that the tool provides the basic framework and code, so the user expands or contracts the code to build a new state tool. Once it is built, all of the above sensitivity analyses can be conducted.

- *Modify the existing Indiana EQIP tool for use by other states*: States have the flexibility to keep the same resource concern definitions but just modify the spatial designation of the maps, or they can introduce new resource concerns or delete some of them. Additionally, if they desire, they can introduce new decision rules by editing the tool.
- *Modify the existing Indiana EQIP tool to use in a new conservation program*: The resource concern description could be different for new conservation programs. The tool could be edited, by keeping the main structure the same, to fit the requirements of the new programs.

11.5 SUMMARY

Each year, thousands of agricultural producers voluntarily submit applications to federal agencies to participate in numerous conservation programs. To select and fund the applications that provide the greatest environmental benefits, the agency personnel developed an application selection process that assigns points and ranks applications based on impacts and costs. This study has illustrated the development of a formal SMCDA model to structure conservation program designs and operation of the SMCDA model in ModelBuilder. To show the SMCDA model and the ArcGIS ModelBuilder's usefulness and flexibility, we structured the 2005 Indiana EQIP and then created a fully functional ModelBuilder application selection tool that functionalizes and automates Indiana NRCS's selection process.

The specification of an SMCDA model helps policy makers and program administrators think systematically and hopefully improve the quality of the results of their decisions. Systematic and structured analysis includes identifying the objectives to achieve the desired program goal and assigning appropriate attributes to attain those objectives in a hierarchical framework; searching out proper and valid methods of measuring attribute effectiveness; eliciting the relative importance of program objectives and attributes; and identifying how to aggregate and collapse performance measurements into a final value that is useful to score and rank applications.

SMCDA allows a transparent approach that increases objectivity and consistency, and generates results that can be repeatable, reviewable, revisable, and easy to understand. As this analysis has demonstrated, both the policy makers and the applicants clearly recognize how the scores are calculated.

ModelBuilder creates a framework that allows users, even inexperienced ones, to easily operate the model and assess the implications of various program elements such as the objectives, attributes, weights, scoring, and decision rules. The tool allows the decision maker to add or delete applications and can include applications from 25 to 800 or more. Also, the user could modify and update the resource concern maps without affecting the tool's operations. Whatever changes are required, it takes a few minutes to complete a comprehensive model run and to automatically update the results based on modifications.

The ModelBuilder application selection tool also provides the opportunity to assess the likely results of a particular program before the first application is submitted. This significantly expands the usefulness of the ModelBuilder tool from scoring, ranking, and selecting applications for funding to examining the possible impacts of different application processes and decision rules.

Finally, a front-end formal model allows us to avoid common pitfalls such as incorrect specification of objectives and attributes and unintended objective weighting. Visual and hierarchical presentations of program objectives and attributes drive policy makers to reanalyze the criteria as well as to clarify the reasons why they are selected and how they contribute to achieve the desired objectives and goals.

The SMCDA model and the ModelBuilder framework allow decisions to be repeated, revised, and reviewed over and over. They include documentation of the process, parameters, requirements, and final results. To create functional application selection processes for other states or for other federal conservation programs, agency administrators only need to update and modify the program component and requirements—objectives, attributes, weights, and decision rules—and tweak the generic ModelBuilder framework for their specific needs.

REFERENCES

Babcock, B.A., P.G. Lakshminarayan, J. Wu, and D. Zilberman. 1997. Targeting tools for the purchase of environmental amenities. *Land Economics* 73 (3):325–339.

Baker, D., D. Bridges, R. Hunter, G. Johnson, J. Krupa, J. Murphy, and K. Sorenson. 2001. *Guidebook to Decision-Making Methods*. Department of Energy. Available at http://kscsma.ksc.nasa.gov/Reliability/Documents/Decision_Making_Guidebook_2002_Dept_of_Energy.pdf (accessed October 16, 2014).

Bertoni, D. 2006. *Agricultural Conservation: USDA Should Improve Its Process for Allocating Funds to States for the Environmental Quality Incentives Program*, United States Government Accountability Office. Available at http://www.gao.gov/assets/260/251620.pdf (accessed October 16, 2014).

Butler, J., D.J. Morrice, and P.W. Mullarkey. 2001. A multiple-attribute utility theory approach to ranking and selection. *Management Science* 47 (6):800–816.

Guikema, S., and M. Milke. 1999. Quantitative decision tools for conservation programme planning: practice, theory and potential. *Environmental Conservation* 26 (3):179–189.

Hajkowicz, S., A. Higgins, K. Williams, D.P. Faith, and M. Burton. 2007. Optimisation and the selection of conservation contracts. *The Australian Journal of Agricultural and Resource Economics* 51 (1):39–56.

Higgins, A.J., S. Hajkowicz, and E. Bui. 2008. A multi-objective model for environmental investment decision making. *Computers and Operations Research* 35 (1):253–266.

Janssen, R. 2001. On the use of multi-criteria analysis in environmental impact assessment in the Netherlands. *Journal of Multi-Criteria Decision Analysis* 10 (2):101–109.

Lahdelma, R., P. Salminen, and J. Hokkanen. 2000. Using multicriteria methods in environmental planning and management. *Environmental Management* 26 (6):595–605.

Lawrence, P.A., J.J. Stone, P. Heilman, and L.J. Lane. 1997. Using measured data and expert opinion in a multiple-objective decision support system for semiarid rangelands. *Transactions of the ASAE* 40 (6):1589–1597.

Linkov, I., A. Varghese, S. Jamil, T. Seager, G. Kiker, and T. Bridges. 2004. Multi-criteria decision analysis: A framework for structuring remedial decisions at contaminated sites. In *Comparative Risk Assessment and Environmental Decision Making*, eds. I. Linkov and A.B. Ramadan, 15–54. Dordrecht: Springer.

Machado, E.A., D.M. Stoms, F.W. Davis, and J. Kreitler. 2006. Prioritizing farmland preservation cost-effectively for multiple objectives. *Journal of Soil and Water Conservation* 61 (5):250–259.

Reichelderfer, K.H., and W.G. Boggess. 1988. Government decision making and program performance: The case of the conservation reserve program. *American Journal of Agricultural Economics* 70 (1):1–11.

Ribaudo, M.O., D.L. Hoag, M. Smith, and R. Heimlich. 2001. Environmental indices and the politics of the conservation reserve program. *Ecological Indicators* 1 (1):11–20.

Sharifi, M.A., and V. Retsios. 2004. Site selection for waste disposal through spatial multiple-criteria decision analysis. *Journal of Telecommunications and Information Technology* 3:1–11.

Smith, P.G.R., and J.B. Theberge. 1987. Evaluating natural areas using multiple criteria: Theory and practice. *Environmental Management* 11 (4):447–460.

Soil and Water Conservation Society and Environmental Defense. 2007. *Environmental Quality Incentives Program (EQIP) Program Assessment*. Available at http://www.swcs.org/documents/filelibrary/EQIP_assessment.pdf (accessed January 9, 2014).

Tecle, A., and L. Duckstein. 1994. Concepts of multicriterion decision making. In *Multicriteria Decision Analysis in Water Resources Management*, eds. J.J. Bogardi and H.P. Nachtnebel, 33–62. Paris: UNESCO Press.

Thinh, N.X., U. Walz, J. Schanze, I. Ferencsik, and A. Goncz. 2004. GIS-based multiple-criteria decision analysis and optimization for land suitability evaluation. Paper presented at Simulation in Umwelt- und Geowissenschaften, Workshop Müncheberg, Aachen Shaker Verlag.

Triantaphyllou, E., and K. Baig. 2005. The impact of aggregating benefit and cost criteria in four MCDA methods. *IEEE Transactions on Engineering Management* 52 (2):213–226.

USDA, NRCS. 2002a. *Conservation Programs Manual*. Washington, DC: USDA.

USDA, NRCS. 2002b. *Risk Assessment for the EQIP Program*. Washington, DC: USDA.

USDA, NRCS. 2006a. *Field Office Technical Guide*. Washington, DC: USDA.

USDA, NRCS. 2006b. *National Conservation Practice Standards*. Washington, DC: USDA.

Yoon, K.P., and C.L. Hwang. 1995. *Multiple-Attribute Decision Making: An Introduction*. Sage University Papers Series, Quantitative Applications in the Social Sciences No. 07-104. Thousand Oaks, CA: Sage Publications.

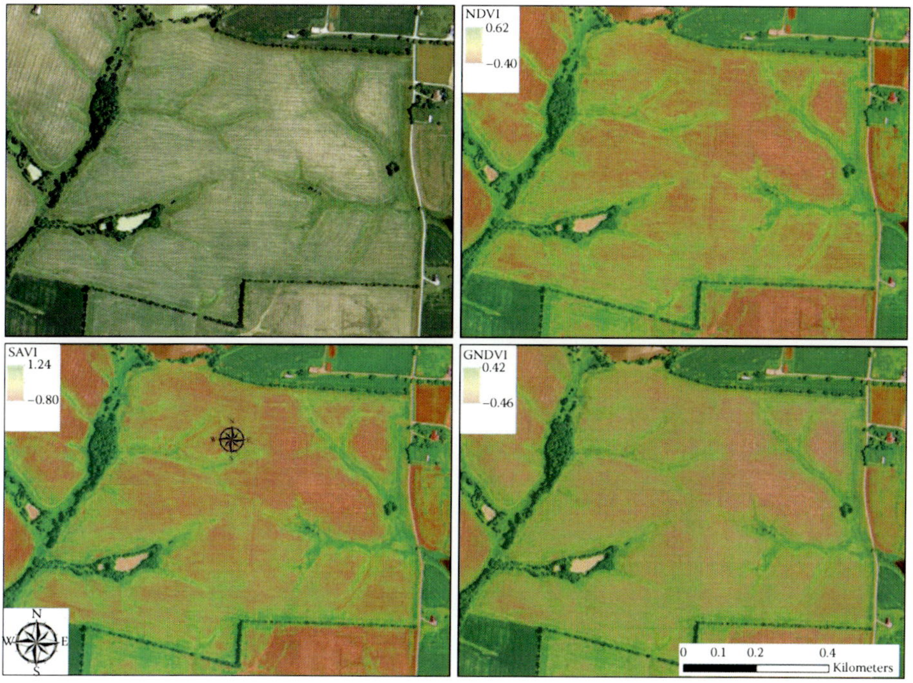

FIGURE 2.12 Example output from three different methods of calculating vegetative index (NDVI, SAVI, and GNDVI) using USGS NAIP imagery in ArcGIS. Detailed instructions are given in the online exercise.

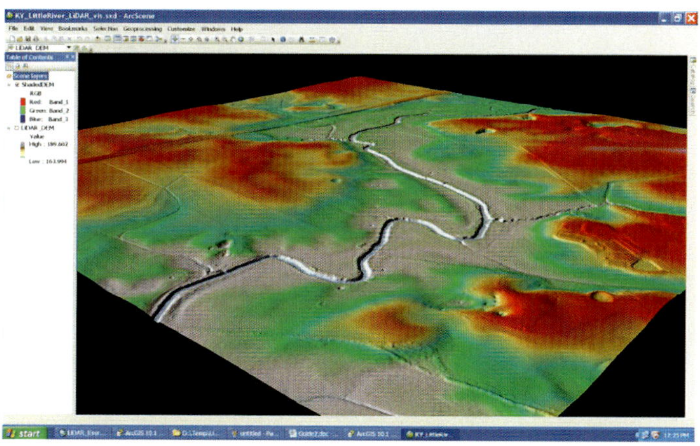

FIGURE 2.14 3-D rendering of a LiDAR-derived hillshade map overlain by elevation contours calculated from the USGS DEM data using ArcGIS. Detailed instructions are given in the online exercise.

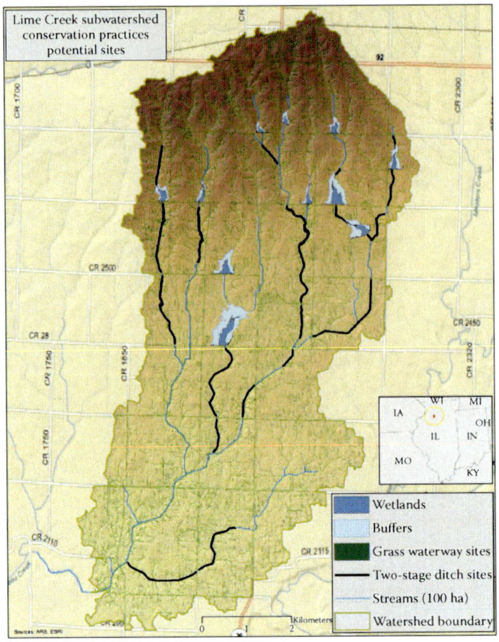

FIGURE 4.4 Map of Lime Creek watershed illustrating a conservation planning scenario that employs three practices to identify potential artificial wetland impoundment areas for shallow flood.

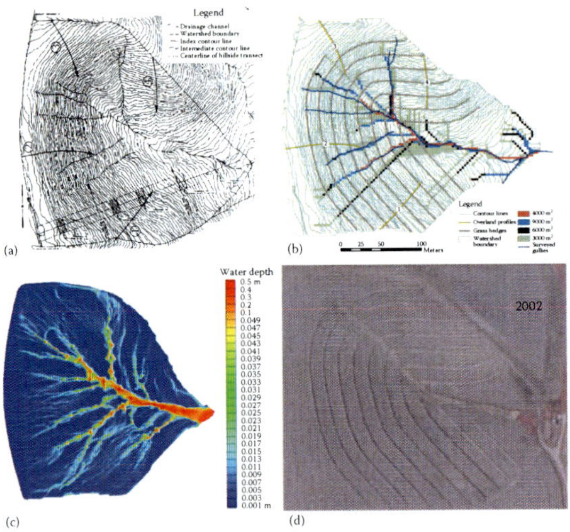

FIGURE 5.2 (a) Topographic map (0.31-m contour interval) of Watershed 11 at Treynor, Iowa, illustrating appropriate RUSLE hillslope flow paths (From Renard, K.G. et al., *Predicting Soil Erosion by Water: A Guide to Conservation Planning with the Revised Universal Soil Loss Equation (RUSLE)*. Agric. Handbook 703, U.S. Department of Agriculture–Agricultural Research Service: Washington, DC, Figure 4-5B, 1997); (b) concentrated flow paths created from a 3-m DEM for alternative minimum contributing areas, indicating that 600 m^2 is the best approximation of the location of concentrated flow areas that end RUSLE hillslope flow paths; (c) water flow depths after 1 h of steady runoff of 50 mm h^{-1} based on a 0.5-m DEM, anisotropic roughness, and a diffusive wave solution (From Vieira, D.A.N., and S.M. Dabney, *Hydrological Processes*, 26:2225–2234, 2012); and (d) aerial photo showing the locations of grass hedges and waterways in 2002.

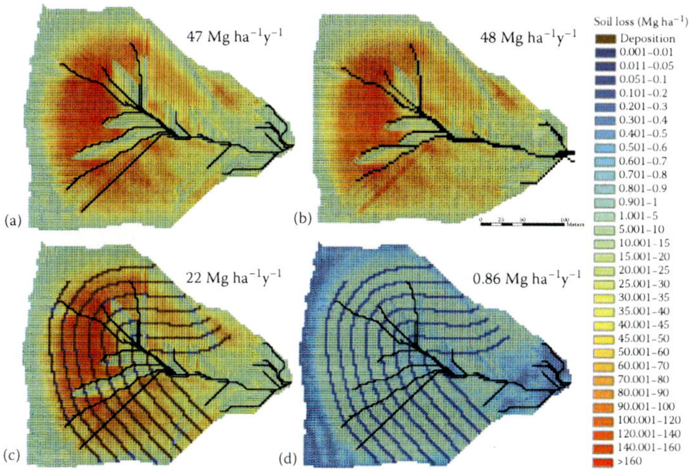

FIGURE 5.3 Flow vectors and patterns of annual RUSLE2 soil loss on a 3-m grid and the field average soil loss for corn yielding 7 Mg ha^{-1} with spring-plow tillage (a, b), with grass hedges (c), or no-till corn with grass hedges (d). Black lines show concentrated flow channels based on 600-m^2 minimum contributing area using a D8 flow-routing algorithm (a, c, d) or using the channel network generation algorithm being developed by Agren, Inc. (b).

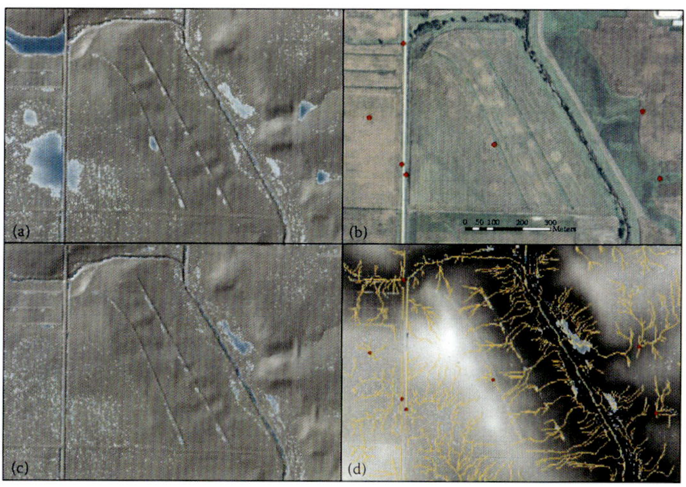

FIGURE 5.4 Shaded relief map of fields in Boone County, Iowa, (41.9315′N, 93.7623′W) showing depths of filled pits (in shades of blue) when flow vectors were determined with a standard D8 algorithm with full pit filling for a 3-m DEM derived from LiDAR (a); color orthophoto showing locations of seven user-selected pixels that were declared "NoData" (b); filled pits resulting from processing the DEM with the seven "NoData" pixels (c); and an elevation map showing a channel network based on 675-m^2 minimum contributing area using a D8 flow-routing algorithm (d).

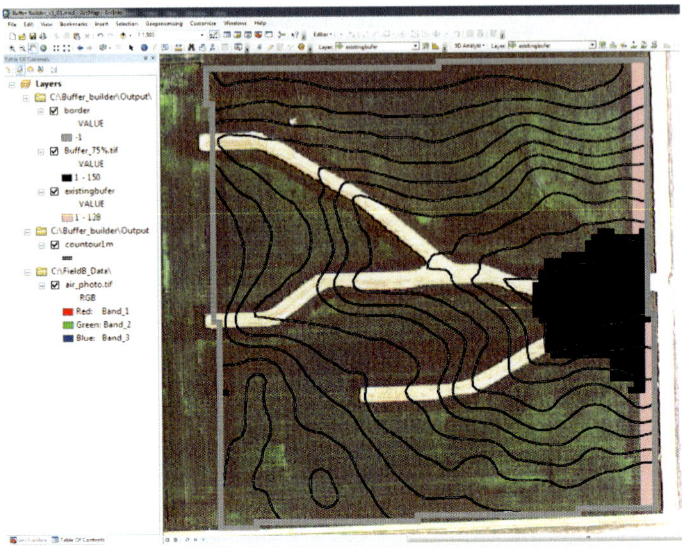

FIGURE 6.5 The display after executing the Buffer Builder Main tool to run both the design and the assessment procedures. The map shows five output layers: designed filter areas, assessed filter polygons, field border raster, and topographic contours on the photo of the field.

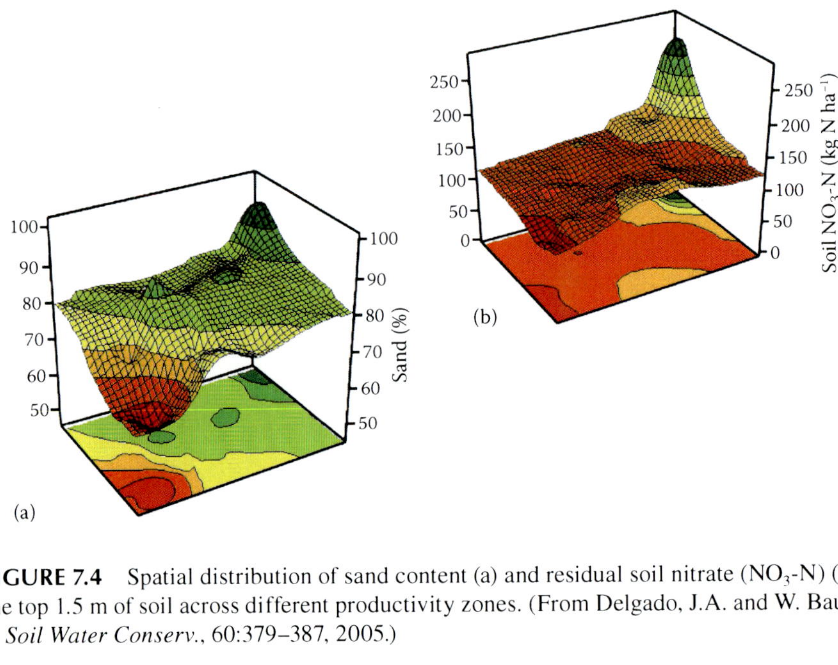

FIGURE 7.4 Spatial distribution of sand content (a) and residual soil nitrate (NO_3-N) (b) in the top 1.5 m of soil across different productivity zones. (From Delgado, J.A. and W. Bausch, *J. Soil Water Conserv.*, 60:379–387, 2005.)

FIGURE 7.5 NLEAP GIS 5.0 Project example showing that the image can be zoomed out so that the entire landscape is visible. Evaluation with color ramping allows identification of the hot spots in the landscape with high risks of nitrogen losses.

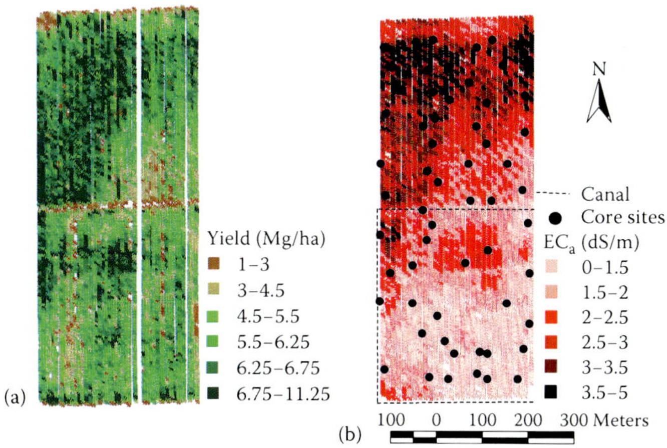

FIGURE 8.4 Maps of (a) cotton yield and (b) EC_a measurements including the locations of the 60 soil core sites. (Modified from Corwin, D.L. et al., *Agronomy Journal* 95(2):352–364, 2003b. With permission.)

FIGURE 8.5 Mobile GPS-based ER equipment showing (a) fixed-array tool bar holding four ER electrodes and (b) a close-up of one of the ER electrodes.

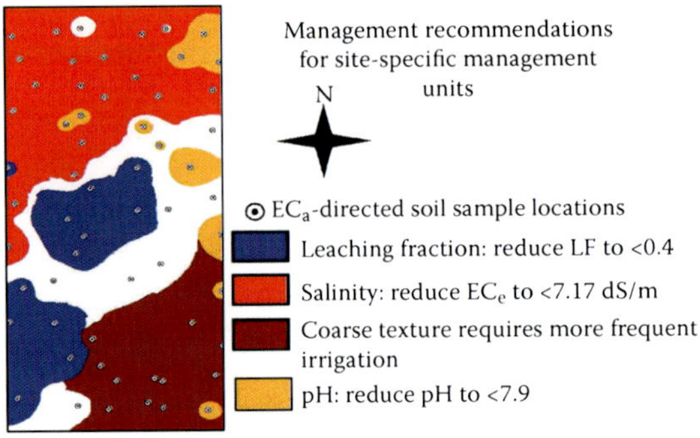

FIGURE 8.9 Site-specific management units for a 32.4-ha cotton field in the Broadview Water District of central California's San Joaquin Valley. Recommendations are associated with the SSMUs for LF, salinity, texture, and pH. (From Corwin, D.L. and S.M. Lesch, *Computers and Electronics in Agriculture*, 46:11–43, 2005. With permission.)

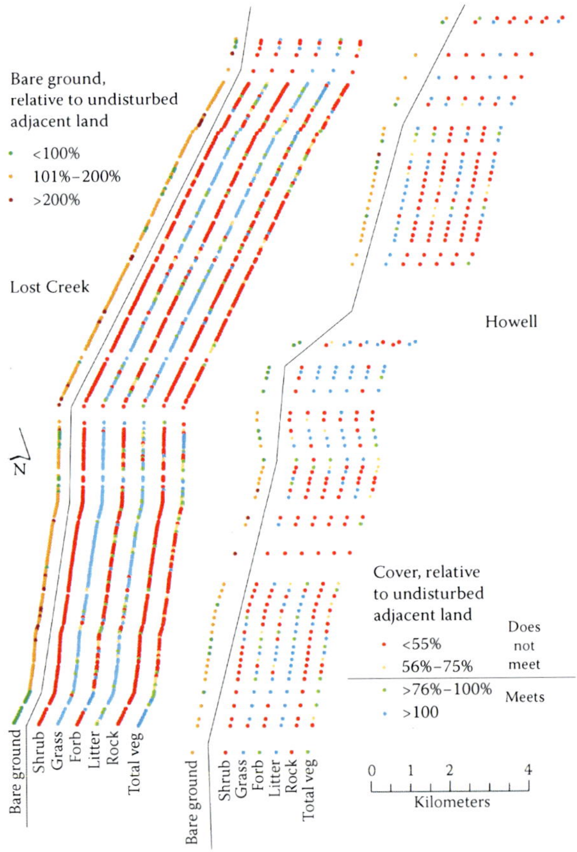

FIGURE 9.2 Cover values for bare ground, shrub, grass, forb, litter, rock, and total vegetation (shrub, forb, and grass) as a percentage of adjacent, undisturbed land cover values. Values were measured from ~5 × 8 m FOV aerial images and plotted to scale, providing a quick and comprehensive view of reclamation progress along two ~35-km stretches of pipeline ROWs in central Wyoming. Cover breaks indicate areas that meet or do not meet the vegetative cover requirements established by the Plans of Development. Skips in the Howell pipeline sequence are the result of technical problems encountered during that survey. (From Booth, D.T. and S.E. Cox, *Environmental Monitoring and Assessment* 158:23–33, 2009.)

FIGURE 9.3 Leafy spurge captured in 10-mm GSD (a) and 1-mm GSD (b) aerial images. The small rectangle shows coverage of the nested 1-mm GSD image. (From Booth, D.T. et al., *Native Plant Journal* 11:327–339, 2010.)

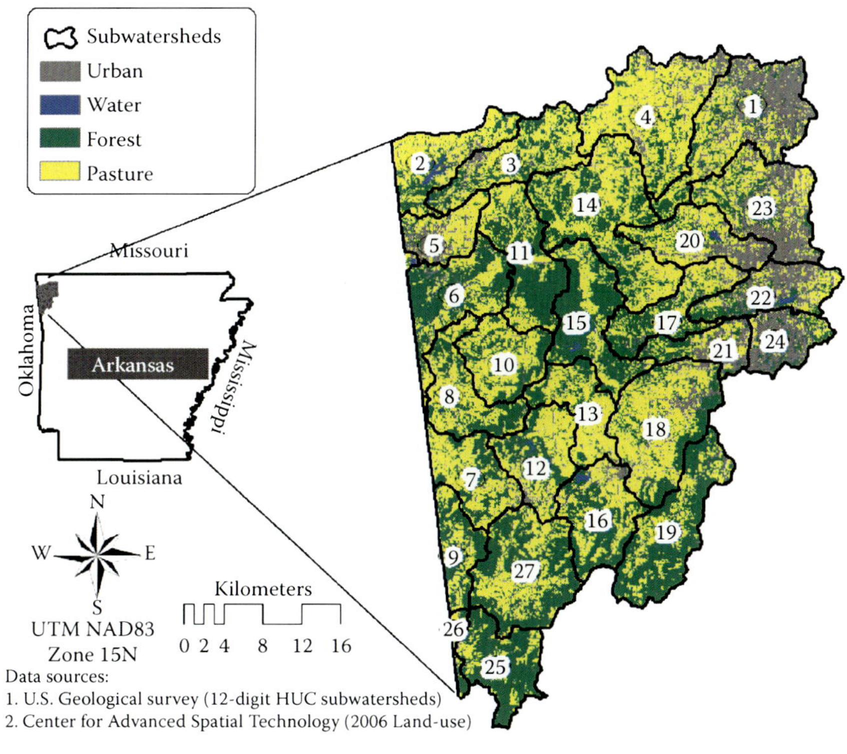

FIGURE 12.2 IRDAA watershed location and 2006 land-use.

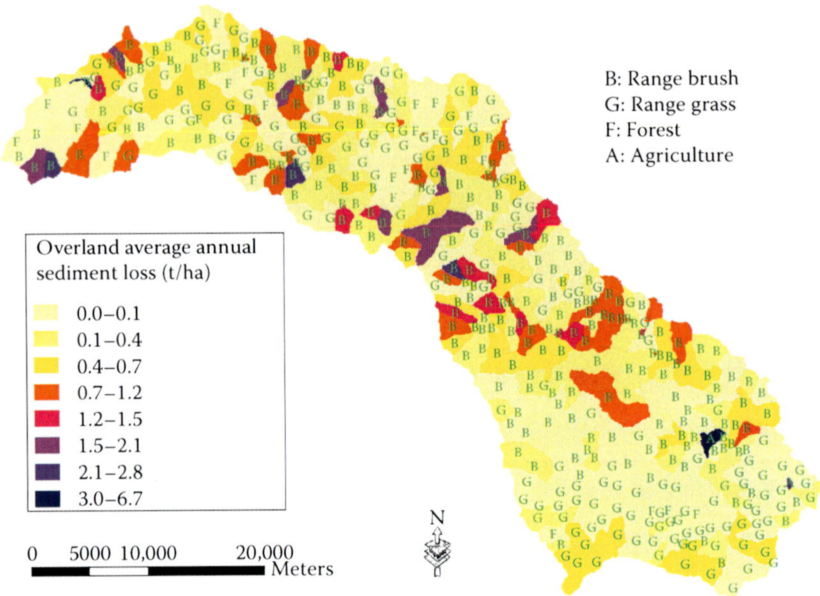

FIGURE 13.5 Average annual sediment losses (1958–2008) from each subarea in the study Cowhouse watershed.

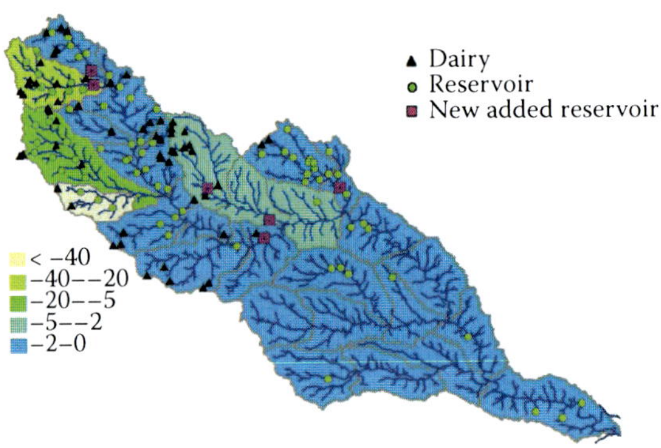

FIGURE 13.10 Simulated total P% change between baseline and the added reservoirs (all manure applied) for each 12-digit watershed in the North Bosque watershed.

12 Integrating Land Use Change Influences in Watershed Model Simulations

*Naresh Pai and Dharmendra Saraswat**

CONTENTS

EXECUTIVE SUMMARY

Watershed models are useful tools to help understand the impact of land use on hydrology. The 2010 release of the Soil and Water Assessment Tool (SWAT) model (SWAT2009) has been provided with a Land Use Update (LUU) module that allows land use to be updated dynamically during simulation runs. In this chapter, we explain the working of the LUU module in SWAT2009 and discuss features of SWAT2009 Land-use Change (SWAT2009_LUC), an interactive tool for developing input files for use by the LUU module. The utility of LUU is demonstrated using a hands-on exercise involving the SWAT2009_LUC tool through a case study conducted in a watershed that is being urbanized in Northwest Arkansas.

* Contact author: dsaraswat@uaex.edu.

KEYWORDS

Hydrologic changes with urbanization, land cover changes, land use changes, Soil and Water Assessment Tool model

12.1 INTRODUCTION

Economic and social conditions in a region are key drivers to the utility of its landscapes. Changes in landscape utility could result in increased urbanization, deforestation, and/or shifts in agricultural practices. These shifts in landscape usage patterns are typically referred to as land-use changes (LUCs). LUCs can have major influences on hydrological, sediment, and nutrient production and transport in a watershed (DeFries and Eshleman 2004; Ahearn et al. 2005; Gitau et al. 2010). Negative consequences may involve flooding, changes in water demand due to urbanization, and loss of ecosystem services due to water quality degradation. In addition, any positive impact gained from implementing conservation practices can be negated by concurrent LUCs (Gitau et al. 2010). Watershed managers and policy makers frequently need to assess the influences of LUCs on water resources but may lack adequate data and information to evaluate this over large spatial and temporal scales. In fact, DeFries and Eshleman (2004) identified LUC as a major emerging issue of the coming century. A literature review of studies dealing with both LUC and climate change has suggested that LUC has had a greater impact on ecological parameters than climate change, contrary to popular perception (Dale 1997).

Mathematical models provide a necessary framework for analyzing interactions between LUCs and water resources because they are comprehensive yet cost-effective. The U.S. Office of Technology Assessment (OTA 1982) reported that the models have the capacity to improve accuracy and to increase effectiveness of information available to managers, decision makers, and scientists. For instance, using a watershed model, Thomas et al. (2009) simulated the ongoing and expected growth of U.S. corn production in the Midwest and reported significant increases in nutrient runoff, erosion, and pesticide losses from agricultural fields. Similarly, Bhaduri et al. (2001) simulated the urbanization phenomenon using the Long Term Hydrologic Impact Analysis (L-THIA) model in an Indiana watershed and reported that an increase in impervious area of only 18% resulted in 80% increase in annual average runoff volume and greater than 50% increases in annual loads of lead, copper, and zinc. Examples such as these substantiate the utility of models in quantifying the hydrological and water quality influences from LUCs.

Among various mathematical models, distributed models have proven useful for a variety of water resource issues (Singh and Frevert 2006). The underlying concept of distributed models is that they divide a watershed into smaller areas to better represent the spatial variation in climate, land use, soil, topography, and management operations. Examples of distributed models include the erosion impact calculator (Williams et al. 1984), precipitation–runoff modeling system (PRMS) (Leavesley et al. 1983), hydrological simulation program—FORTRAN (Bicknell et al. 1997), soil and water assessment tool (SWAT) (Arnold et al. 1998), European Hydrologic System (MIKE-SHE) (Bathurst 1986), and Modelo de Erosão Fisico e Distribuido

(Nunes et al. 2005). The focus of this chapter is the SWAT model, developed by the USDA Agricultural Research Services (USDA-ARS) (Arnold et al. 1998) and used worldwide (Gassman et al. 2007; Tuppad et al. 2011).

Land use and vegetation cover, which are important inputs in distributed models, are frequently estimated with remotely sensed data and geographical information systems (GISs) (Vieux 2004). Thematically classified maps of a watershed, depicting land use/land cover (LULC) at a particular instant during the study period, have proven successful for supplying land use spatial distribution information to several watershed models (Singh and Woolhiser 2002), including SWAT. However, LULC maps, like most other input datasets, could also be a source of model errors. Inaccuracies in LULC data can result from different stages of the production cycle including, but not limited to, multispectral imagery that is used for LULC development, image classification techniques, and errors in ground-truth data used for training the classifier (Jensen 1996). LULC errors, resulting from spatial inaccuracies, can propagate to distributed model simulations; this error has been quantified by several studies (see, for example, Cotter et al. 2003; Eckhardt et al. 2003; Miller et al. 2007; Pai and Saraswat 2013).

While it is important to improve the spatial accuracy of LULC maps to reduce model errors, several model applications indicate the need to represent land use information accurately in the temporal scale. For instance, several LUC studies were conducted by replacing the LULC information in a calibrated/validated SWAT model with alternative scenarios (e.g., Lenhart et al. 2003; Chaplot et al. 2004; Wilson and Weng 2011). However, results reported from those approaches could be far removed from actual land uses since the simulation reproduction was driven primarily by initial model parameterization obtained during calibration (Lenhart et al. 2003). In contrast, Miller et al. (2002) used four SWAT models, each building upon a single land use map from different time periods (1973, 1986, 1992, and 1997), to quantify the influence of LUC in San Pedro Basin, Mexico. Although this approach provided temporally unique simulations, it is time-consuming and simplistic in implementation. Other studies have shown that ecosystem responses to changes in land use patterns are complex and nonlinear (Li et al. 2007; Ghaffari et al. 2010) and call for better integration of LUC influences in distributed model simulations.

In this chapter, we would like to engage the reader in the changes that have been made to the SWAT watershed model and discuss the role of a supporting tool for integrating concurrent LUC influences in model simulations. To explore the material further, we provide a case study in an Arkansas watershed, demonstrating the applicability of the new land use-related SWAT model feature. Although the results of this case study are specific to the SWAT model, it is our hope that the approach being advanced and the facts asserted would invite advancements toward an integrated LUC feature in the framework of other watershed models.

12.2 METHODS

12.2.1 Model Description

SWAT (Arnold et al. 1998) is a watershed-scale model that operates on a daily time step and simulates the long-term impacts of various management activities on the quantity and quality of water. This GIS-based model has been used globally at various spatial scales ranging from plots (Veith et al. 2008) to continents (Schuol et al. 2008). The model is being actively supported by USDA-ARS while being developed and applied in both academia and industry. The model algorithms, interfaces, and several supporting tools are freely available for download from the SWAT website (http://swat.tamu.edu).

Within the taxonomy of watershed models, the SWAT model falls under the semiphysical, semidistributed, continuous, long-term, hydrologic, and water-quality model category. Its semidistributed properties are particularly relevant in this study. The model divides a watershed into subwatersheds, and subwatersheds are further divided into homogeneous, lumped landmasses known as hydrological response units (HRUs). Each HRU is delineated (and distinguished from other HRUs) based on a specific land use, soil, and slope category. Therefore, a greater number of land use, soil, and slope categories in a watershed result in a larger number of HRUs.

The SWAT model simulates watershed processes, such as those related to rainfall–runoff, sediment, and chemical transport, initially at the HRU scale. The information required to simulate these watershed processes is obtained through a combination of built-in databases and user-supplied input data. To support the task of preparing input data, SWAT modelers use graphical user interface tools (or software) that are integrated with GIS such as ArcGIS (ESRI, Redland, California) or MapWindow (http://www.mapwindow.org). These integrated SWAT–GIS tools, such as ArcSWAT (for ArcGIS; see Winchell et al. 2008) or MWSWAT (for MapWindow; see George and Leon 2007), have simplified and expedited the processing of watershed-scale datasets into a SWAT-compatible format.

12.2.2 Land-Use Update in SWAT2009/SWAT2012

Several watershed-scale remote sensing and GIS layers, such as the digital elevation model (DEM), LULC, and soil data, inform the SWAT model about the characteristics of a watershed. The SWAT–GIS tools mentioned previously extract and process such information at watershed, subwatershed, and HRU spatial scales. This information is then consolidated in the form of parameter values, which are written in text files and stored in the TxtInOut folder of the SWAT model. The SWAT algorithm, compiled as a SWAT2009.exe executable file, uses these parameter values to simulate watershed processes on a daily basis.

Among numerous model parameters, a particular interest in this study is the fractional area (FR) of HRU within each subwatershed, which is represented by a variable named HRU_FR in the HRU general input file. It is useful to understand the process by which the HRU_FR value is derived by the SWAT. Using a raster-based operation, the SWAT–GIS interface calculates the number of cells that have the unique combination of a particular land use, soil, and slope within a subwatershed and divides it by the total number of cells within a subwatershed. This calculation results in the HRU_FR value, which is defined as the FR of the total subwatershed contained in the HRU. The value of HRU_FR ranges from 0 to 1, with higher values indicating larger occupation in the subwatershed. For instance, HRU_FR of 0.1 for a particular HRU indicates that it occupies 10% of the total subwatershed area. Thus, the HRU_FR value of an HRU and its land use, in combination, inform the model about the existing land use distribution. Note that before SWAT2009, the HRU_FR was considered a constant value within the model, maintaining an essentially static land use distribution during the simulations.

In January 2010, when the 2009 version of the SWAT model (SWAT2009) was released, the HRU_FR was redefined as a variable parameter. In other words, users can now update this parameter as many times as necessary during the simulation period. This simple feature in SWAT2009, called land use update (LUU), provides a powerful method to vary the simulated land use distribution dynamically during model simulation (Arnold et al. 2010). More importantly, it ensures that model responses such as the flow of water and pollutant transport in a watershed are a function of its changing land use.

The data requirements for the LUU feature in SWAT2009 include two files: one containing the dates on which land use needs to be updated (lup.dat) and the second containing the area of each HRU on update dates (fnam.dat; one file per LULC layer). The formatting requirement for these files is available in the SWAT2009 input/output manual (see Arnold et al. 2011, Chapter 37, p. 515). When these files are placed in the TxtInOut folder of the SWAT model, the LUU is activated automatically. The availability of temporal land use layers during a study period represents an opportunity to activate the LUU feature and to study historic LUC impacts. However, the number of HRUs in a SWAT model could often run into thousands, whereas the HRU topology is fragmented in nature, as shown by Pai et al. (2012). Mapping a regular LULC grid against numerous and fragmented HRUs, to calculate the HRU_FR parameter, represents a computational challenge. Thus, although the new feature is useful, its intensive data requirement could limit its usage.

12.2.3 Automation of LUU Parameterization

In support of the LUU feature in SWAT2009, we developed an automated geospatial technique to calculate the HRU_FR using an LULC raster. We simplify the explanation of our geospatial technique by considering a hypothetical rectangular subwatershed consisting of five rows and six columns. Figure 12.1 illustrates the process by which the original HRU layer is updated in response to an input LULC raster for this hypothetical subwatershed. The HRU layer for this subwatershed has three HRUs, while the corresponding input LULC layer also has three land use categories. Note that this input LULC is different from the one that was used for the initial HRU delineation.

Our GIS technique first scans through the text files in the TxtInOut folder of the SWAT model to identify the land use category associated with each HRU. This land use category, identified by SWAT during initial HRU delineation, is considered its base LULC. For instance, in Figure 12.1, the hypothetical base LULC categories for HRU ID 1, 2, and 3 are 3, 2, and 1, respectively. Thereafter, our technique performs a raster overlay operation between the HRU layer and a new user input LULC raster. The purpose of this overlay operation is to identify corresponding cells within the HRU raster that do not contain its base LULC. For instance, in Figure 12.1, a single cell in HRU ID 1 (left dotted box) and five cells in HRU ID 3 (right dotted box) do

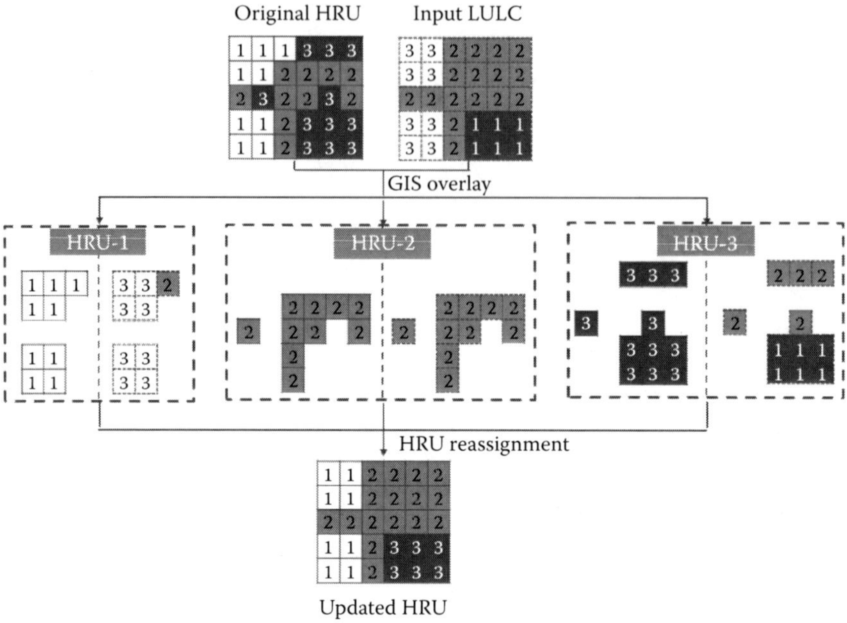

FIGURE 12.1 Schematic of the geospatial technique used to develop an updated HRU layer based on input LULC layer.

not conform to the base definition of its respective HRU and hence need to be reclassified. We call these nonconforming HRU cells, since they do not conform to the original land use designation of their respective HRUs. Next, the technique searches for an HRU that has a similar base LULC (LULC 2 in this case) and soil definition, and reclassifies the nonconforming HRU cells. In Figure 12.1, six nonconforming cells were assigned an HRU ID of 2 to create an updated HRU layer. The technique then calculates the HRU_FR for this updated HRU layer. These HRU_FR represent information about the changed LULC in the watershed. The updated HRU_FR is written in a text file (fnam.dat) per the syntax requirements and then copied to the TxtInOut folder of the SWAT model to activate the LUU feature in SWAT2009.

This automatic technique has been incorporated in a desktop tool called SWAT2009_LUC (Pai and Saraswat 2011b) to make it available to a larger user base. Although SWAT2009_LUC was programmed using the proprietary MATLAB®* (The Mathworks, Natick, Massachusetts) language, it was compiled as a standalone software so that it can be used on any Microsoft® Windows® computer. The SWAT2009_LUC user manual (Pai and Saraswat 2011a) provides information about the tool development and functionalities. The tool can be obtained by downloading from the SWAT website http://swat.tamu.edu/software/links-to-related-software/ or by contacting the authors directly.

* MATLAB® is a registered trademark of The MathWorks, Inc. For product information, please contact: The MathWorks, Inc., 3 Apple Hill Drive, Natick, MA 01760-2098, USA. Tel: 508 647 7000; Fax: 508 647 7001. E-mail: info@mathworks.com; Web: www.mathworks.com.

12.2.4 LUU Evaluation: A Case Study

The LUU feature in SWAT 2009 and our automated technique discussed earlier were evaluated for the Illinois River Drainage Area in Arkansas (IRDAA; Figure 12.2) watershed. The IRDAA is an eight-digit hydrologic unit code watershed, with a drainage area of 1963 km^2 and located in the Northwest Arkansas region. The region has been experiencing growth at a fast pace due to increased job opportunities in retail and food production industries. Additionally, urbanization has been active for the past decade in this region in general (Gitau et al. 2010), and the IRDAA watershed in particular (Haggard 2010), giving rise to several water quality concerns (Saraswat and Pai 2011). Thus, the IRDAA watershed provides a suitable study area to test the SWAT2009_LUC tool and the LUU module.

SWAT model development for this watershed has been reported extensively by Pai et al. (2011). In this study, the emphasis is on applying the LUU feature and studying the resulting changes in watershed responses. Therefore, the IRDAA SWAT model was simulated with and without the LUU feature, and percentage change in surface runoff and groundwater was studied at the subwatershed scale using the following formula:

$$\text{Percentage change} = \frac{(\text{output}_{\text{without-LUU}} - \text{output}_{\text{with-LUU}})}{\text{output}_{\text{with-LUU}}} \times 100 \qquad (12.1)$$

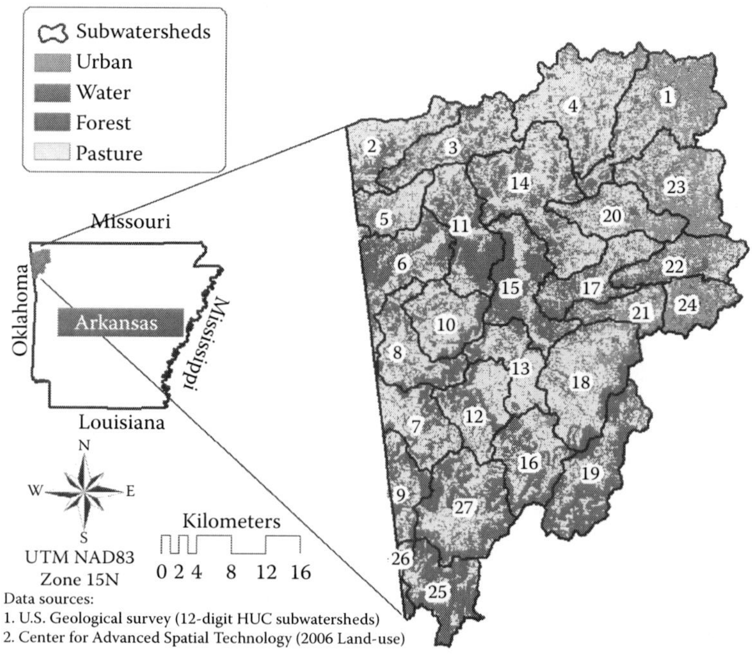

FIGURE 12.2 **(See color insert.)** IRDAA watershed location and 2006 land use.

LULC layers from 1999, 2004, and 2006 developed by the Center for Advanced Spatial Technologies at the University of Arkansas were processed using SWAT2009_LUC and used for the with-LUU scenario. The without-LUU scenario used only a single land use layer from 2006. There were two objectives of this evaluation: (1) to test the functionalities of the SWAT2009_LUC tool and (2) to verify the sensitivity of the SWAT model to LUCs in the IRDAA watershed.

12.3 RESULTS AND DISCUSSION

12.3.1 LUCs in IRDAA

When simulating LUC scenarios, land use layers are often obtained from different agencies that may use different classification schemes for LULC categories. Depending on the classification scheme used, the number of land use categories can differ between LULC rasters. However, the LUU feature in SWAT2009 allows users to update the area of only existing HRUs, which have predefined land uses. New land uses cannot be incorporated once the HRUs have been delineated. It was therefore essential to postprocess the LULC layers of the IRDAA watershed so that the land use categories were similar. Postprocessing involved merging detailed land use categories, for instance, oak, cedar, and bald cypress tree varieties, to the generic forest category when detailed categories were not available in all LULC layers. Future work is required in this area to allow modelers to incorporate new and emerging land uses within the watershed model framework.

Figure 12.3a shows the historical trend of land use distribution within the IRDAA watershed. The land use distribution at the watershed scale generally showed increasing urban areas and decreasing pasture areas during the study period (1999–2008). This LUC, however, was not uniform across the watersheds. Subwatersheds 1, 21, 22, 23, and 24 had greater changes in the urban and pasture areas compared to other watersheds (Figure 12.3b). These subwatersheds are likely to be more impacted due to the LUCs, and hence, it is essential that watershed models capture the LUC at its finest scale. The distributed nature of the SWAT model and the implementation of the LUU module at the HRU scale make it useful to evaluate such nonuniform LUCs.

12.3.2 Evaluation of Automated Approach for LUU Parameterization

The LUU module in SWAT2009 requires the area of every HRU each time the land use needs to be updated. Three land use geospatial datasets from 1999, 2004, and 2006 were processed using the SWAT2009_LUC tool. The programming for this tool was done in MATLAB, which enabled efficient, matrix-based manipulations through its Mapping Toolbox. This was leveraged in the SWAT2009_LUC tool by organizing HRU and land use information into a grid format. The size of the grid used was 7040 rows and 4630 columns. This grid size was a function of the original HRU layer created by ArcSWAT, which in turn is a function of the DEM layer that is used for watershed delineation. The tool was utilized to create LUU module input files for the IRDAA watershed consisting of 1126 HRUs, which were delineated using the 2006 land use layer.

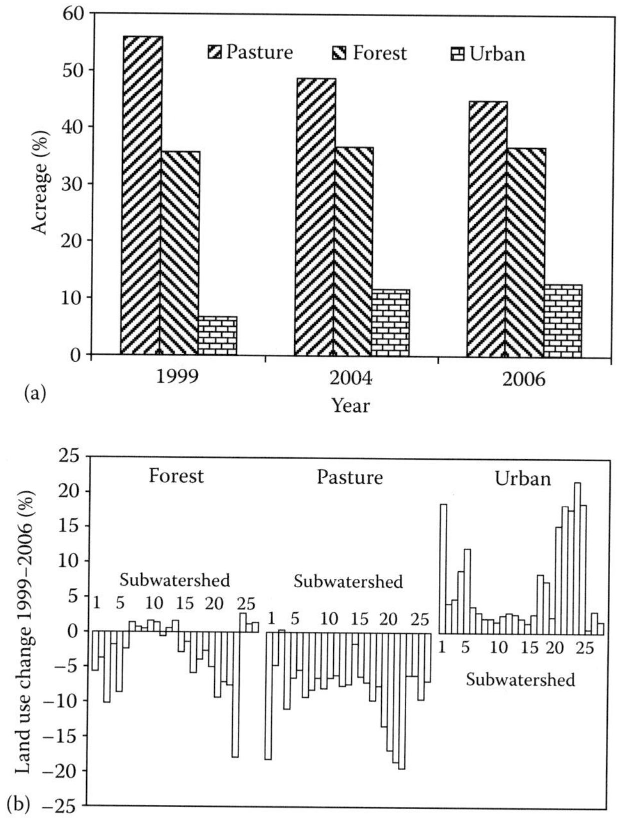

FIGURE 12.3 LUC in the IRDAA watershed: (a) watershed scale; (b) subwatershed scale. (From Pai, N. and D. Saraswat. *Trans. ASABE* 54(5):1649–1658, 2011.)

With all the necessary input layers and information fed into SWAT2009_LUC, the processing time for the three land use layers was approximately 5 min on a desktop computer with a 64-bit processor and 8 GB of random access memory (RAM). The tool automatically copied the required files (lup.dat, file1.dat, file2.dat, and file3.dat) to the TxtInOut folder of the SWAT model: this activated the LUU module. The SWAT2009_LUC website offers this tool for both 32- and 64-bit computers running a Microsoft Windows operating system.

12.3.3 LUC Impacts

A case study was conducted to test the effectiveness of the LUU module and the functionality of the SWAT2009_LUC tool. The purpose of this case study was not to reaffirm hydrological impacts of urbanization in the IRDAA watershed. The noticeable effects of urbanization, from a hydrological standpoint, are the alterations in transport pathways of water in a watershed. Consequently, increases in volume and velocity of surface water and decreases in groundwater as a response to urbanization are expected and have already been widely reported (see Dale 1997 for a literature review). Rather, this case study aims to highlight the change in SWAT model outputs when LUCs are integrated within simulations.

Turning the LUU module off (i.e., single land use layer simulation) and on (i.e., multiple land use layer simulation) sequentially allowed observation of the impact of LUC on SWAT output. Figure 12.4 shows the effect of integrating multiple land use layers on SWAT output at a subwatershed level. We found that once the LUU module was turned off, annual average surface runoff generally increased (Figure 12.4a) while groundwater generally decreased (Figure 12.4b), as compared to simulation output obtained while the LUU module was turned on. Specifically, the percentage change ranged from a 0.5% decrease (subwatershed 18) to a 12.8% increase (subwatershed 22) for surface runoff and from a 1.4% increase (subwatershed 3) to a 15.3% decrease (subwatershed 23) for groundwater (Figure 12.4).

With the LUU module turned off, the model simulates a constant and relatively high urban area from 2006 throughout the study period. For instance, on a subwatershed basis, the percentage increase in urban area from 1999 to 2006 ranged from 0.5% to 21.8%. The higher surface runoff and the lower groundwater simulated when the LUU module was deactivated are a direct effect of using a single land use layer from 2006. In contrast, activating the LUU module resulted in a gradual changing of land use from pervious to impervious surfaces from 1999 to 2008, which accounted for lower overall surface runoff and higher groundwater simulations.

Higher surface runoff and lower groundwater are typically associated with greater impervious areas (Tong et al. 2009), which were seen in the 2006 layer. Consequently, temporal subwatershed-scale surface runoff was likely overpredicted. This effect was greater for subwatersheds showing the largest increases in urban areas where, correspondingly, the highest groundwater increases and evapotranspiration and surface runoff decreases were exhibited. The larger groundwater flow simulated when the LUU module was activated is perhaps a better reflection

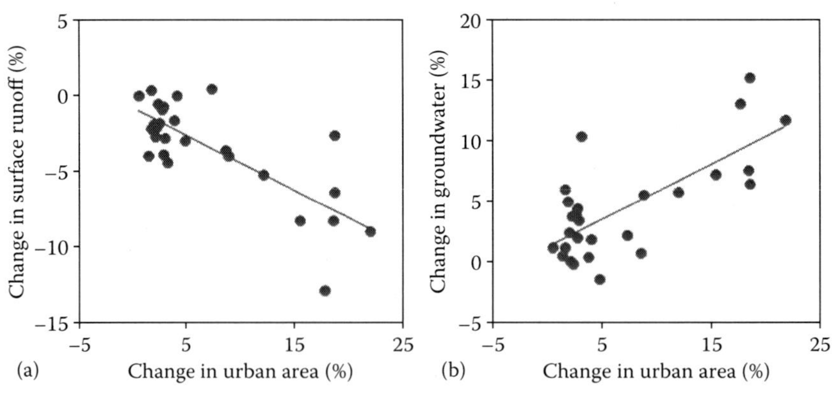

FIGURE 12.4 Effect of urbanization on two SWAT subwatershed-scale output: (a) changes in surface runoff output; (b) changes in groundwater output.

of temporal variation of land uses in IRDAA. The results also demonstrate that the LUU module imparts sensitivity to the model against changing land uses in the watershed. However, this sensitivity may not be spatially uniform; subwatersheds with greater urbanization exhibited greater impacts on hydrological processes. By using the SWAT2009_LUC tool and activating the LUU module in the SWAT, temporal land uses can be quickly input into the SWAT model, and the impact of LUCs can be better appraised.

Several recent SWAT model applications demonstrate its usability in identifying pollutant hotspots in a watershed so that they could be prioritized for management action (Tripathi et al. 2003; White et al. 2009; Pai et al. 2011). In such applications, model outputs are compared between either HRUs or subwatersheds to identify those that contribute disproportionately. For such intersubwatershed/HRU comparison projects, the distributed nature of the SWAT model and this new ability to integrate spatially nonuniform LUC influences are essential. In addition, studies have also documented the sensitivity of the SWAT model to accurate land use input (Heathman et al. 2009). Other researchers observed that small errors in impervious land surfaces could have a substantial effect on the uncertainty of runoff modeling results (Stuebe and Johnson 1990; Endreny et al. 2003; White and Chaubey 2005). Hence, improved spatially distributed model responses can be expected from using the SWAT2009_LUC tool.

Improvements in spatially distributed responses may not be visible for watershed models that are calibrated at a single gage in a watershed with little LUC. In other words, the benefits of integrating LUC influences within model simulations is dependent on the intensity and scale of LUCs. Results from this study showed that the model sensitivity to LUU module activation is not spatially uniform. Subwatersheds with greater LUCs (such as in subwatersheds 22, 23, and 24) exhibited an average of 13% higher groundwater and 8% lower surface runoff when the LUU module was activated. This is likely to improve the temporal predictive ability of the model since it is now a function of LUCs. Unfortunately, we did not have a long-term measured dataset from the urbanizing subwatersheds of the IRDAA watershed to verify if the activation of the LUU module resulted in better predictions. Future analyses could build on this work by quantifying the improvements in model responses using observed datasets.

12.4 SUMMARY

Watershed models are regarded as a primary tool for understanding the long-term influences of LUC on hydrology and water quality. However, model simulation depends on the spatial and temporal accuracy of its input datasets. Studying the integration of multiple temporal LULC datasets in the SWAT model was the objective of this study. Watershed responses at the outlet of each subwatershed of IRDAA were simulated with and without the newly released LUU module. The LUU module allows modelers to update the land use distribution within the model dynamically during the simulations. Without the LUU module, the IRDAA SWAT model used information from a single land use layer from 2006, which represented urbanized conditions (0.5% to 59.8% at the subwatershed scale), throughout the study period. In contrast, when the LUU module was activated, the model could leverage information about gradually urbanizing variable conditions in the watershed between 1999 and 2008. Results from this study showed that the surface runoff could be overpredicted by up to 13%, and groundwater could be underpredicted by up to 15% at the subwatershed scale when only a single LULC is used.

In addition to studying the LUC influences, the chapter also introduced the SWAT2009_LUC geospatial tool. SWAT2009_LUC provides a user-friendly graphical user interface environment for SWAT modelers to process multitemporal land use layers of a study area. The user can interactively provide locations of input data through the tool. The tool is stand-alone (independent of any proprietary software) and fast (<5 min for processing three land use layers for the eight-digit IRDAA watershed) because it leverages the efficiency of matrix-based raster calculations in its geospatial technique (see Pai and Saraswat 2011b for more details). The tool seamlessly interacts with existing SWAT model projects to extract necessary information and automatically activates the LUU feature. A demonstration video on the SWAT2009_LUC tool is available on the website (http://baegrisk.ddns.uark.edu/SWAT_Model_Tools/SWAT2009_LUC/). The tool is expected to assist SWAT modelers in evaluating the LUC influences in their watersheds.

12.5 TUTORIAL

A tutorial consisting of step-by-step instructions for using SWAT2009_LUC is provided on the CRC Press website at http://www.crcpress.com/prod/isbn/9781429867228. The purpose of this tutorial is to familiarize users with the capabilities of the SWAT2009_LUC tool. An example dataset is also available in the online materials for this book from the CRC website at http://www.crcpress.com/product/isbn/9781439867228. An example dataset can also be downloaded from the SWAT2009_LUC website.

ACKNOWLEDGMENTS

Financial support for this study was provided by the Arkansas Natural Resources Commission through the Clean Water Act 319(h) grant funding program. We thank Dr. Raghavan Srinivasan for answering many questions related to HRU delineation in ArcSWAT. Dr. Jeff Arnold, Ms. Nancy Sammons, and Dr. Mike White helped us to better understand the LUU module.

REFERENCES

Ahearn, D.S., R.W. Sheibley, R.A. Dahlgren, M. Anderson, J. Johnson and K.W. Tate. 2005. Land use and land cover influence on water quality in the last free-flowing river draining the western Sierra Nevada, California. *J. Hydrol.* 313:234–247.

Arnold, J.G., P.W. Gassman and M.J. White. 2010. New developments in the SWAT ecohydrology model. In *Proceedings of 21st Century Watershed Technology: Improving Water Quality and Environment*. St. Joseph, MI: ASABE.

Arnold, J.G., J.R. Kiniry, R. Srinivasan, J.R. Williams, E.B. Haney and S.L. Neitsch. 2011. *Soil and Water Assessment Tool Input/Output File Documentation Version 2009*. Texas Water Resources Institute Technical Report No. 365. College Station, TX: Texas A&M University.

Arnold, J.G., R. Srinivasan, R.S. Muttiah and J.R. Williams. 1998. Large-area hydrologic modeling and assessment: Part I. Model development. *J. Am. Water Resour. Assoc.* 34(1):73–89.

Bathurst, J.C. 1986. Physically-based distributed modelling of an upland catchment using the Système Hydrologique Européen. *J. Hydrol.* 87(1–2):79–102.

Bhaduri, B., M. Minner, S. Tatalovich and J. Harbor. 2001. Long-term hydrologic impact of urbanization: A tale of two models. *J. Water Resour. Plan. Manage.* 127(1):13–19.

Bicknell, B.R., J.C. Imhoff, A.S. Donigian and R.C. Johanson. 1997. *Hydrological Simulation Program—FORTRAN (HSPF), User's Manual for Release 11. EPA 600/R-97/080.* Athens, GA: United States Environmental Protection Agency.

Chaplot, V.A., A. Saleh, D.B. Jaynes and J. Arnold. 2004. Predicting water, sediment, and NO3-N loads under scenarios of land use and management practices in a flat watershed. *Water Air Soil Pollut.* 154(1–4):271–293.

Cotter, A.S., I. Chaubey, T.A. Costello, T.S. Soerens and M.A. Nelson. 2003. Water quality model output uncertainty as affected by spatial resolution of input data. *J. Am. Water Resour. Assoc.* 39(4):977–986.

Dale, V.H. 1997. The relationship between land-use change and climate change. *Ecol. Appl.* 7(3):753–769.

DeFries, R. and K. Eshleman. 2004. Land-use change and hydrologic processes: A major focus for the future. *Hydrol. Proc.* 18(11):2183–2186.

Eckhardt, K., L. Breuer and H.G. Frede. 2003. Parameter uncertainty and the significance of simulated land use change effects. *J. Hydrol.* 273(1–4):164–176.

Endreny, T.A., C. Somerlot and J.M. Hassett. 2003. Hydrograph sensitivity to estimates of map impervious cover: A WinHSPF BASINS case study. *Hydrol. Proc.* 17(5):1019–1034.

Gassman, P.W., M.R. Reyes, C.H. Green and J.G. Arnold. 2007. The soil and water assessment tool: Historical development, applications, and future research directions. *Trans. ASABE* 50(4):1211–1250.

George, C. and L. Leon. 2007. WaterBase: SWAT in an open source GIS. *Open Hydrol. J.* 1:19–24.

Ghaffari, G., S. Keesstra, J. Ghodousi and H. Ahmadi. 2010. SWAT-simulated hydrological impact of land use change in the Zanjanrood basin, Northwest Iran. *Hydrol. Proc.* 24(7):892–903.

Gitau, M., I. Chaubey, E. Gbur, J. Pennington and B. Gorham. 2010. Impacts of land use change and best management practice implementation in a Conservation Effects Assessment Project watershed: Northwest Arkansas. *J. Soil Water Conserv.* 65(6):353–368.

Haggard, B.E. 2010. Phosphorus concentrations, loads, and sources within the Illinois River drainage area, northwest Arkansas, 1997–2008. *J. Environ. Qual.* 39(6):2113–2120.

Heathman, G.C., J.C.I. Ascough and M. Larose. 2009. Soil and Water Assessment Tool evaluation of soil and land use geographic information system data sets on simulated stream flow. *J. Soil Water Conserv.* 64(1):17–32.

Jensen, J.R. 1996. *Introductory Digital Image Processing*, 2nd edn. Upper Saddle River, NJ: Prentice-Hall.

Leavesley, G.H., R.W. Lichty, B.M. Troutman and L.G. Saindon. 1983. *Precipitation–Runoff Modeling System: User's Manual.* Water Resource Investigation Report 83-4328. Denver, CO: U.S. Geological Survey.

Lenhart, T., N. Fohrer and H.G. Frede. 2003. Effects of land use changes on the nutrient balance in mesoscale catchments. *Phys. Chem. Earth* 28(34):1301–1309.

Li, K.Y., M.T. Coe, N. Ramankutty and R. De Jong. 2007. Modeling the hydrological impact of land-use change in West Africa. *J. Hydrol.* 337(3–4):258–268.

Miller, S.N., D.P. Guertin and D.C. Goodrich. 2007. Hydrologic modeling uncertainty resulting from land cover misclassification. *J. Am. Water Resour. Assoc.* 43(4):1065–1075.

Miller, S.N., W.G. Kepner, M.H. Mehaffey, M. Hernandez, R.C. Miller, D.C. Goodrich, K. Devonald, D.T. Heggem and W.P. Miller. 2002. Integrating landscape assessment and hydrologic modeling for land cover change analysis. *J. Am. Water Resour. Assoc.* 38(4):915–929.

Nunes, J.P., G.N. Vieira, J. Seixas, P. Goncalves and N. Carvalhais. 2005. Evaluating the MEFIDIS model for runoff and soil erosion prediction during rainfall events. *Catena* 61(1):210–228.

Office of Technology Assessment (OTA). 1982. Chapter 1: Summary, issues, and options. In *Use of Model for Water Resource Management, Planning, and Policy*. Washington, DC: U.S. Government Printing Office.

Pai, N. and D. Saraswat. 2011a. *SWAT2009_LUC User Manual*. Fayetteville, AR: University of Arkansas, Department of Biological and Agricultural Engineering. Available at http://baegrisk.ddns.uark.edu/SWAT_Model_Tools/SWAT2009_LUC/docs/SWAT2009_LUC_UserMan.pdf (accessed July 20, 2012).

Pai, N. and D. Saraswat. 2011b. SWAT2009_LUC: A tool to activate the land use change module in SWAT 2009. *Trans. ASABE* 54(5):1649–1658.

Pai, N. and D. Saraswat. 2013. Impact of land use land cover categorical uncertainty on SWAT hydrological modeling. *Trans. ASABE* 56(4):1387–1397.

Pai, N., D. Saraswat and M. Daniels. 2011. Identifying priority subwatersheds in the Illinois River drainage area in Arkansas watershed using a distributed modeling approach. *Trans. ASABE* 54(6):2181–2196.

Pai, N., D. Saraswat and R. Srinivasan. 2012. Field_SWAT: A tool for mapping SWAT output to field boundaries. *Comput. Geosci.* 40:175–184.

Saraswat, D. and N. Pai. 2011. Spatially distributed hydrologic modeling in Illinois River Drainage Area in Arkansas using SWAT. In *Soil Hydrology, Land Use, and Agriculture: Measurement and Modelling*, ed. M.K. Shukla, 196–210. Oxfordshire, UK: CABI International.

Schuol, J., K.C. Abbaspour, R. Srinivasan and H. Yang. 2008. Estimation of freshwater availability in the West African sub-continent using the SWAT hydrologic model. *J. Hydrol.* 352(1):30–49.

Singh, V.P. and D.K. Frevert. 2006. *Watershed Models*. Boca Raton, FL: Taylor & Francis.

Singh, V.P. and D.A. Woolhiser. 2002. Mathematical modeling of watershed hydrology. *J. Hydrol. Eng.* 7(4):270–292.

Stuebe, M.M. and D.M. Johnston. 1990. Runoff volume estimation using GIS techniques. *J. Am. Water Resour. Assoc.* 26(4):611–620.

Thomas, M.A., B.A. Engel and I. Chaubey. 2009. Water quality impacts of corn production to meet biofuel demands. *J. Environ. Eng.* 135(11):1123–1135.

Tong, S.T.Y., A.J. Liu and J.A. Goodrich. 2009. Assessing the water quality impacts of future land use changes in an urbanising watershed. *Civ. Eng. Environ. Syst.* 26(1):3–18.

Tripathi, M.P., R.K. Panda and N.S. Raghuwanshi. 2003. Identification and prioritization of critical sub-watersheds for soil conservation management using the SWAT model. *Biosyst. Eng.* 85(3):365–379.

Tuppad, P., K.R. Douglas-Mankin, T. Lee, R. Srinivasan and J.G. Arnold. 2011. Soil and Watershed Assessment Tool (SWAT) hydrologic/water quality model: Extended capability and wider adoption. *Trans. ASABE* 54(5):1677–1684.

Veith, T.L., J.G. Arnold and A.N. Sharpley. 2008. Modeling a small, northeastern watershed with detailed, field-level data. *Trans. ASABE* 51(2):471–483.

Vieux, B.E. 2004. *Distributed Hydrologic Modeling Using GIS*. Norwell, MA: Kluwer Academic Publishers.

White, K.L. and I. Chaubey. 2005. Sensitivity analysis, calibration, and validations for a multi-site and multivariable SWAT model. *J. Am. Water Resour. Assoc.* 41(5):1077–1089.

White, M.J., S.H. Stoodley, S.J. Phillips, D.E. Storm and P.R. Busteed. 2009. Evaluating non-point source critical source area contributions at the watershed scale. *J. Environ. Qual.* 38(4):1654–1663.

Williams, J.R., C.A. Jones and P.T. Dyke. 1984. A modeling approach to determining the relationship between erosion and soil productivity. *Trans. ASAE* 27(1):129–144.

Wilson, C.O. and Q. Weng. 2011. Simulating the impacts of future land use and climate changes on surface water quality in the Des Plaines River watershed, Chicago Metropolitan Statistical Area, Illinois. *Sci. Total Environ.* 409:4387–4405.

Winchell, M.R., R. Srinivasan, M. Diluzio and J. Arnold. 2008. *ArcSWAT User's Guide*. Temple, TX: Blackland Research Center, Texas Agricultural Experiment Station.

13 Combining Landscape Segmentation and a Agroecosystem Simulation Model

Xiuying Wang, Pushpa Tuppad, and Jimmy R. Williams

CONTENTS

EXECUTIVE SUMMARY

Conservation practices are often implemented based on experience and field conditions. More informed conservation planning using simulation models allows targeting placement of practices to critical areas. Simulation models can be used to simulate the effectiveness of various conservation practices over long time periods and over large areas, thereby providing feedback to future conservation planning. This eliminates ineffective conservation practices and expands the temporal and spatial scales of conservation measures. The Agricultural Policy/Environmental eXtender (APEX) model has been developed to assess a wide variety of agricultural water resource, water quality, and other environmental problems. The major APEX model components and available APEX geographic information system (GIS)–based interfaces that support APEX applications are described in this chapter. Case studies that describe and discuss the model use for conservation planning are provided to further explore the potential of this simulation model to enhance conservation planning.

KEYWORDS

Agricultural Policy/Environmental eXtender (APEX), conservation practice, hydrology, nonpoint source pollution, sediment

13.1 INTRODUCTION

Numerous hydrologic and environmental models and tools have been developed to complement field studies and assess conservation practices. These tools differ in the way they represent watershed processes and in the ability to represent spatial and temporal variability throughout study areas. One of these models is the Agricultural Policy Environmental Extender (APEX) model (Williams and Izaurralde 2006). It is a flexible and dynamic tool that evaluates a wide array of management strategies applied to crop, pasture, and grazing lands. APEX extends the ability of the field-scale model—Environmental Policy Impact Climate (Williams et al. 1984; Williams 1995; Izaurralde et al. 2006)—for simulating complex farm or watershed landscape processes and management practices. APEX is capable of estimating the long-term sustainability of land management with respect to erosion (wind, sheet, and channel), economics, water supply, water quality, soil quality, plant competition, weather, and pests for agricultural lands. APEX routes water, sediment, nutrients, and pesticides across landscapes through channels, floodplains, and reservoirs to the watershed outlet. The routing algorithms account for filter strips, terraces, and waterways.

Simulation-aided conservation planning is advantageous for conservation and program evaluation. The primary emphasis of soil and water conservation planning is to identify critical areas with high levels of runoff and nonpoint source pollutants. Different landscape positions have different surface and subsurface geometries and soil properties; therefore, these positions differentially affect hydrologic and chemical processes. Accounting for these differences and different land uses and cultural practices across landscapes can greatly improve modeling accuracy and guide conservation planning throughout landscapes to maximum environmental efficacy.

APEX has been applied for evaluating the effectiveness of conservation practices and for conducting scenario analyses (e.g., Wang et al. 2008, 2009; Yin et al. 2009; Mudgal et al. 2010; Tuppad et al. 2010). The most significant application of APEX is the ongoing National Cropland Assessment component of the Conservation Effects Assessment Project (CEAP) (Duriancik et al. 2008; USDA-NRCS 2009). A comprehensive review of the APEX model applications can be found in Gassman et al. (2010). The APEX model parameterization of inputs for different conservation practices and the determination of sediment delivery ratios for the Upper Mississippi River Basin (UMRB) for the CEAP project are reported in Wang et al. (2011a). Other recent APEX applications include analyses of the effects of agroforestry buffers on reducing runoff and sediment losses from small-grazed pasture watersheds in north-central Missouri (Kumar et al. 2011) and the effects of bioenergy cropping systems on pollutant losses from representative soils in three Iowa counties (Powers et al. 2011). The objectives of this study are to briefly describe the major components of the APEX model, to summarize geographical information system (GIS) interfaces that support APEX applications for conservation planning, and to demonstrate APEX applications with case studies and an appended hands-on exercise.

13.2 METHODS

13.2.1 Model Description

The APEX model was developed for use in whole farm and/or watershed management. Management capabilities include irrigation, drainage, furrow diking, buffer strips, terraces, waterways, fertilization, forest management, manure management, lagoons, reservoirs, crop rotation and selection, pesticide application, grazing, and tillage. The model operates with daily time steps. Farms or watersheds can be subdivided into hydrologically connected subareas based on terrain attributes, fields, soil types, landscape positions, or any other desirable configurations. The major components in APEX include climate, hydrology, erosion/sedimentation, crop growth, carbon cycling, nutrient cycling, and routing components. A brief overview of these components is provided here. A detailed, more theoretical description of APEX can be found in Williams and Izaurralde (2006).

13.2.1.1 Climate

Daily precipitation, maximum and minimum air temperature, and solar radiation are the required weather variables for APEX simulations. Wind speed and relative humidity are also required if the Penman methods (Penman 1948) are used to estimate potential evaporation. Wind speeds are needed when wind-induced erosion or dust emission and distribution are to be simulated. These weather variables can be either entered by the user or simulated at run time. For weather simulation, monthly weather statistics are required. Precipitation is generated in the model based on a first-order Markov chain model developed by Nicks (1974). Precipitation can also be generated spatially for watershed applications covering larger areas and/or encompassing regions with steep rainfall gradients. Air temperature and solar radiation are generated in the model using a multivariate generation approach described by Richardson (1981). Wind generation in APEX is based on the Wind Erosion Continuous Simulation (WECS) model (Potter et al. 1998).

13.2.1.2 Hydrology

Daily runoff volume, peak runoff rate, subsurface flow, percolation below the soil profile, evapotranspiration, and snowmelt are considered in the APEX hydrology component. Rainfall interception by plant canopy is considered, and the excess falls to the soil surface. If snow is present, it may be melted as a function of the daily maximum air temperature and the snow pack temperature to estimate the total daily water in the area of interest. Runoff can be calculated using either a modification of the Natural Resources Conservation Service (NRCS) curve number method (Mockus 1969; USDA-NRCS 2004) or the Green and Ampt infiltration equation (Green and Ampt 1911).

Peak runoff rate can be estimated using either the modified rational formula (Williams 1995) or the SCS TR-55 method (USDA-NRCS 1986). The subsurface flow component computes vertical and horizontal subsurface flow simultaneously using storage routing and pipe flow equations. Vertical and horizontal flows are partitioned as a function of the vertical flow travel time and the horizontal travel time. Horizontal flow is partitioned between quick return flow and subsurface flow based on the ratio of upland slope length versus reach channel length. The vertical percolation flows to groundwater storage, which is further partitioned into return flow and deep percolation. Return flow is added to channel flow from the subarea, while it is assumed that deep percolation is lost from the system. The groundwater component partitions flow between deep percolation and return flow using the groundwater storage residence time and a partitioning coefficient.

Five options are provided in APEX for estimating potential evaporation: Hargreaves (Hargreaves and Samani 1985), Penman (1948), Priestley–Taylor (Priestley and Taylor 1972), Penman–Monteith (Monteith 1965), and Baier–Robertson (Baier and Robertson 1965). Potential soil water evaporation is estimated as a function of potential evaporation and leaf area index. Actual soil water evaporation is estimated by using exponential functions of soil depth and water content. Plant water evaporation is simulated as a linear function of potential evaporation and leaf area index.

13.2.1.3 Soil Erosion/Sedimentation

Eight options are provided in APEX for simulating water-induced erosion caused by runoff from rainfall and irrigation: the Universal Soil Loss Equation (USLE) method (Wischmeier and Smith 1978); the Onstad–Foster (AOF) modification of the USLE (Onstad and Foster 1975); the Modified Universal Soil Loss Equation (MUSLE) method (Williams 1975); three MUSLE variants described by Williams (1995), which are referred to as MUST (theoretical version), MUSS (small watershed version), and MUSI (approach that uses input coefficients); the Revised Universal Soil Loss Equation (RUSLE) method (Renard et al. 1997); and the Revised Universal Soil Loss Equation—Version 2 (RUSLE2) (USDA-ARS 2005).

Wind erosion is calculated with the WECS (Potter et al. 1998), which requires the daily distribution of wind speed. The approach estimates potential wind erosion for a smooth bare soil by integrating the erosion equation through a day using the wind speed distribution. The potential erosion is then adjusted according to soil properties, surface roughness, vegetative cover, and distance across the field in the wind direction.

13.2.1.4 Crop Growth

The plant growth and competition component of APEX simulates crop rotations and other cropping/vegetation systems such as cover crops, double cropping, plant–weed competition, pastures, and tree growth. Each crop has a unique set of crop parameters, which are packaged in the model databases. Phenological development of the crop is based on daily heat unit accumulation. The daily growth may be affected by atmospheric CO_2 concentration and physiological stresses. Annual crops grow from planting date to harvest date or until the accumulated heat units equal the user-specified potential heat units for the crop. Perennial crops maintain their root systems throughout the year, but they may become dormant after frost. They resume growth when the average daily air temperature exceeds their base temperature. APEX also considers mixed plant stands competing for light, water, and nutrients.

13.2.1.5 Carbon and Nitrogen Cycling

APEX incorporates carbon and nitrogen algorithms similar to the Century model (Parton et al. 1994; Vitousek et al. 1994) to distribute C and N across soil layers into several pools: metabolic litter; structural litter; and active, slow, and passive humus (Izaurralde et al. 2006, 2007). The soil carbon sequestration is estimated as a function of climatic conditions, soil properties, and management practices.

The components of the N cycle that are simulated include atmospheric N inputs, fertilizer and manure N applications, crop N uptake, mineralization, immobilization, nitrification, denitrification, ammonia volatilization, organic N transport on sediment, and nitrate–nitrogen (NO_3–N) losses through leaching, surface runoff, lateral subsurface flow, and tile flow. Denitrification is a function of temperature and water content (Williams 1995) with the requirement of anaerobic conditions and a carbon source. Nitrification, the conversion of ammonia N to NO_3–N, is estimated based on the first-order kinetic rate equation of Reddy et al. (1979). Atmospheric emissions of N gases from the soil profile simulated in APEX include N_2 and nitrous oxide (N_2O) and ammonia volatilization. Volatilization is estimated simultaneously with nitrification. The organic N loss is estimated using a modified loading function (Williams and Hann 1978), considering sediment yield, organic N loss in the soil surface, and an enrichment ratio. The soluble N loss is estimated by considering the change in concentration (Williams 1995). The concentration in a soil layer decreases exponentially as a function of flow volume. A list of parameters related to these components may be adjusted to improve results (Wang et al. 2011b).

13.2.1.6 Phosphorus

APEX estimates soluble P runoff loss as a function of the concentration of labile P in the top soil layer, runoff volume, and a linear adsorption isotherm. Sediment transport of P is estimated with a modified loading function originally developed by McElroy et al. (1976). The P mineralization model is a modification of the Production of Arid Pastures Limited by Rainfall and Nitrogen (PAPRAN) mineralization model (Seligman and van Keulen 1981).

Mineralization from the fresh organic P pool is estimated as the product of the mineralization rate and the fresh organic P content. Mineralization of organic P associated with humus is estimated for each soil layer as a function of soil water content, temperature, and bulk density. Mineral P is transferred among three pools: labile, active mineral, and stable mineral.

13.2.1.7 Routing Component

APEX has two options for routing water through channels and flood plains: a daily time step average flow method and a short time interval complete flood-routing method. The complete flood-routing approach simulates the dynamic stream flow, whereas the daily time step method can only estimate the daily water yield. Sediment is routed through the channel and floodplain separately. The sediment-routing equation is a variation of Bagnold's (1977) sediment transport equation, which estimates the transport concentration capacity as a function of velocity. The organic forms of N and P and adsorbed pesticide are transported by sediment and are routed using an enrichment ratio approach. The enrichment ratio is estimated as the ratio of the mean sediment particle size distribution of the outflow to that of the inflow. Organic N and P mineralization in the channels is not considered because, in general, the travel time is short. Mineral nutrient and soluble pesticide losses occur only if flow is lost within the reach. The pesticide-routing approach is the same as described for nutrients.

13.2.2 GIS Interfaces for APEX

APEX, as a spatially distributed hydrological and cropping systems model, has the ability to simulate key landscape processes and capture the land use/soil/management variability by segmenting the study area into hydrologically connected landscape units called *subareas*. GIS interfaces have been developed for APEX to facilitate landscape segmentation for modeling purposes and to develop model input and output management. GIS interfaces for APEX together with several major APEX applications for conservation planning are summarized and discussed in the following to highlight the capabilities and strengths of the APEX model.

13.2.2.1 AVSWAT–APEX Subbuilder

Di Luzio et al. (2004) described an ArcView GIS platform for the Soil and Water Assessment Tool (SWAT; see Arnold et al. 1998) called AVSWAT. AVSWAT automates watershed delineation and hydrologic network identification, calculation of parameters that describe subbasin geometric and topographic characteristics, channel dimensions, and land use, soils, and slope area distributions. Before any APEX-specific GIS-based interfaces were developed, the AVSWAT watershed delineation and land use/soil characterization functionalities and the corresponding output files had been used in an APEX subbuilder program (Williams et al. 2004) to build an

APEX subarea file (Wang et al. 2009; Tuppad et al. 2010). The file contains subarea information, indicating which soil and operation schedule files will be used for each subarea, subarea catchment characteristics, routing instructions, and channel geometry of routing reaches.

13.2.2.2 SWAT–APEX

The AVSWAT interface was further expanded to automate APEX inputs and perform nested APEX simulations on the field, whole-farm, or small watershed scale within a SWAT watershed application through an innovative SWAT–APEX program (SWAPP) (Saleh and Gallego 2007). The SWAPP program is initiated with SWAT GIS input data layers created by the AVSWAT interface for the respective watershed of interest. The user specifies the type of land uses to be simulated by APEX rather than by SWAT. Then, all the required APEX data files for selected land uses from the SWAT format are transferred to the APEX format and simulated in APEX. The next step is to process the combined simulation results of SWAT and APEX. An enhanced version of SWAPP, called Comprehensive Economic Environmental Optimization Tool (CEEOT)–SWAPP (Saleh et al. 2008), expands the interface between a farm-level economic model and APEX and/or SWAT; therefore, it is able to estimate net farm returns and other economic indicators for representative farms.

13.2.2.3 WinAPEX–GIS

WinAPEX (Steglich and Williams 2008) is a Windows interface developed to provide APEX users with a user-friendly environment to construct APEX input data and provide editing functions. A combined Field Hydro Tool created within ArcGIS 9.2 (Kemanian et al. 2009) and the WinAPEX modeling system called WinAPEX–GIS has been developed to build large watershed applications with a landscape positioning function. The data inputs and model outputs were loaded into an Access database that allowed for rapid editing, mapping, and building a large number of scenarios. GIS tools are used to delineate a watershed into subbasins based on the watershed digital elevation model (DEM) and separate the landscape of each subbasin into two positions, accomplished using a GIS algorithm that performs several functions:

1. Calculate the position of each DEM cell by assigning a percentage distance between the stream line and the ridge line of each subbasin
2. Calculate the change in slope for each of the cells
3. Identify the cells that had the most rapid change in slope, found in the lower 40% of the subbasin
4. Assign the cells below the maximum slope change as lower landscape positions and the cells above that line to the upper landscape position

The rule set required that all upper landscape positions flow into or through a lower landscape position before entering the stream channel. The lower landscape position is assumed to have a drainage channel that receives water from the immediately adjacent upper landscape position.

13.2.2.4 ArcAPEX

ArcAPEX is an ArcGIS-based user interface designed to automate the input parameterization of APEX (Tuppad et al. 2009). ArcAPEX is an extension to the ArcGIS software package. The interface has been developed for use with ArcGIS versions 9.2 and 9.3. ArcAPEX uses topographic, weather, land use, and soil spatial datasets to facilitate model input. The subarea boundaries and the hydrologic connectivity between subareas are based on DEM or by user-predefined subarea boundaries and streams that are closely associated with specific agricultural field boundaries. The interface has been designed to create either a stand-alone APEX project or an APEX–SWAT integrated project. The overall modeling support is similar to that in SWAPP. The APEX–SWAT integrated application requires user identification of a preexisting SWAT project dataset. It enables the development of a watershed-scale model that incorporates multiple scales into the simulation. Therefore, APEX is implemented for more detailed simulation of farms or small subwatersheds with complex agronomic systems. Alternatively, SWAT is implemented for larger subwatersheds characterized by simple agricultural systems and nonagricultural landscapes, as well as for integrating constituent (runoff, sediment, nutrient, and pesticide) contributions from all subwatersheds and simulating in-stream channel processes.

13.3 APPLICATIONS

13.3.1 Shoal Creek Watershed

The Shoal Creek watershed (22.5 km^2) is located in Coryell County, Texas (Figure 13.1). It is part of the U.S. Army's Fort Hood military reservation area. Military maneuvers damage vegetation and disturb land conditions, which result in vegetation loss, soil exposure and erosion, runoff channelization, and gully system development. Historically, the watershed was covered with mixed-grass prairie and small hillock ranges. The watershed received Best Management Practices (BMPs), including the implementation of contour soil ripping in 2001 and gully plugs during 2002 to 2004 (Figure 13.1). The APEX model was calibrated and validated for the pre-BMP and post-BMP conditions in the Shoal Creek watershed using daily runoff and sediment yield. The calibrated model was used for scenario analysis to quantify the BMP benefits for military training lands (Wang et al. 2009). The average annual precipitation during the study period (1997–2005) was 760 mm.

The AVSWAT interface of the SWAT2005 version was used to delineate the watershed and estimate catchment characteristics for preparing the APEX subarea configuration file. The interface delineates the watershed into subwatersheds based on the DEM and a user-defined critical source threshold area for stream definition. The watershed was subdivided into 183 subareas (Figure 13.2) ranging in size from 0.17 to 110 ha with the consideration of the contour ripping and gully plug positions, as shown in Figure 13.1. The stream threshold area of 0.5 ha was used to identify more streams so that additional outlets could be added at the locations where gully plugs and contour ripping were implemented. Extensive outlet editing was conducted to delete other unnecessary outlets generated due to the small stream threshold area used. Some gully plugs were grouped together in one

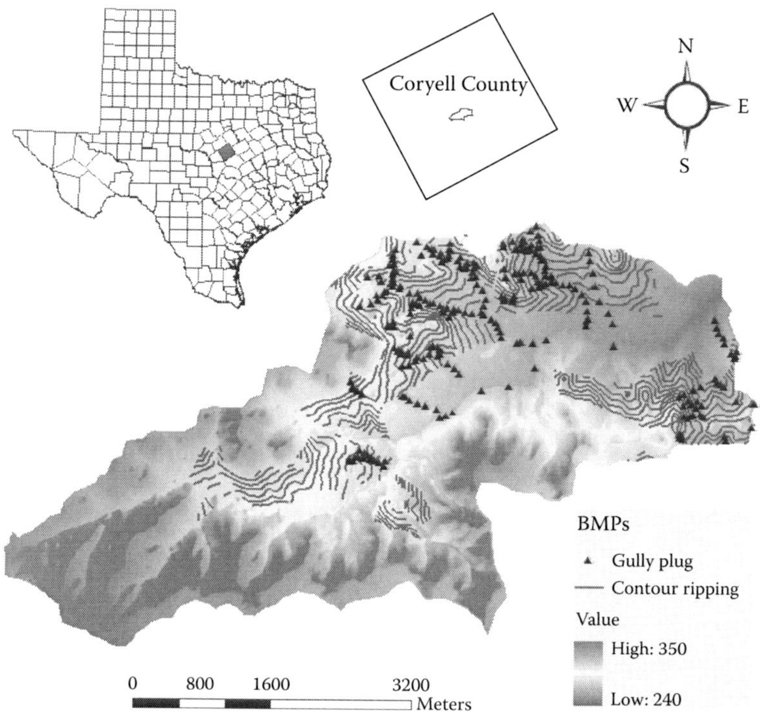

FIGURE 13.1 Location of the Shoal Creek watershed in Central Texas, DEM (5 × 5 m), and installed conservation practices. (From Wang, X. et al., *Trans. ASABE* 52(4):1181–1192, 2009.)

subarea where it was either impossible to add an outlet or it would result in a subarea that was too small.

The effectiveness of conservation practices on runoff and sediment yield was evaluated by comparing responses between a preconservation scenario and the current baseline. Changes in runoff and sediment yield between the scenario and measured values under conservation conditions provided the percentage of reductions due to implemented conservation practices in the watershed.

These deep ripping and gully plugs were placed based on experience and field conditions. The combined benefits of deep ripping and gully plugs were quantified as a reduction of 47% in runoff and 87% in sediment yield at the watershed outlet. Runoff reduction for these subareas with contour ripping practice ranged between 40% and 46% and between 54% and 62% in sediment reduction when compared with the corresponding no conservation condition. Subareas with gully plugs have 88% to 95% reduction in sediment.

13.3.2 Cowhouse Creek Watershed Study

The Cowhouse Creek is located in north-central Texas (Figure 13.3). It is a tributary of the Brazos River, most of which is rangeland within the Lampasas Cut Plain. From the headwaters in northeast Mills County, the creek flows southeast for approximately 90 mi through Hamilton, Coryell, and Bell counties and drains into Belton Lake. The

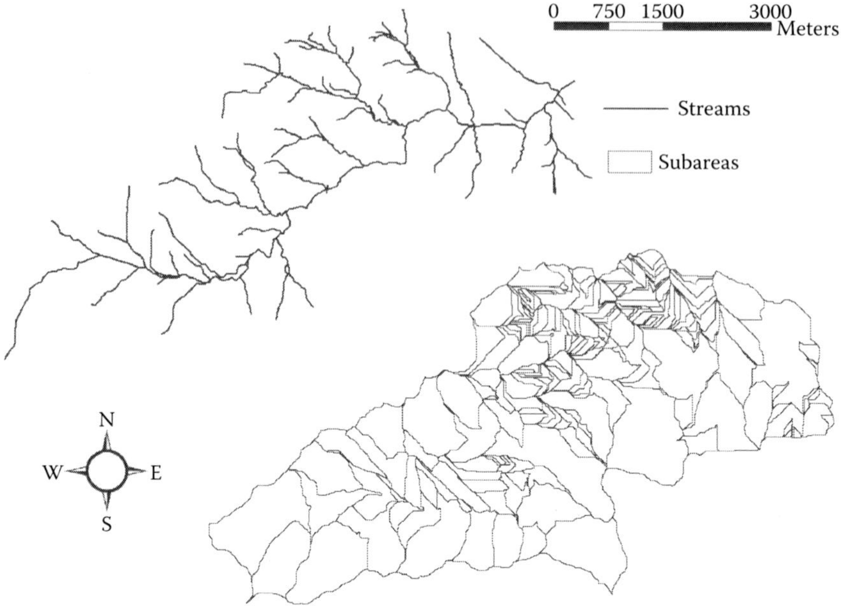

FIGURE 13.2 Stream network and the associated 183 subareas delineated from the 5 × 5 m DEM for the Shoal Creek watershed. (From Wang, X. et al., *Trans. ASABE* 52(4):1181–1192, 2009.)

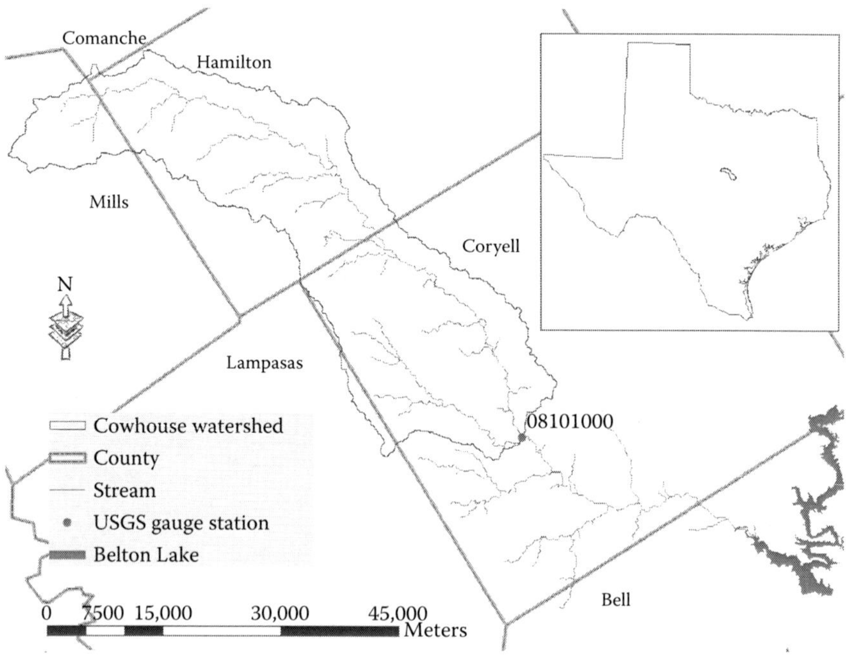

FIGURE 13.3 Location of the Cowhouse Creek in north-central Texas.

study area is the watershed area upstream of the U.S. Geological Survey (USGS) gauge site 08101000 (Figure 13.3), which has a drainage area of 1178 km^2. Long-term annual precipitation averages 775 mm. The major soil series include Nuff (silty–clayey), Doss (silty–clayey), Topsey (clayey–loamy), and Brackett (clayey–loamy), followed by Evant (silty–clayey) and Eckrant (clayey–skeletal, smectitic, thermic Lithic Haplustolls).

A DEM of 10 × 10 m resolution was used to establish the topographic characteristics of the watershed using the ArcAPEX interface (Tuppad et al. 2009). The watershed was divided into 445 subareas (Figure 13.4). The USDA-NRCS land use/land cover data and SSURGO soil data were used to define the subarea land use and soil characteristics.

The watershed is predominantly brush range (62%) and rangeland (34%). Cropping systems were chosen that represented vegetation community and management practices in the area. The brush range was assigned grass and mesquite trees. The grasses on the brush range and rangeland were grazed at a stocking rate of 10 ha/head. Heavy woodlands (4%) were simulated by a cover of deciduous trees with no harvesting of biomass. Agriculture land was less than 0.1% with continuous corn seeded with a planter in April and harvested in September. The tillage system consisted of one field cultivation operation and disk tillage before planting for seedbed preparation. Fertilizers were broadcast at the rates of 65 kg N ha^{-1} before planting corn.

The monthly record of stream flow at the USGS gauge station 08101000 (Figure 13.3) was used to calibrate and validate the hydrology component of the APEX model. The Blackland Research and Extension Center began collecting water quality and quantity data for this watershed in 1998. The monthly sediment yields were used to calibrate and validate the erosion/sedimentation component of the APEX model.

Rangeland management scenarios of brush control were analyzed using the calibrated APEX model. Long-term scenario simulation results of sediment yields (1951–2008) were compared with those from the baseline condition. Three brush control scenarios with the removal of mesquites were established: from all range brush areas, from areas with Evant soil, and from areas with Evant or Eckrant soils. These practices were simulated with all inputs held consistent with the baseline, except the removal of mesquite. The benefits of conservation practices on sediment yield were reported both at the subarea level (overland processes) and at the watershed outlet (which includes overland contribution and routing of the constituent through the stream network within the watershed).

The long-term simulation was used to target critical areas. Although the predicted average annual sediment yield at the watershed outlet is 0.23 t ha^{-1} from 1958 to 2008 (0.32 t ha^{-1} from 1998 to 2008; see Figure 13.5), the predicted overland sediment losses ranged from 0.0 to 6.7 t ha^{-1}. The high sediment losses were from the agriculture lands and some range brush lands. Because the agriculture land accounts for less than 0.1% of the total land area, the critical areas are range brush lands. There are 249 out of 445 subareas simulated as range brush. Figure 13.6 plots the overland sediment losses for each range brush subarea grouped by soils. On average, the subareas with Evant soil have relatively higher sediment losses, followed by areas with Eckrant soil, among the range brush lands. These two soils are classified as hydrologic soil group D. Therefore, they have high runoff potential. For this study area, in addition to the two soils, Pidcoke, Tarpley, and Real soils also belong to hydrologic soil group D. A mapping of the surface runoff overlay with soils confirms that these soils have higher surface runoff. However, the Evant and Eckrant soils have relatively higher soil

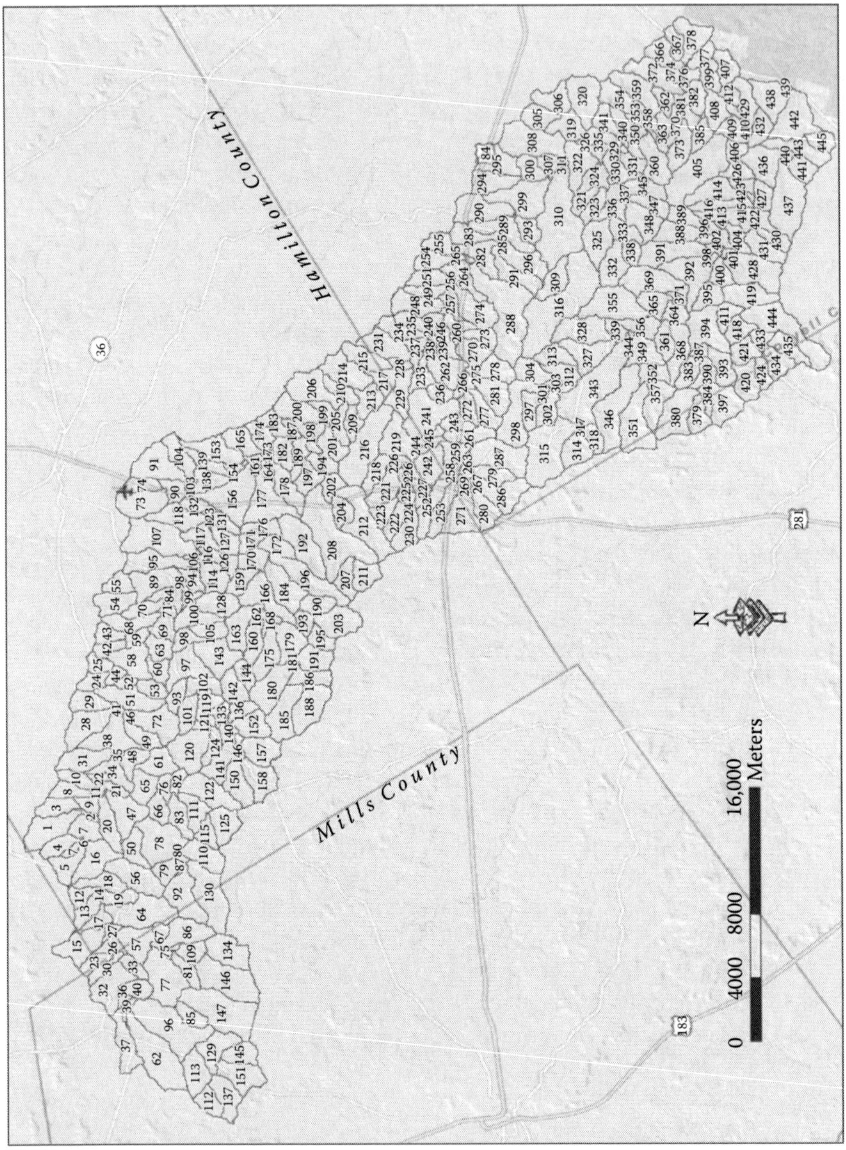

FIGURE 13.4 Subarea delineation of the Cowhouse watershed upstream of the USGS station 08101000.

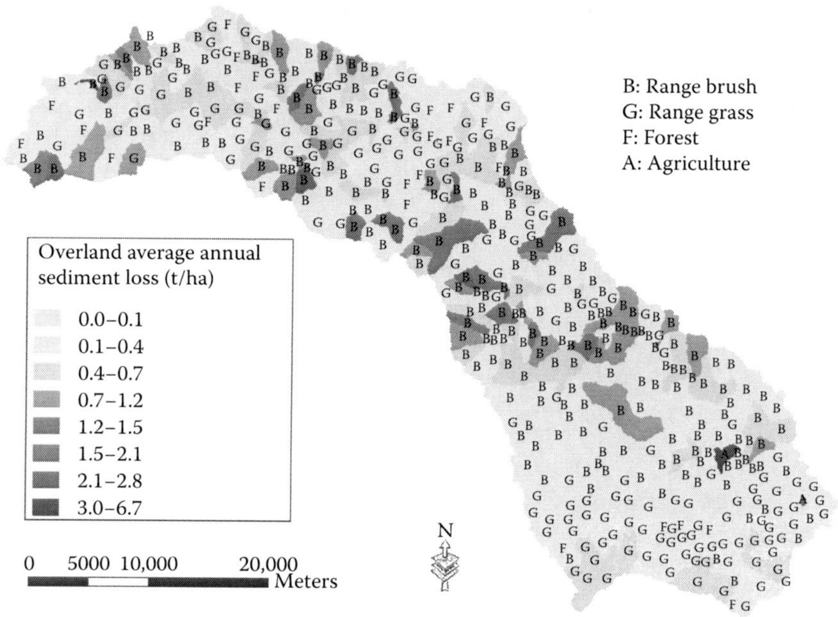

FIGURE 13.5 **(See color insert.)** Average annual sediment losses (1958–2008) from each subarea in the study Cowhouse watershed.

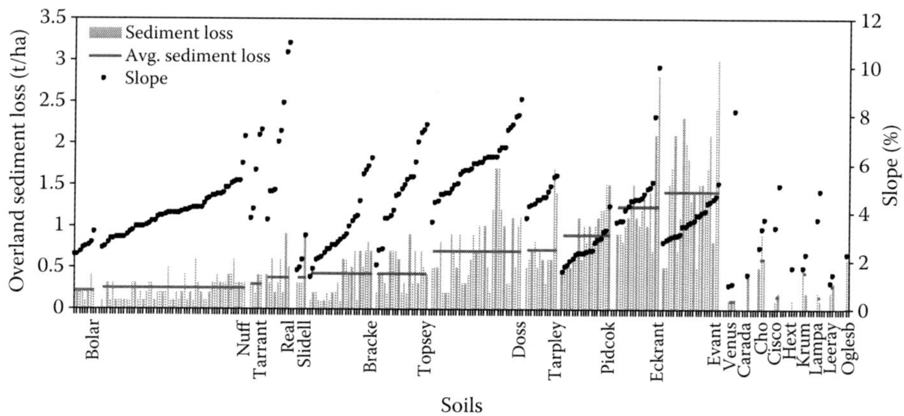

FIGURE 13.6 Average annual sediment losses (1958–2008) from each range brush subarea in the study Cowhouse watershed.

erodibility factors, which leads to the fact that subareas with Evant and Eckrant soils have relatively higher sediment losses among the range brush lands. Therefore, the scenario analysis was conducted for the conversion of range brush on Evant and Eckrant soils (18% of the range brush area) to range grasses. This was compared to the scenario in which all the range brush was removed (Table 13.1). The reductions at the watershed outlet were less compared to the reductions predicted at the subarea level (see overland

TABLE 13.1
APEX-Predicted Effects of Rangeland Management Practices (1951–2008)

Scenarios	Treated Area	Sediment Loading to Watershed Outlet t/year	Sediment Loading to Watershed Outlet Reduction (%)	Average Overland Sediment Loss Reduction for Treated Areas (%)
Baseline		34,060	—	—
Range brush on Evant soil to range grass	6%	31,719	−6.9	−58.8
Range brush on Evant and Eckrant soils to range grass	11%	29,840	−12.4	−53.8
All range brush to range grass	63%	24,769	−27.3	−48.5

sediment loss reduction for treated areas in Table 13.1). The most effective treatment to reduce overland sediment loss is the conversion of range brush to range grass on Evant soil areas, which reduced sediment by 58.8% on average at the subarea level.

13.3.3 North Bosque Watershed

The North Bosque watershed (3227 km^2) in Central Texas is a Special Emphasis watershed in the CEAP watershed studies (Richardson et al. 2008). The watershed flows from Erath County through Hamilton, Bosque, and McLennan Counties to create Lake Waco (Figure 13.7). The long-term average annual precipitation for this watershed is 842 mm. The land use was 63.5% native pastures/range, 14.2% improved pasture, 10.4% mesquite/cedar mix, 6.4% deciduous trees (including wetlands), 4.3% cropland, 1% urban, and 0.2% dairies and milk sheds. The goal of the CEAP watershed study is to identify conservation practices to reduce the phosphorous-loading levels flowing from the North Bosque watershed into Lake Waco (Dyke et al. 2010).

The WinAPEX–GIS interface was used in this study (Table 13.2), with the 10-m elevation data from the Texas Natural Resource and Information Systems used to develop the detailed watershed data. The watershed was divided into small subunits to let one subarea represent one field or pasture with the upper and lower landscape positions identified. Figure 13.8 shows the upper and lower landscape positions within one 12-digit watershed in the study watershed, delineated by the USGS using a nationwide system that is based on surface hydrologic features.

Cropping systems were chosen that represented crop and management practices in the area: waste application areas have Coastal Bermuda; pasture and range areas have combinations of grasses and shrubs that included mesquite trees, Little Bluestem grass, and side oats grama grass. Croplands have three rotations: (1) continuous corn, (2) corn/sorghum, and (3) wheat/sorghum.

APEX has been calibrated and validated based on available data at the Hico monitoring station (Figure 13.7) from 1993 to July 1998 (Wang et al. 2011b). The watershed area above Hico is 926 km^2. It contains 41 of the 61 dairies in the North Bosque (Figure 13.9). There are two primary types of reservoirs found in the study area: those constructed by the NRCS as flood prevention reservoirs and private water

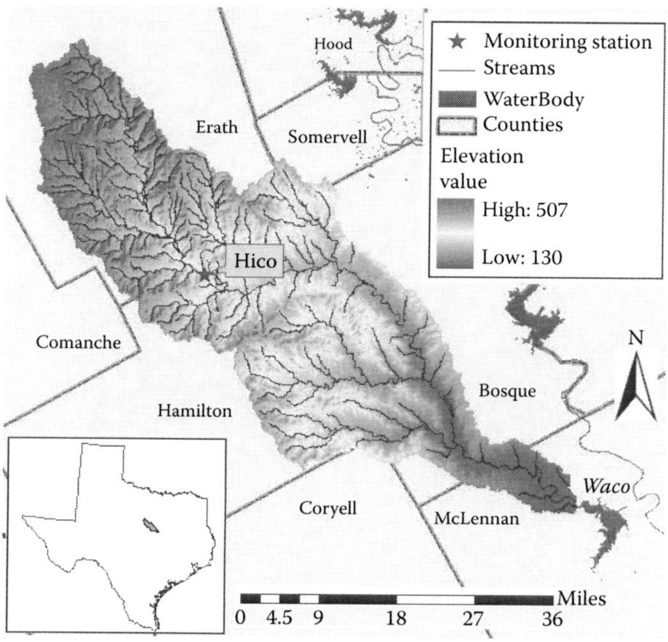

FIGURE 13.7 Locations of the North Bosque watershed in Texas.

TABLE 13.2
GIS-Assisted APEX Applications for Conservation Planning in Central Texas

Watershed	Drainage Area (km²)	Number of Subareas	Conservation Practice/ Planning	GIS Interface
Shoal Creek	23	183	Gully plug Contour ripping	AVSWAT–APEX subbuilder
Cowhouse Creek	1178	445	Range brush to range grass	ArcAPEX
North Bosque	3227	15,088	Nutrient management Manure transfer Adding reservoirs Modified buffer Pasture/hay land planting	WinAPEX–GIS

supply reservoirs built by other organizations. The calibrated model was applied to evaluate the effects of various conservation practices on three levels: farm level, subbasin level (12-digit watershed), and watershed level (North Bosque). The major conservation practices simulated were nutrient management, manure transfer, reservoirs, modified buffer, and pasture/hay land planting.

The exercise of dividing the watershed into very small subbasins for modeling purposes provided a clear demonstration of the ability of targeting a small percentage of the total land area to make a significant improvement in watershed health and

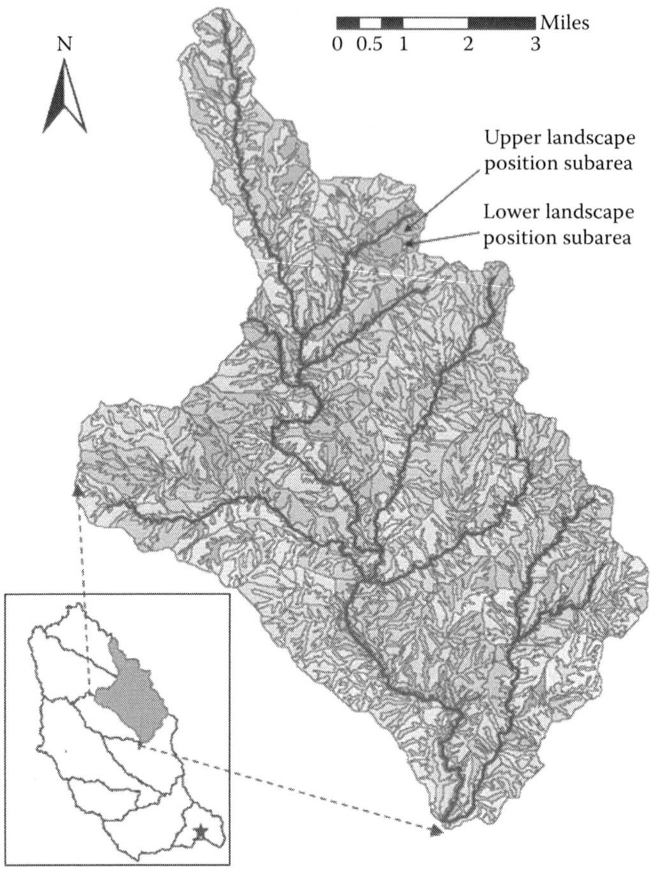

FIGURE 13.8 Subarea delineation considering the upper and lower landscape positions within one 12-digit hydrologic unit code (HUC) in the study watershed.

nutrient loadings into large lakes such as Lake Waco. The majority of the practices of this study targeted the waste application fields near the dairies in the northern part of the watershed. These areas accounted for less than 14% of the total land area drained by the Bosque River.

Changing the placement of the manure on the field and removing the manure for composting reduced the amount of nutrients leaving the waste application areas. At the mouth of the North Bosque, phosphorus loadings for various scenarios range from decreases of 29%, when multiple practices were applied to the watershed, to increases of 21%, when reservoirs were removed from the watershed (Dyke et al. 2010). The multiple practices include the following: 50% of the manure produced in the watershed was applied onto the waste application fields, and the remainder of the manure was hauled off to locations outside the watershed; lower landscape management with manure is applied on upper landscapes only; and six new reservoirs were added.

Phosphorus, nitrogen, and sediment leaving the farm were decreased in all application areas where conservation buffers or pasture planting was applied. Field-level

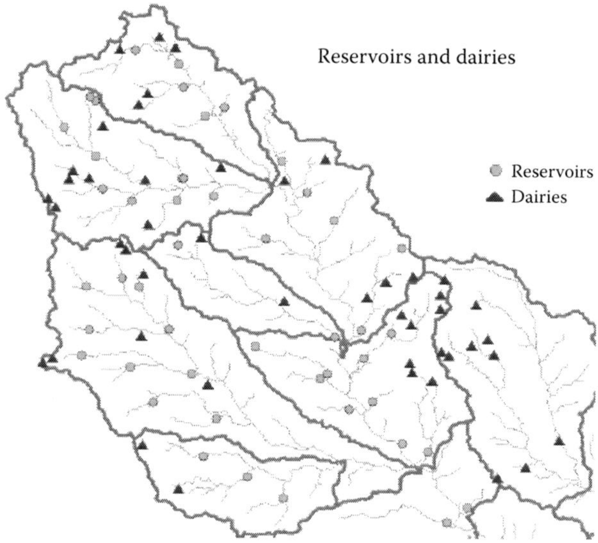

FIGURE 13.9 Locations of the dairies and the reservoirs above the Hico monitoring station.

phosphorus reductions ranged from less than 1% to over 50% for manure application areas. Nitrogen reductions ranged from less than 1% to over 30% for these areas. Farm-level sediment reductions ranged from less than 1% to over 70% for waste application areas. At the watershed level, as more conservation practices were applied, scenarios showed progressive improvement of watershed health. The larger improvements came when six new reservoirs were added (Figure 13.10). These reservoir locations were chosen to provide buffering of unprotected subwatersheds that

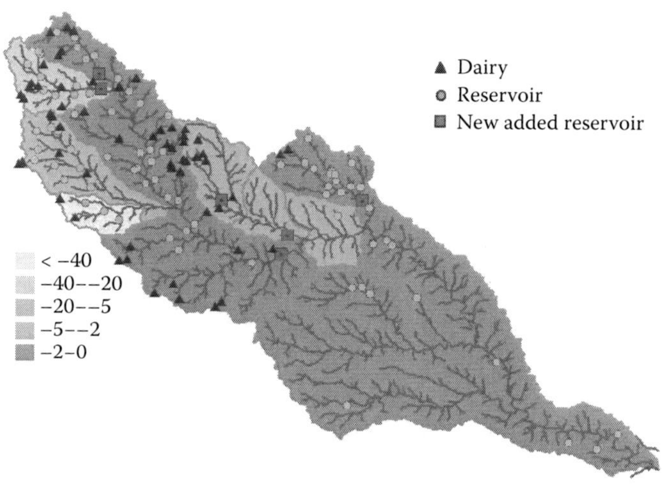

FIGURE 13.10 **(See color insert.)** Simulated total P% change between the baseline and the added reservoirs (all manure applied) for each 12-digit watershed in the North Bosque watershed.

contained dairies in the upper reaches from the mainstreams as water flowed into Lake Waco. The addition of these reservoirs reduced both phosphorus and nitrogen loads leaving the North Bosque by around 29%. This translated to around 15% improvement in water quality entering Lake Waco.

13.4 DISCUSSION AND CONCLUSIONS

The three case studies used different GIS interfaces to facilitate the preparation of model input files. The watershed drainage areas range from 23 to 3227 km^2. The process of watershed delineation, stream network identification, and study area partitioning into smaller units was accomplished through a GIS tool and utilized DEMs. Fine-resolution DEM and detailed land use and soil maps provide a better base for generating accurate estimations to capture significant topographic, land use, and soil variability within the watershed. For the Shoal Creek watershed, the DEM of 5 × 5-m was used, and the deep ripping and gully plug locations were considered during watershed delineation, so that small units were delineated to capture the gully plugs and the contour-ripping areas. The study by Wang et al. (2009) was focused on evaluating the effectiveness of conservation practices that are currently in place. A more informed conservation planning in regard to the placement of practices can be done by ranking the sediment losses from all the 183 subareas, so conservation practices can be specifically placed on targeted areas that produce high sediment losses. Some conservation practices that are not very effective or that have high establishment or maintenance costs may be eliminated to allow more resources for more promising conservation practices.

The use of simulation models as a decision support system for the management of conservation practices and assessment of their quantitative benefits at the watershed level prior to implementation is beginning to gain acceptance. For the rangeland Cowhouse Creek watershed, the APEX model was used to simulate the effectiveness of rangeland management over a long time period and a large area. It expands the temporal and spatial scales, targets critical areas, and provides feedback to future conservation planning, which will eliminate ineffective conservation practice placements. Utilizing the approach provides land managers the ability to test land management considerations prior to allocation of resources, thus reducing some of the uncertainty and risk associated with conservation practices. In doing so, land managers have a valuable tool that will provide guidance as to what practices are feasible for specific locations within the landscape and some predictive capacity to evaluate the costs and benefits of various applications.

The separation of the upper and lower position areas in the landscape allows alternative treatment. The North Bosque study implies that (1) conservation practices that cause divergence of the runoff water over the landscape, slowing the channelization of the water, will improve the quality of the water that eventually reaches the stream; (2) the removal or haul-off of a portion of manure from the basin has a significant impact on the nutrient loads reaching the streams in the watershed; and (3) the careful placement of a small number of new reservoirs in a watershed that protect previously unprotected regions of the watershed that contribute nutrient loadings to the stream can significantly improve the water quality in downstream water supplies.

This modeling process also provides land managers with the ability to test practices at specific locations within a watershed as a means for effectively siting practices on the landscape, where the highest return in reduction of erosion is predicted.

Given these results, the modeled conservation practices have been effective in reducing nonpoint source pollution at all levels. The model capabilities will be of great value in aiding natural resource management agencies in optimizing the use of limited conservation resources to provide the highest impact within communities where land management practices are being proposed.

13.5 HANDS-ON EXERCISE

The objectives of this exercise are (1) to set up an APEX project and (2) to familiarize the user with the capabilities of APEX through hands-on exercise in ArcAPEX.

To start the exercise, the user first needs to install the ArcAPEX ArcGIS extension, which is a graphical user interface for APEX. The software and the data used in this exercise are available from the CRC website: http://www.crcpress.com/product/isbn/9781439867228. The ArcAPEX ArcGIS extension may also be downloaded from ftp://ftp.brc.tamus.edu/pub/outgoing/srin/apex/. The extension is for ArcGIS 9.3. In order to install ArcAPEX, one must have the following: (1) Microsoft Windows XP or Windows 2000; (2) ArcGIS 9.3 with SP1 (build 1850); (3) ArcGIS Spatial Analyst 9.3; (4) ArcGIS Dot Net support (usually found in C:\ProgramFiles\ArcGIS\DotNet); (5) Microsoft.Net Framework 2.0; and (6) Adobe Acrobat Reader version 8 or higher.

A step-by-step guide is provided on the CRC Press website. A detailed ArcAPEX training manual is available from the CRC website: http://www.crcpress.com/product/isbn/9781439867228. The following are the key procedures necessary for modeling using APEX:

- Create the APEX project.
- Delineate the designated watershed for modeling.
- Define land use/soil/slope data grids.
- Determine the distribution of subareas based on land use and soil data.
- Define rainfall, temperature, and other weather data.
- Write the APEX input files—requires access to data on soil, weather, land cover, plant growth, fertilizer and pesticide use, tillage, and urban activities.
- Edit the input files—if necessary.
- Setup and run APEX—requires information on simulation period, PET estimation method, and other options.
- View the APEX output (usually followed by calibration/validation and/or conservation planning).

After the exercise, users should try to address the following questions:

1. Which subarea has the highest sediment loss and/or nutrient losses? Can you explain why?
2. What conservation practices might help to reduce sediment and/or nutrient losses? Try them out and make interpretations for your conservation planning. You'll want to calculate the edge-of-field and watershed effects. You

may refer to additional scenarios available from the CRC website: http://www.crcpress.com/product/isbn/9781439867228, such as adding a grassed buffer strip to targeted subareas and manure management, etc.

ACKNOWLEDGMENTS

Copyright permission granted for figures used in the article are acknowledged for ASABE for Wang et al. (2009) and ASA • CSSA • SSSA for Wang et al. (2011b). Acknowledgment is also given to R. Srinivasan and Evelyn Steglich for using the ArcAPEX training dataset and manual material in the online exercise.

REFERENCES

Arnold, J.G., R. Srinivasan, R.S. Muttiah, and J.R. Williams. 1998. Large area hydrologic modeling and assessment part I: Model development. *J. Am. Water Resour. Assoc.* 34:73–89.

Bagnold, R.A. 1977. Bed-load transport by natural rivers. *Water Resour. Res.* 13:303–312.

Baier, W., and G.W. Robertson. 1965. Estimation of latent evaporation from simple weather observations. *Can. J. Plant Sci.* 45:276–284.

Di Luzio, M., R. Srinivasan, and J.G. Arnold. 2004. A GIS-coupled hydrological model system for the watershed assessment of agricultural nonpoint and point sources of pollution. *Trans. GIS* 8(1):113–136.

Duriancik, L.F., D. Bucks, J.P. Dobrowski, T. Drewes, S.D. Eckles, L. Jolley, R.L. Kellogg, D. Lund, J.R. Makuch, M.P. O'Neill, C.A. Rewa, M.R. Walbridge, R. Parry, and M.A. Weltz. 2008. The first five years of the Conservation Effects Assessment Project. *J. Soil Water Conserv.* 63(6):185A–197A.

Dyke, P., X. Wang, J.R. Williams, and T. Dybala. 2010. *Modeling Nutrient Loads and Management Operations in the North Bosque Watershed.* Temple, TX: Blackland Research and Extension Center. Available at http://blackland.tamu.edu/files/2012/09/The-North-Bosque-11-08-2010-For-NRCS-Final.pdf (accessed January 10, 2014).

Gassman, P.W., J.R. Williams, X. Wang, A. Saleh, E. Osei, L.M. Hauck, R.C. Izaurralde, and J.D. Flowers. 2010. The Agricultural Policy Environmental eXtender (APEX) model: An emerging tool for landscape and watershed environmental analyses. *Trans. ASABE* 53(3):711–740.

Green, W.H., and G.A. Ampt. 1911. Studies on soil physics: 1. Flow of air and water through soils. *J. Agric. Sci.* 4:1–24.

Hargreaves, G.H., and Z.A. Samani. 1985. Reference crop evapotranspiration from temperature. *Appl. Eng. Agric.* 1:96–99.

Izaurralde, R.C., J.R. Williams, W.B. McGill, N.J. Rosenberg, and M.C. Quiroga Jakas. 2006. Simulating soil C dynamics with EPIC: Model description and testing against long-term data. *Ecol. Model.* 192:362–384.

Izaurralde, R.C., J.R. Williams, W.M. Post, A.M. Thomson, W.B. McGill, L.B. Owens, and R. Lal. 2007. Long-term modeling of soil C erosion and sequestration at the small watershed scale. *Clim. Change* 80:73–90.

Kemanian, A.R., P. Duckworth, and J.R. Williams. 2009. A spatially distributed modeling approach for precision conservation and agroecosystem design. In *Proceedings of the International Farming Systems Design Conference*, 153–154, Monterey, CA.

Kumar, S., R.P. Udawatta, S.H. Anderson, and A. Mugdal. 2011. APEX model simulation of runoff and sediment losses for grazed pasture watersheds with agroforestry buffers. *Agroforest. Syst.* 83(1):51–62.

McElroy, A.D., S.Y. Chiu, J.W. Nebgen, A. Aleti, and F.W. Bennett. 1976. *Loading Functions for Assessment of Water Pollution from Nonpoint Sources*. EPA 600/2-76-151. Washington, DC: U.S. Environmental Protection Agency.

Mockus, V. 1969. Hydrologic soil–cover complexes. In *SCS National Engineering Handbook. Section 4. Hydrology*, 10.1–10.24. Washington, DC: USDA-Soil Conservation Service.

Monteith, J.L. 1965. Evaporation and environment. *Symp. Soc. Exp. Biol.* 19:205–234.

Mudgal, A., C. Baffaut, S.H. Anderson, E.J. Sadler, and A.L. Thompson. 2010. APEX model assessment of variable landscapes on runoff and dissolved herbicides. *Trans. ASABE* 53(4):1047–1058.

Nicks, A.D. 1974. Stochastic generation of the occurrence, pattern, and location of maximum amount of daily rainfall. Misc. Publ. No. 1275. In *Proceedings Symposium on Statistical Hydrology*, 154–171. Washington, DC: USDA.

Onstad, C.A., and G.R. Foster. 1975. Erosion modeling on a watershed. *Trans. ASAE* 18:288–292.

Parton, W.J., D.S. Schimel, D.S. Ojima, and C.V. Cole. 1994. A general model for soil organic matter dynamics: Sensitivity to litter chemistry, texture and management. In *Quantitative Modeling of Soil-Forming Processes*, eds. R.B. Bryant and R.W. Arnold, 147–167. Special Publication No. 39. Madison, WI: Soil Science Society of America.

Penman, H.L. 1948. Natural evaporation from open water, bare soil, and grass. *Proc. Royal Soc. London A* 193(1032):120–145.

Potter, K.N., J.R. Williams, F.J. Larney, and M.S. Bullock. 1998. Evaluation of EPIC's wind erosion submodel using data from southern Alberta. *Can J. Soil Sci.* 78:485–492.

Powers, S.E., J.C. Ascough II, R.G. Nelson, and G.R. Larocque. 2011. Modeling water and soil quality environmental impacts associated with bioenergy crop production and biomass removal in the Midwest USA. *Ecol. Model.* 14:2430–2447.

Priestley, C.H.B., and R.J. Taylor. 1972. On the assessment of surface heat flux and evaporation using large-scale parameters. *Monthly Weather Rev.* 100:81–92.

Reddy, K.R., R. Khaleel, M.R. Overcash, and P.W. Westerman. 1979. A nonpoint source model for land areas receiving animal wastes: II. Ammonia volatilization. *Trans. ASABE* 22:1398–1404.

Renard, K.G., G.R. Foster, G.A. Weesies, D.K. McCool, and D.C. Yoder. 1997. Predicting soil erosion by water: A guide to conservation planning with the revised universal soil loss equation (RUSLE). *Agric. Handb. 703*. Washington, DC: USDA-ARS.

Richardson, C.W. 1981. Stochastic simulation of daily precipitation, temperature, and solar radiation. *Water Resour. Res.* 17:182–190.

Richardson, C.W., D.A. Bucks, and E.J. Sadler. 2008. The conservation effects assessment project benchmark watersheds: Synthesis of preliminary findings. *J. Soil Water Conserv.* 63(6):590–604.

Saleh, A., and O. Gallego. 2007. Application of SWAT and APEX using the SWAPP (SWAT–APEX) program for the Upper North Bosque River watershed in Texas. *Trans. ASABE* 50(4):1177–1187.

Saleh, A., E. Osei, and O. Gallego. 2008. Use of CEEOT–SWAPP modeling system for targeting and evaluating environmental pollutants. In *Proceedings 21st Century Watershed Technology Conference: Improving Water Quality and Environment*, eds. E.W. Tollner and A. Saleh. St. Joseph, Mich.: ASABE.

Seligman, N.G., and H. van Keulen. 1981. PAPRAN: A simulation model of annual pasture production limited by rainfall and nitrogen. In *Proceedings, Workshop: Simulation of Nitrogen Behaviour of Soil–Plant Systems*, eds. M.J. Frissel and J.A. van Veen, 192–221. Wageningen, The Netherlands: Centre for Agricultural Publishing and Documentation.

Steglich, E.M., and J.R. Williams. 2008. *Agricultural Policy/Environmental Extender Model: User's Manual. Version 0604 DOS and WinAPEX Interface*. BREC Report 2008-16. Temple, TX: Texas A&M University, Texas AgriLIFE Research, Blackland Research and Extension Center. Available at http://apex.tamu.edu/media/34652/apex-user-manual.pdf (accessed January 10, 2014).

Tuppad, P., C. Santhi, X. Wang, J.R. Williams, R. Srinivasan, and P.H. Gowda. 2010. Simulation of conservation practices using the APEX model. *Appl. Eng. Agric.* 26(5): 779–794.

Tuppad, P., M.F. Winchall, X. Wang, R. Srinivasan, and J.R. Williams. 2009. ARCAPEX: ARCGIS interface for Agricultural Policy Environmental Extender (APEX) hydrology/water quality model. *Int. Agric. Eng. J.* 18(1–2):59–71.

USDA-ARS. 2005. *Revised Universal Soil Loss Equation 2: Overview of RUSLE2*. Oxford, MS: USDA-ARS National Sedimentation Laboratory, Watershed Research Physical Processes Unit. Available at http://www.ars.usda.gov/Research/docs.htm?docid=6010 (accessed January 10, 2014).

USDA-NRCS. 1986. *Urban Hydrology for Small Watersheds*. Technical Release 55 (TR-55). Washington, DC: USDA-NRCS. Available at http://www.cset.sp.utoledo.edu/~nkissoff/pdf/CIVE-3520/Modified-tr55.pdf (accessed October 8, 2014).

USDA-NRCS. 2004. Estimation of direct runoff from storm rainfall. In *NRCS National Engineering Handbook, Part 630, Hydrology*, 10.1–10.22. Washington, DC: USDA-NRCS. Available at http://directives.sc.egov.usda.gov/OpenNonWebContent.aspx?content=17752.wba (accessed January 10, 2014).

USDA-NRCS. 2009. *Conservation Effects Assessment Project (CEAP)*. Washington, DC: USDA National Resources Conservation Service. Available at http://www.nrcs.usda.gov/wps/portal/nrcs/main/national/technical/nra/ceap/ (accessed January 10, 2014).

Vitousek, P.M., D.R. Turner, W.J. Parton, and R.L. Sanford. 1994. Litter decomposition on the Mauna Loa environmental matrix, Hawaii: Patterns, mechanisms, and models. *Ecology* 75:418–429.

Wang, X., P.W. Gassman, J.R. Williams, S.R. Potter, and A.R. Kemanian. 2008. Modeling the impacts of soil management practices on runoff, sediment yield, maize productivity, and soil organic carbon using APEX. *Soil Tillage Res.* 101(1–2):78–88.

Wang, X., D.W. Hoffman, J.E. Wolfe, J.R. Williams, and W.E. Fox. 2009. Modeling the effectiveness of conservation practices at Shoal Creek watershed, Texas using APEX. *Trans. ASABE* 52(4):1181–1192.

Wang, X., N. Kannan, C. Santhi, S.R. Potter, J.R. Williams, and J.G. Arnold. 2011a. Integrating APEX output for cultivated cropland with SWAT simulation for regional modeling. *Trans. ASABE* 54(4):1281–1298.

Wang, X., A. Kemanian, and J.R. Williams. 2011b. Special features of the EPIC and APEX modeling package and procedures for parameterization, calibration, validation, and applications. In *Methods of Introducing System Models into Agricultural Research*, eds. L.R. Ahuja and L. Ma, 177–208. Advances in Agricultural Systems Modeling 2. Madison, WI: American Society of Agronomy (ASA)—Crop Science Society of America (CSSA)—Soil Science Society of America (SSSA).

Williams, J.R. 1975. *Sediment Yield Prediction with Universal Equation Using Runoff Energy Factor*. ARS-S-40. Washington, DC: USDA-ARS.

Williams, J.R. 1995. The EPIC model. In *Computer Models of Watershed Hydrology*, ed. V.P. Singh, 909–1000. Highlands Ranch, CO: Water Resources Publications.

Williams, J.R., and R.W. Hann. 1978. *Optimal Operation of Large Agricultural Watersheds with Water Quality Constraints*. Technical Report No. 96. College Station, TX: Texas Water Resources Institute, Texas A&M University.

Williams, J.R., and R.C. Izaurralde. 2006. The APEX model. In *Watershed Models*, eds. V.P. Singh, and D.K. Frevert, 437–482. Boca Raton, FL: CRC Press.

Williams, J.R., C.A. Jones, and P.T. Dyke. 1984. A modeling approach to determining the relationship between erosion and soil productivity. *Trans. ASAE* 27(1):129–144.

Williams, J.R., E. Wang, A. Meinardus, W.L. Harman, M. Siemers, and J.D. Atwood. 2004. *APEX Users Guide. V.2110*. Temple, TX: Texas A&M University, Texas Agricultural Extension Service, Texas Agricultural Experiment Station, Blacklands Research Center.

Wischmeier, W.H., and D.D. Smith. 1978. Predicting rainfall erosion losses, a guide to conservation planning. *USDA Agric. Handbook 537*. Washington, DC: USDA.

Yin, L., X. Wang, J. Pan, and P.W. Gassman. 2009. Evaluation of APEX for daily runoff and sediment yield from three plots in the Middle Huaihe River Watershed, China. *Trans. ASABE* 52(6):1833–1845.

14 Spatial Economics Decision-Making Guide for Conservation Reserve Program Enrollment

Carl R. Dillon, Jordan M. Shockley, and Joe D. Luck

CONTENTS

EXECUTIVE SUMMARY

Yield maps, geographic information system (GIS), and economic analysis tools can be used to improve the profitability of Conservation Reserve Program (CRP) land enrollment decisions. This chapter provides a theoretical basis for such an approach, a step-by-step guide, and a case study. The economic advantages of using precision agriculture tools for selecting land for CRP enrollment to maximize profitability will be demonstrated in the case study. This study will also describe how this approach could be utilized at different levels of analysis, from simple to more intricate.

KEYWORDS

Economic break-even criterion, geographic information system (GIS), precision Conservation Reserve Program (CRP) enrollment

14.1 INTRODUCTION

The decision of what, if any, farmland to enroll in the Conservation Reserve Program (CRP) is important to farmers. The CRP is a rental payment and cost share program offered through the US Department of Agriculture (USDA) and is administered by the Farm Service Agency to compensate farmers for removing land from production. Eligible land for CRP enrollment includes cropland, including field margins, which have been planted in 4 of the past 6 years, as well as certain marginal pastureland. The benefits provided by CRP include reduced erosion, improved environmental well-being, and wildlife preservation. These attributes are examples of what economists would call market failures in that they represent benefits that are not reflected in the price of agricultural commodities. In other words, the free market does not generally pay farmers for positive externalities such as environmental improvements. As such, there is a market failure that can be rectified through CRP wherein society (the government) pays farmers for enhanced environmental stewardship. Land enrolled in CRP provides farmers with economic incentive and a potential for greater profitability than they may have if they were to produce crops in these areas.

Precision agriculture (e.g., yield maps), geographical information systems (GISs), and economic tools can improve the profitability of decisions regarding land enrollment in CRP. This concept was originally proposed by Stull et al. (2001, 2004) and has been further expanded to include wildlife conservation (McConnell et al. 2010) and whole-farm economic optimization (Dillon and Shockley 2010). While the potential is great, guidelines are needed for practitioners to apply these techniques on-farm.

Prior research regarding spatial analyses has demonstrated the logic for establishing variable width conservation vegetative filters based on water flow (Dosskey et al. 2005). Precision spatial information has also been shown to have the potential to reduce nitrate leaching (Delgado and Bausch 2005). Precision conservation may have the potential to improve the environment. Additional environmental benefits of CRP participation include enhancements in soil productivity (Ribaudo et al. 1989). The economic value of environmental improvements considering soil productivity and wildlife has been estimated (Hansen 2007). Policy changes continually incorporate a growing understanding of the consequences of CRP. Therefore, a decision-making guide for CRP enrollment must demonstrate adequate flexibility to modification under these dynamic implementation conditions. Such is the intent of the decision-making tool to be presented here.

The economic elements of CRP inherently give rise to factors that complicate the decision-making process for farmers who are considering enrollment. The lag of CRP payment levels relative to increasing rental rates continues to be an issue today just as it was during the rising crop prices presented in a 1998 study by Cooper and Osburn. Components of contract design and their economic implications for CRP enrollment have been investigated (Wu and Guan 2009). Enhancements such as reduced contract length might serve as a mechanism to encourage farmers to enroll in CRP when facing a lagging, low CRP payment level relative to expected profitability from continued production. Uncertainty in general and decision irreversibility in

CRP participation have also been recognized and studied from an economic perspective (Isik and Yang 2004). While facing some forms of uncertainty, it should be recognized that the CRP has inherently displayed benefits to farmers beyond the opportunity for increased profitability. One example of such benefit is that a farmer receives a constant CRP payment as an alternative to the uncertain yields and market prices associated with production on CRP-eligible land. The multitude of considerations in CRP enrollment provide justification for a straightforward decision-making tool that combines economic criteria with precision spatial data. Consequently, the objectives of this study are to provide a theoretical basis for a precision CRP enrollment decision-making tool, to convey the application procedure for such a tool, and to demonstrate its use with a case study. While this study can provide insights, with modification, to many areas of CRP enrollment, the focus here is upon continuous (not subject to competitive bidding) CRP enrollment of grassed filter strips.

14.2 METHODS

Farmers frequently make choices among competing alternatives that are based on economic consequences resulting from the underlying production environment. Consequently, it is not surprising that production economics decision-making models often can serve as helpful guides in making these choices. The methods will follow the development of a production economic decision-making tool by examining four elements. First, the underlying economic theory is established as a basic framework. This serves as a base for the development of a more practical and useful planning tool—the partial budget and the subsequent break-even analysis. Spatial elements that are inherent in precision agriculture and GIS are then utilized to develop net returns maps. These GIS data (yield maps or net returns maps) are then combined with economic criteria to create spatially depicted decision guides to help delineate areas for potential CRP enrollment.

14.2.1 Output Substitution Theory

A farmer considering CRP enrollment essentially is choosing between the production of two outputs—environmental improvement or crop yields. The first output of environmental enhancement is not normally a market good but in this case includes corresponding governmental payment for CRP participation. The alternative output is a crop (e.g., corn or soybeans) with potential net returns from the production and marketing of that commodity. Therefore, the decision to enroll in CRP implies an opportunity cost of foregoing the production of agricultural commodities on the farmland in question. Consequently, a classical output substitution (enterprise combination or crop mix) is an appropriate theoretical framework for making the decision of what land to enroll in CRP.

In an output–output (or output substitution) model, the physical trade-offs between different outputs or products can be defined. This relationship is commonly referred to as a production possibilities frontier or a production possibilities curve (PPC). The PPC depicts the possible combinations of efficient production of outputs (i.e.,

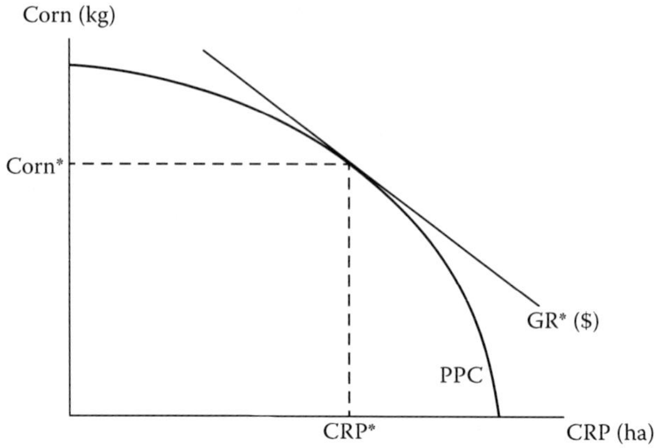

FIGURE 14.1 Graphical depiction of optimal precision CRP enrollment selection, where CRP is Conservation Reserve Program participation (ha); Corn is corn production (total kg for the farm); GR is gross revenue (total $ for the farm); PPC is production possibility curve; and * indicates the economically optimal level.

commodities) for a given set of constant resources or inputs (factors of production) available to the producer. This may be represented in many different ways including graphically as shown in the example in Figure 14.1. This simple two-output model includes two alternatives of crop production, specifically corn, and the enrollment in CRP as germane for our purposes here. The maximum level of corn production corresponds to no CRP enrollment, while some corn production is possible under maximum CRP enrollment as not all land is eligible for the program. While the physical possibilities are herein represented, the decision of what to produce between the two alternatives requires adding economic data. Specifically, the farmer's goal in this case is to maximize gross revenue (GR). Thus, the effective prices for the two alternatives are multiplied by the amount of the two goods produced and combined to account for total GR. Thus, the producer maximizes GR to a point that is still physically possible, resulting in the optimal output mix (Figure 14.1). Graphically, this is the tangency point between the GR line and the PPC, which requires equivalency of the slopes of both. Mathematically, this is where the marginal rate of transformation (the slope of the PPC that depicts the rate at which one output can be transformed into the other) is equal to the output price ratio (the slope of the GR line). Depending on the curvature of the PPC and the effective output prices, the economically optimal level of CRP enrollment can be no enrollment, maximum eligible enrollment, or some level in between.

14.2.2 Partial Budgeting and Break-Even Analysis

While useful as a theoretical framework, the underlying economic output substitution model may be embodied in a more useable economic decision-making tool as a simple partial budget. A partial budget is a farm management planning guide that summarizes the changes in profits associated with a proposed change to a whole-farm operation. Consequently, it is an appropriate and practical tool for the consideration of CRP enrollment. It offers the advantage of providing an analytical framework for making decisions by focusing solely on the aspects of the farm plan that are affected by the proposed alteration. Negative impacts on profitability (additional costs and reduced revenues) are subtracted from positive changes in profitability (additional revenues and reduced costs) to determine a net change in profit. Changes that would result in a lower expected profit (net change is negative) are disregarded as unfavorable. If net change in profitability is zero, then the decision maker would be indifferent to making the proposed change.

For the example at hand, the costs of CRP enrollment include additional costs of grassy strip establishment and maintenance as well as reduced revenue from foregone crop sales. The benefits of CRP enrollment include additional revenue from government payment and reduced costs from crop production. The costs are subtracted from the benefits of CRP to establish the net change in profits. An example partial budget for CRP enrollment may be found in Table 14.1. A positive change in profit is indicated, inferring that a switch from corn production to CRP enrollment would be profitable in this instance. While an average yield for all CRP-eligible land could be used, precision conservation concepts would allow for GIS to be applied for a more accurate assessment of the profitability of CRP enrollment. Thus, the use of spatially recorded yield data is desirable. Naturally, the development of a partial budget for every point on a farm field is unattractive. Fortunately, the dilemma is easily rectified by the calculation of the break-even yield required to justify CRP enrollment, as developed by Stull et al. (2004), or through the use of spatially dependent net returns estimations, as discussed in Section 14.2.3.

We note from the partial budget that the farmer is economically indifferent to CRP enrollment at the point where there is a zero net change in profits (wherein marginal benefits are equal to marginal costs from CRP enrollment). Specifically, this is

$$CRP + VC - EST - PY = 0 \tag{14.1}$$

where CRP = CRP payments ($/ha or $/ac) including cost share for vegetative buffer strip establishment and maintenance; VC = variable cost ($/ha or $/ac) of crop

TABLE 14.1
Partial Budget Example for CRP Enrollment Decision

Additional costs:	Additional revenue:
EST = $15	CRP payment = $375
Reduced revenue:	Reduced costs:
PY = $0.16 × 4700 = $752	VC = $450
Total: $767	Total: $825
Net change in profits = $825 − $767 = +$58	

Note: Budget is per year per unit of land area (ha).

production removed due to enrollment; EST = annualized establishment and maintenance costs (\$/ha or \$/ac) for vegetative buffer strip; P = crop price (\$/kg or \$/bu); and Y = crop yield (kg/ha or bu/ac).

By solving this equation for yield, one can determine the break-even yield or the yield at which the farmer neither makes nor loses money as a result of CRP enrollment. This gives

$$(CRP + VC - EST)/P = Y \tag{14.2}$$

This is a powerful concept in that the break-even yield can then be compared spatially on a yield map to highlight profitable versus unprofitable areas of the field for CRP enrollment. Specifically, for areas of the field where the expected crop yields are greater than this break-even yield level, expected profits would be greater by remaining in crop production. For lower-yielding areas of the field, the CRP enrollment would provide greater expected profitability than the alternative of remaining in crop production.

14.2.3 Net Returns Mapping

A yield map can serve as an excellent tool in farm management decision making and can be expanded into a net returns map, which visually depicts the profits (more correctly, net returns above specified costs) on a spatial basis. At the simplest level, this can be straightforward accounting of crop price multiplied by the spatially dependent yield (GR by location) less specified costs. Specified costs could be total costs reflecting both operating costs (such as seed, fertilizer, and fuel) and ownership costs (such as machinery depreciation and interest on machinery investment) per unit of land area (e.g., hectare or acre). Alternatively, the specified cost might represent only operating costs or some select costs of relevance. Conceptually, the estimation of components within a net returns calculation could become more accurate and involved as desired with supporting data. For example, variable rate nitrogen application would necessitate the spatially dependent costs of nitrogen as applied. Likewise, costs that are yield-dependent (e.g., grain drying) could conceivably be reflected spatially. Furthermore, it may be possible to consider variation in machinery performance across the field (such as for headlands or changes in speed due to slope) and the corresponding changes in machinery costs. Such a net returns map has considerable possibilities in farm management applications including the identification of less profitable areas of the field. Where CRP-eligible areas of the field have net returns that are inferior to CRP payments, a farm manager should consider enrollment. One advantage of using net returns maps for delineating CRP-eligible areas is that an average across multiple years of crop rotation is permitted through a common unit (dollars). This contrasts with yield maps wherein averaging different crop yields (such as corn and soybean) across years is not relevant to this decision. Ultimately, the inclusion of economic criteria is critical in the decision-making process whether that step be done as part of the spatial depiction directly (i.e., net return maps) or as a means of establishing a break-even yield criterion for use with a yield map.

14.2.4 CRP Enrollment Delineation

The analytical framework of economic theory, economic planning tools, and GIS can conceptually be merged as a guide for the CRP enrollment decision. Notably, a desirable feature of the analytical framework lies in its power to be implemented at different levels. On a relatively simple level, a producer could get a precision agriculture consultant to make a yield map highlighting areas above and below the break-even yield that is calculated using the formula developed above. On a more intricate level, a producer might develop his or her own CRP enrollment recommendation map. As an example procedure for this, the case study discussed in Section 14.3 provides a stepwise procedure for doing so. Nonetheless, the potential still exists for expanding the concept to incorporate complexities such as farmer risk attitude and impacts of varying forms of land acquisition (ownership versus rental), as evident in Stull et al. (2001).

The optimal level and location of precision CRP enrollment are indeed affected by the farmer's attitude toward risk and the form of land tenure (Stull et al. 2001). As a constant source of income that is not weather-dependent, CRP reduces the risk borne by the farmer relative to the alternative of producing a crop that has fluctuating yields. Thus, greater levels of aversion to risk by farmers may lead to more land being enrolled in CRP. A more risk-averse farmer may choose to enroll land in CRP to avoid yield variability associated with production even if that land may be slightly more profitable while in production. Land tenure also has impacts on CRP enrollment. Some tenant farmers have approached land owners to encourage CRP enrollment for low-yielding areas. The idea is that the land owner can get the CRP payments, while the tenant can avoid paying rent on that land enrolled in CRP. Land tenure naturally affects the optimal percentage of CRP enrollment from the farmer's perspective. Perhaps not surprisingly, more land is enrolled in CRP when the farmer is also the owner of the land. Meanwhile, a producer farming land rented under a crop share arrangement faces reduced benefits and would therefore want to enroll less land in CRP (Stull et al. 2001). While crop share rental arrangements embody a sharing of production risk between the land owner and the farmer tenant, the potential for CRP enrollment to further reduce risk is still evident. This means farmers, whether land owners or tenants, with greater aversion to risk would still wish to enroll more eligible land in CRP than their risk-neutral counterparts.

14.3 CASE STUDY

With the theoretical basis for a precision CRP enrollment decision-making tool now established, attention can be turned to other study objectives. These objectives are to convey the application procedure for such a tool and to demonstrate the use of such a tool with a case study. This is accomplished through a description of the case study used, a presentation of the stepwise procedure used for the precision CRP enrollment decision-making tool, and a discussion of the resulting CRP enrollment recommendations.

14.3.1 Background

The economic decision-making process often entails gathering underlying physical data, collecting relevant economic data, and applying the appropriate economic decision-making criteria. The case of CRP enrollment delineation is no exception. The physical data required are embodied in the yield maps for the area being considered for CRP enrollment. As an example application of the decision-making tool, the case study of corn yield data for a cooperating farmer in Shelby County, Kentucky, is used. The corn is planted in a 2-year rotation with double-cropped soybean and winter wheat production under no-till conditions, common to Kentucky row crop agriculture. The field being considered is 11.51 ha (28.44 ac) and is largely composed of a silt loam soil with a gently rolling topography. Raw crop yield monitor data were filtered for measurement error and inappropriate anomalies in the data using the same criteria of Stull et al. (2004). Specifically, harvester data standards include filtering for usable combine speed (allowing a range of 63.5 to 355.6 cm or 25 to 140 in./s), moisture level (10–25%), and mass flow (less than 170.25 kg or 75 lb./s). For this example, the filtered crop yields resulted in a mean of 7998.45 kg/ha (127.43 bu/ac) with a spatial standard deviation of 1791.38 kg/ha (28.54 bu/ac).

With the physical yield map data established, the question arises as to what economic data are needed. Crop prices, crop production costs, and CRP-related data regarding payment levels and vegetative buffer strip establishment and maintenance costs are required. A 2011 expected median corn price of $0.1677/kg ($4.26/bu), less basis and hauling costs, was used for this study (World Agricultural Outlook Board 2011). Corn production costs from Halich (2011) were used, but excluded cash rental charges, resulting in $928.99/ha ($376.11/ac). Similarly, the annualized establishment and maintenance costs associated with a grassy vegetative filter strip was taken from Stull et al. (2004) and updated to 2010 using the producer price index to account for inflation resulting in a level of $30.90/ha ($12.51/ac). Notably, 50% of these establishment and maintenance costs are offset as part of the cost share elements of CRP, and such is reflected in subsequent analysis. The CRP payment beyond this cost share is at a level of $363.09/ha ($147.00/ac) for the case at hand. Upon collecting the physical and economic data, the following step in the decision-making process of applying the economic decision rule is conducted as outlined in Section 14.3.2.

14.3.2 Stepwise Procedure

The existence of two alternate procedures is embodied with the analytical framework. One is a simpler approach whereby a break-even yield for CRP enrollment indifference is calculated. Then, a yield map is created to highlight areas above and below this calculated yield in developing CRP enrollment recommendations. The calculated break-even CRP enrollment yield for the economic data used herein is 7615.57 kg/ha (121.33 bu/ac). Alternatively, a more advanced procedure involves the decision maker using GIS software for a more thorough and detailed analysis. Under this more advanced procedure, a case study example of CRP enrollment delineation may be determined in a series of steps. While specific elements that are unique to the particular software package are inevitable (as demonstrated in the detailed tutorial on the online materials accompanying this book), a discussion of the general stepwise procedure is helpful. There are seven basic steps involved in conducting the CRP enrollment decision guide used herein:

Step 1: Develop a filtered yield map.
Step 2: Gather economic data.
Step 3: Develop a map depicting net returns above selected costs by calculating expected net returns under continued production.
Step 4: Delineate the CRP-eligible regions of the field.
Step 5: Calculate the expected net returns under CRP enrollment.
Step 6: Select the maximum expected net returns option between CRP enrollment or continued production by location.
Step 7: Compare and contrast results.

Recall that appropriate analysis necessitates good input data; therefore, the yield map should be appropriately filtered for erroneous data. The procedure used for this example case study was discussed in Section 14.3.1 as was the economic data. With the first two steps accomplished, development of a net returns map (loosely called a "profit map") may now be undertaken. While a more involved procedure accounting for spatially dependent costs may arise, a straightforward calculation of gross returns less costs of production was adequate here. Furthermore, a single crop year was used in this example, but averaging across multiple years is desirable when the data are available. Delineation of CRP-eligible areas for the example at hand involves including blue line stream data and creating a buffer around the environmentally sensitive area. Specifically, this delineation depicts the areas of the field that are within 45.72 m (150 ft.) of a blue line stream (which is defined as a stream that flows most of the year or all year round). The expected net returns under CRP enrollment for this example is \$347.64/ha (\$140.75/ac). This in turn can be compared on a spatial basis to the net returns map using a computerized maximum function in GIS software to visually depict an ideal CRP enrollment recommendation map. Notably, this would fail to capture the realistic elements of logistics in that it would represent desired enrollment on a cell-by-cell or point basis, while in practice one would enroll a contiguous strip of land in CRP. Nonetheless, it serves as a useful guide. The final step involves comparing and contrasting the economic results from such options as no CRP enrollment (continued production), complete CRP enrollment of all eligible land, and precision-guided partial enrollment of land in CRP. The economic results for the case at hand are presented in Section 14.3.3. While the generalities presented above provide a more universal conceptual framework for conducting this analysis, a more detailed step-by-step example could prove instructional. Consequently, a tutorial with the unique elements of ArcGIS software using ArcMap v10.0 is available in the online materials accompanying this book. The tutorial provides a thorough, detailed application of this process for a specific case study using the data that are also included in the online materials for this chapter. The files may be used to extract shape files for the yield map and blue line streams used in the tutorial.

14.3.3 Results

The results for the three decisions of complete CRP enrollment, no CRP enrollment, and precision CRP enrollment are shown in Table 14.2. Enrollment of the entire area eligible for CRP land in a vegetative filter strip results in a net return of $595.73. Continued production on this 1.71 ha (4.23 ac) would provide an expected net return of $704.64 based on the 1-year yield map. Notably, sampling multiple years of data would be more desirable. At an average of 7994.69 kg/ha (127.37 bu/ac), there is little difference in the mean yield for this particular CRP-eligible strip and the mean yield for the field as a whole (7998.45 kg/ha or 127.43 bu/ac). Net returns of $857.14 result from using the precision CRP enrollment strategy for an increase of $152.50 over continued crop production. Under the precision CRP strategy, 0.59 ha of land would be enrolled for a vegetative filter strip, while the remaining 1.14 ha would be used for crop production. This serves as verification that the economically guided decision, by design, results in the greatest economic outcome. Furthermore, it could be argued that a farmer would be otherwise disinclined economically from enrolling any land in CRP, as continued production is preferred to entire CRP participation. This provides partial evidence that delineation for precision CRP enrollment might enhance environmental benefits by encouraging producers to participate in CRP on a portion of their eligible land when otherwise they might choose not to participate. For this case study, precision enrollment resulted in 0.59 ha (1.43 ac) or approximately 34% of the eligible land being recommended for enrollment. The mean yield on the remaining 1.14 ha (2.80 ac) eligible for enrollment but remaining in production increased to 8992.06 kg/ha (143.26 bu/ac), further enhancing the economic return.

The precision CRP recommendation map is displayed in Figure 14.2. It should be noted that this serves as a decision-making guide and still requires modification. Specifically, it is impractical to enroll a single cell in the middle of the eligible area. However, it is not a requirement that all eligible land be enrolled in order to participate in CRP. Thus, the numerical results provided herein are estimates that would change as the farmer excludes and includes adjacent cells to develop a manageable contiguous vegetative CRP filter strip. Recall that these results could incorporate expanded considerations such as attitudes toward risk or varying land tenure arrangements that potentially affect the recommended CRP enrollment area.

TABLE 14.2
Case Study Results for Alternative CRP Enrollment Decisions

	All Eligible CRP	No CRP	Precision CRP
Expected net returns ($)	595.73	704.64	857.14
Land in CRP (ha)	1.73	0	0.59
Land in production (ha)	0	1.73	1.14
Yield (kg/ha)	0	7994.69	8992.06

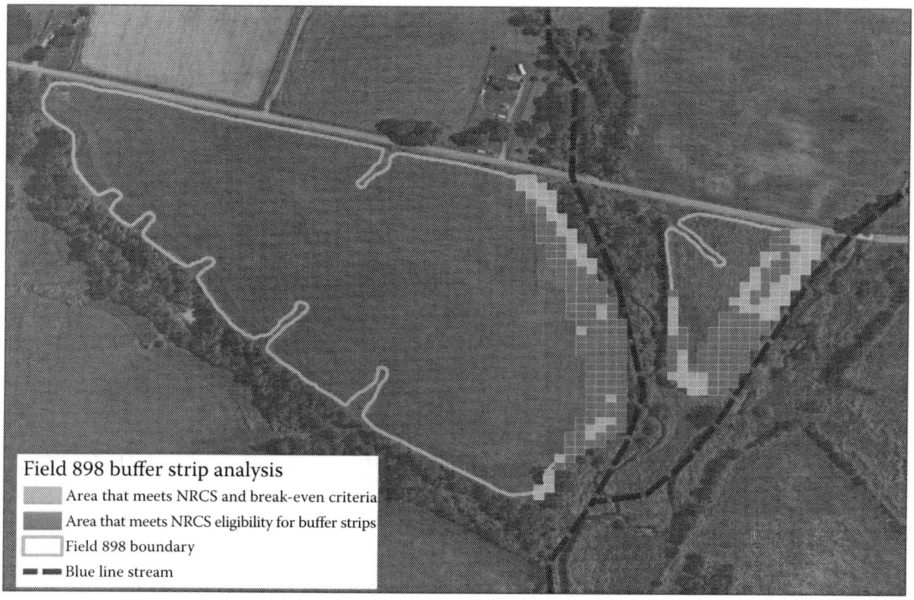

FIGURE 14.2 Precision CRP recommendation map.

14.4 SUMMARY

The melding of economic analysis with yield maps offers opportunities for enhanced profitability by improving enrollment in the CRP. The purpose of this chapter was to provide guidelines for delineating such precision CRP recommendations. Introductory background material was presented that first established the basics of CRP and prior research of relevance. The theoretical underpinnings for a methodological framework for developing such a recommendation map were discussed briefly. A stepwise procedure was presented that emphasized the potential for a simple or more advanced approach. The calculation of a break-even yield allows for a relatively simple-to-use method. A more detailed procedure permitted expected economic comparisons among no CRP enrollment, precision CRP selective enrollment, and the enrollment of all land eligible. A case study provided evidence that precision CRP enrollment can be an optimal enrollment strategy. Precision CRP enrollment displayed net returns that were almost 22% greater than the next best alternative of foregoing CRP participation for this case study. Therefore, precision CRP enrollment might possibly encourage environmental improvement by persuading producers to enroll some land that might not otherwise be put into CRP. The potential for expanding the analysis to include such issues as land tenure arrangement and risk attitude of the decision maker was also discussed.

The contributions of this chapter are fourfold. Firstly, demonstration of a detailed guide to making precision CRP enrollment decisions encourages the adoption of the strategy. Secondly, the flexibility of the decision-making tool offers a range of opportunities from simple (but at a cost of hiring a consultant to develop a break-even map) to more involved (but with enhanced information and possibly less cost). Thirdly, evidence that the precision CRP strategy afforded by the decision-making tool leads to greater profitability was provided. Finally, logic was offered that the precision CRP strategy can lead to environmental improvements through increased enrollment over the binary choice of either enrolling all eligible land or not enrolling any land.

ACKNOWLEDGMENTS

We especially thank Mike Ellis and Worth and Dee Ellis Farms for their willingness to share data for the case study used here as well as for their tireless support of the College of Agriculture at the University of Kentucky. We also acknowledge the helpful review from the students from the University of Kentucky BAE 599 Precision Agriculture course of spring 2011 as well as reviewers and editors. We also convey our heartfelt appreciation to those who assisted with the development of the tutorial: Teri Dowdy with Biosystems and Agricultural Engineering, University of Kentucky, and Dr. John Fulton, associate professor of Agricultural Engineering, Ohio State University.

REFERENCES

Cooper, J.C., and C.T. Osburn. 1998. The effect of rental rates on the extension of Conservation Reserve Program contracts. *American Journal of Agricultural Economics*. 80(1):184–194.

Delgado, J.A., and W.C. Bausch. 2005. Potential use of precision conservation techniques to reduce nitrate leaching in irrigated crops. *Journal of Soil and Water Conservation*. 60(6):379–387.

Dillon, C.R., and J.M. Shockley. 2010. Precision management for enhancing farmer net returns with the Conservation Reserve Program. *Proceedings of the 10th International Conference on Precision Agriculture*, Denver, CO, July 18–20. Electronic (CD) Proceedings.

Dosskey, M.G., D.E. Eisenhauer, and M.J. Helmers. 2005. Establishing conservation buffers using precision information. *Journal of Soil and Water Conservation*. 60(6):349–354.

Halich, G. 2011. *Corn and Soybean Budgets 2011*. Lexington, KY: University of Kentucky Cooperative Extension Service.

Hansen, L. 2007. Conservation Reserve Program: Environmental benefits update. *Agricultural and Resource Economics Review*. 36(2):267–280.

Isik, M., and W. Yang. 2004. An analysis of the effects of uncertainty and irreversibility on farmer participation in the Conservation Reserve Program. *Journal of Agricultural and Resource Economics*. 29(2):242–259.

McConnell, M.D., L.W. Burger, Jr., and W. Givens. 2010. Precision conservation: Using precision agriculture technology to optimize conservation and profitability in agricultural landscapes. *Proceedings of the 10th International Conference on Precision Agriculture*, Denver, CO, July 18–20. Electronic (CD) Proceedings.

Ribaudo, M.O., S. Piper, G.D. Schaible, L.L. Langner, and D. Colacicco. 1989. CRP: What economic benefits? *Journal of Soil and Water Conservation*. 44(5):421–424.

Stull, J.D., C.R. Dillon, S.G. Isaacs, and S.A. Shearer. 2004. Using precision agriculture technology for economically optimal strategic decisions: The case of CRP filter strip enrollment. *Journal of Sustainable Agriculture*. 24(4):79–96.

Stull, J.D., C.R. Dillon, S.A. Shearer, S.G. Isaacs, and S. Riggins. 2001. Use of geographically referenced grain yield data for location and enrollment of vegetative buffer strips in the Conservation Reserve Program. Paper presented at the *2001 American Society of Agricultural Engineers Annual International Meeting*, Sacramento, CA, July 30–August 1, 2001. ASAE Paper No. 011028. St. Joseph, MI: ASAE.

World Agricultural Outlook Board. 2011. *World Agricultural Supply and Demand Estimates*. Technical Bulletin No. WASDE-463. Washington, DC: U.S. Department of Agriculture.

Wu, F., and Z. Guan. 2009. Contract designs and participation in the Conservation Reserve Program in the era of biofuel production. Selected paper presented at the *27th International Conference of Agricultural Economists*, Beijing, China, August 16–22, 2009.

Index

Page numbers followed by f and t indicate figures and tables, respectively.

D

E

F

M

S

U

V

W

Y

Z